Environmental Science and Engineering

The ultimate goal of this series is to contribute to the protection of our environment, which calls for both profound research and the ongoing development of solutions and measurements by experts in the field. Accordingly, the series promotes not only a deeper understanding of environmental processes and the evaluation of management strategies, but also design and technology aimed at improving environmental quality. Books focusing on the former are published in the subseries Environmental Science, those focusing on the latter in the subseries Environmental Engineering.

Keng Yuen Foo · Rajesh Banu · Lai Kuan Lee ·
Jia Wen
Editors

Water Nexus Approach for Sustainable Development

Selected Papers of the 2024 6th International Conference on Environmental Prevention and Pollution Control Technologies (EPPCT 2024)

Springer

Editors
Keng Yuen Foo
River Engineering and Urban Drainage Research Centre (REDAC)
Universiti Sains Malaysia
Nibong Tebal, Malaysia

Lai Kuan Lee
School of Industrial Technology
Universiti Sains Malaysia
Gelugor, Malaysia

Rajesh Banu
Central University of Tamilnadu
Thiruvarur, Tamil Nadu, India

Jia Wen
College of Environmental Science and Engineering
Hunan University
Changsha, China

ISSN 1863-5520 ISSN 1863-5539 (electronic)
Environmental Science and Engineering
ISBN 978-3-031-89237-0 ISBN 978-3-031-89238-7 (eBook)
https://doi.org/10.1007/978-3-031-89238-7

This Springer imprint is published by the registered company Springer Nature Switzerland AG
The registered company address is: Gewerbestrasse 11, 6330 Cham, Switzerland

Committee Members

Conference Chair

Prof. Yasin Orooji, Zhejiang Normal University, China

Technical Program Committee Chair

Prof. Manish Kumar, University of Petroleum and Energy Studies (UPES), India

Organizing Committee Chair

Assoc. Prof. Rongkui Su, Central South University of Forestry and Technology, China

Publication Chairs

Prof. Foo Keng Yuen, Universiti Sains Malaysia (USM), Malaysia
Assoc. Prof. Abdul-Sattar Nizami, Government College University Lahore, Pakistan
Assoc. Prof. Jia Wen, Hunan University, China

Committee Members

Prof. Abdul Rauf, Nanjing University of Information Science and Technology, China
Prof. El-Sayed Salama, Lanzhou University, China
Prof. Dr. Vinayak Adimule, Angadi Institute of Technology and Management, India
Prof. Tarun Kumar Lohani, Arba Minch University, Ethiopia
Assoc. Prof. Yunqing LIU, YiLi Normal University, China
Assoc. Prof. Amar Razzaq, Huanggang Normal University, China
Assoc. Prof. Jia Wen, Hunan University, China
Assoc. Prof. Rongkui Su, Central South University of Forestry and Technology, China
Assoc. Prof. Lai Kuan Lee, Universiti Sains Malaysia, Malaysia
Assoc. Prof. J. Rajesh Banu, Central University of Tamil Nadu, India
Asst. Prof. Nirban Laskar, Mizoram University, Aizawl, India
Asst. Researcher, Zhirong Zhao, North China Electric Power University, China
Dr. Fathiah Mohamed Zuki, University Malaya, Malaysia
Dr. Abdullah Bilal Öztürk, Yildiz Technical University, Turkey
Dr. Saeed Shojaei, Yazd University, Iran

Preface

This book entitled "Water Nexus Approach for Sustainable Development—Selected Papers of the 2024 6th International Conference on Environmental Prevention and Pollution Control Technologies (EPPCT 2024)" provides rigorous discussions, case studies, and recent developments in the emerging areas of water management. The readers would benefit from enhancing their knowledge and skills in the domain areas. Also, this Proceedings may help the readers in developing new and innovative ideas.

From April 26 to 28, 2024, in Kunming, China, the academic feast EPPCT 2024 gathered a diverse array of famous scholars, experts, engineers and technicians as well as technology developers at home and abroad to discuss the latest research results and development trends in water management, pollution control and environmental protection. Experts and scholars from international institutions such as Universiti Sains Malaysia (USM), Malaysia, gathered with those from China's Tongji University, Hebei University of Science and Technology, Zhejiang Normal University, Hunan University, Central South University of Forestry and Technology, etc., to perform enlightening keynote speeches and share their scientific research results and practical experiences in water management, pollution control and environmental protection and other disciplines. They highlighted the importance of integrating new technologies with traditional approaches to achieve greater environmental sustainability. In addition, many elites from the business community also participated in the conference enthusiastically, bringing a wealth of practical cases and cutting-edge thinking to the conference.

Apart from the wonderful academic reports and discussions, this conference also included many high-quality academic contributions, forming a rich collection of papers. These papers cover a wide range of technological achievements in natural environment protection and pollution treatment, encompassing researches on natural environment preservation, water pollution control, risk assessment, policies, nexus approach and their related domains, with both theoretical discussions and practical analyses, fully demonstrating the latest research dynamics and academic level in the field.

We would like to express our sincere gratitude and thanks to all the participants for their hard work and contributions and also acknowledge the conference organizers for their careful preparations and thoughtful arrangements, which provided a good international platform for academic exchanges and cooperation among the participants.

This book seeks to offer valuable insights and share successful experiences among scholars, engineers, and environmental professionals from the world's leading universities, research centers, and governmental and industrial departments engaged in environmental management, catering especially to regions and countries grappling with on-going wastewater pollution issues.

Nibong Tebal, Malaysia — Keng Yuen Foo
Thiruvarur, India — Rajesh Banu
Gelugor, Malaysia — Lai Kuan Lee
Changsha, China — Jia Wen

Contents

Part I
Water Resources Management and Monitoring

Chapter 1
Research and Application of Heavy Metal Wastewater Treatment by Constructed Wetland

Haitao Tian, Ruqiong Qin, and Wenting Yang

Abstract Through the study of the mechanism of removing heavy metal pollutants by constructed wetland, the selection methods of wetland plants and substrates were explored for the types of heavy metal pollutants. Combined with the research results of related process design, a constructed wetland system with Leersia hexandra Swartz as wetland plants and a mixture of red soil and zeolite as substrates was constructed. The design parameters of the constructed wetland system and the operating parameters such as hydraulic loading and influent concentration were studied. The research results can provide some reference for the technical development and application of constructed wetland treatment of heavy metal wastewater.

Keywords Constructed wetland · Heavy metal wastewater · Design parameters · Operation parameters

1.1 Introduction

Heavy metal wastewater is mainly generated from chemical industry, metallurgy, mining, electronics, instrumentation, machinery manufacturing and other industries. For instance, the production processes of electroplating, leather, fertilizer and paper in the chemical industry will discharge heavy metal wastewater [1]. The pollutants in heavy metal wastewater, such as Cd, Cr, Mn, Ni, As, will cause serious harm to both human health and the environment. Therefore, the heavy metal wastewater should be treated effectively and meet the corresponding discharge standards. The common treatment methods of heavy metal wastewater include physical method, chemical method, physical–chemical method and biological method. The first three methods can transform the heavy metal substances in the wastewater into insoluble substances by precipitation or floating removal, or concentrate and separate

H. Tian (✉) · R. Qin · W. Yang
School of Equipment and Environment, Guangxi Polytechnic of Construction, Nanning, China
e-mail: tianhaitao110@126.com

K. Y. Foo et al. (eds.), *Water Nexus Approach for Sustainable Development*,
Environmental Science and Engineering, https://doi.org/10.1007/978-3-031-89238-7_1

heavy metal substances. The biological method mainly uses plants and microorganisms to remove heavy metals from the wastewater [2]. Constructed wetland is an eco-friendly, low-carbon biological wastewater treatment technology. As an organic whole of plants, microorganisms and substrates, constructed wetlands can play the role in coordination, integrating plant absorption, microbial degradation, filtration, co-sedimentation, ion exchange and other treatment functions. At the same time, constructed wetlands have the advantages of low investment, low operation energy consumption, and less management workload. Therefore, constructed wetlands have good application prospects in the treatment of heavy metal wastewater.

1.2 Mechanism of Removing Heavy Metal Pollutants by Constructed Wetland

The treatment of heavy metal wastewater by constructed wetland is mainly through the synergistic action of plants, microorganisms and substrates. The functions of these three components in the treatment of heavy metal wastewater are shown in Table 1.1.

1.3 Technology Design of Constructed Wetland

According to the characteristics of water flow, the types of constructed wetlands are mainly divided into surface flow constructed wetlands, horizontal subsurface flow constructed wetlands, and vertical subsurface flow constructed wetlands. The technology design of constructed wetland includes the selection of plants, the choice of substrate, and the selection of design parameters, etc.

1.3.1 Selection of Plants

The selection of plants for constructed wetlands requires two main considerations. Firstly, the plants should have strong enrichment and transport ability. This means that they can absorb heavy metal pollutants in large quantities and transport them from roots to stems and leaves, removing heavy metal pollutants from the constructed wetland substrates and extending the service life of the substrates. Secondly, plants need to have strong tolerance and be able to grow normally in heavy metal wastewater. Based on the above two aspects, heavy metal hyperaccumulating plants have a significant advantage. They have higher enrichment and transport coefficient for corresponding heavy metals, and have much higher heavy metal tolerance than ordinary plants, so they can grow in high-concentration heavy metal wastewater. The

Table 1.1 Components and functions of constructed wetland

Component	Type	The role in heavy metal wastewater treatment
Plants	Aquatic plants, hygrophytes, etc.	(1) Extraction: plants absorb heavy metals and enrich them in stems and leaves, and heavy metals are removed from constructed wetlands by harvesting [3] (2) Filtration: plant roots remove heavy metals from wastewater by absorbing or depositing [4] (3) Volatilization: plant volatilize of heavy metals into the atmosphere, so as to remove the heavy metals in the wastewater [5] (4) Fixation: plants remove heavy metals from wastewater by immobilizing them to soil or humus (5) Provide oxygen for the degradation of pollutants (6) Provide attachment sites for microorganisms (7) Maintain the stability of the system
Microorganisms	Bacteria, fungi, etc.	(1) Absorption: microorganisms absorb heavy metals as nutrients to sustain themselves (2) Precipitation: microorganisms release proteins to convert heavy metals into insoluble substances and precipitate them (3) Catalysis: microorganisms can enhance the effect of plant adsorption and enrichment of heavy metals
Substrates	Soil, sand, gravel, ceramisite, etc.	(1) Exchange adsorption: the cation adsorbed on the soil colloid surface is replaced by heavy metal ions, so as to adsorb the heavy metals (2) Obligate adsorption: The soil colloidal surface and heavy metal ions are adsorbed by obligate force (3) Provide the foundation for the growth of plants and microorganisms (4) Buffer effect: reduce the toxicity of heavy metals to plants and microorganisms

Table 1.2 The common heavy metal hyperaccumulating plants and their characteristics

Type of metal	Super-enriched plants	Enrichment coefficient	Transport coefficient
Cr	*Leersia hexandra Swartz* [6]	56.83(leaf)	11.59(leaf/root + stem)
Cd	*Viola baoshanensis* [7]	5.26(aboveground part)	2.22(aboveground part)
Cd	*Aster subulatus Michx* [8]	14.3(stem), 13.7(leaf)	1.21(stem), 1.16(leaf)
As	*Pteris vittata L* [9]	7.42(pinna)	3.93(pinna)
As	*Pteris nervosa* [10]	6.62(aboveground part)	2.60(aboveground part)
Zn	*Sedum alfredii H* [11]	1.94(aboveground part)	–
Mn	*Phytolacca acinosa Roxb.* [12]	0.25(leaf)	23.77(leaf), 3.56(stem)

Note enrichment coefficient = plant heavy metal content/soil heavy metal content; transport coefficient = plant ground heavy metal content / root heavy metal content. The enrichment coefficient and transport coefficient in the table are the maximum data of plants under field growth conditions in the corresponding literature

common heavy metal hyperaccumulating plants and their characteristics are shown in Table 1.2.

1.3.2 Selection of Substrate

Constructed wetland substrate should have strong adsorption capacity to remove a large number of heavy metal pollutants in wastewater. As a place for plants and microorganisms to cling, the substrate should be conducive to the growth of plants and the stimulation of microbial activity, and indirectly improve the treatment effect of constructed wetlands. Constructed wetland substrate should also have a high hydraulic conductivity to avoid substrate blockage resulting in overland flood. Constructed wetland substrates are also targeted for the treatment of different types of heavy metal wastewater, and the comparison of common substrates and treatment effects is shown in Table 1.3.

1.3.3 Selection of Design Parameters of Constructed Wetland

At present, the existing domestic industry standards and local standards on the design of constructed wetlands are mainly to treat urban domestic sewage. The constructed wetlands with heavy metal wastewater as the treatment object can be designed by referring to its design principle and combining its own characteristics. Based on the industry standard *Technical specification for wastewater natural treatment project*

Table 1.3 Comparison of common substrates and treatment effects

Types of wastewater	Wetland plant	Types of constructed wetland	Heavy metal treatment effect
Cr-containing wastewater	*Leersia hexandra Swartz*	Vertical subsurface flow	Peat > rice soil > red soil > sand [13]
Cd-containing wastewater	*Phragmites australis, Typha angustifolia, Myriophyllum verticillatum*	Surface flow	Fly ash haydite > clinoptilolite > soil [14]
As-containing wastewater	*Pteris vittata L*	Vertical subsurface flow	Manganese sand > ceramsite [15]

CJJ/T54-2017 issued by the Ministry of Housing and Urban–Rural Development, the relevant design parameters are summarized in Table 1.4.

Table 1.4 Design parameters of constructed wetland

Design parameter	Types of constructed wetland		
	Surface flow	Horizontal subsurface flow	Vertical subsurface flow
Surface load, removal rate, hydraulic loading, hydraulic retention time	It should be determined by experiment or operation experience of constructed wetlands under similar conditions		
Hydraulic gradient	< 0.5%	0.5–1%	< 0.5%
Length–width ratio	> 3:1	3:1–10:1	1:1–3:1
Substrate aggregate	–	Influent: 25–15 mm Principal region: 4–8 mm Effluent: 10–15 mm	Principal region: 2–5 mm Transition region: 5–10 mm Catchment region: 10–15 mm
Substrate depth	–	Influent: 0.6–1.2 m Principal region: 0.6–1.2 m Effluent: 0.6–1.2 m	Principal region: 0.8–1.2 m Transition region: 0.2–0.3 m Catchment region: 0.2–0.3 m
Over height	0.3–0.6 m	0.3 m	0.3 m

1.4 Research on the Application of Constructed Wetland to Treat Heavy Metal Wastewater

1.4.1 Construction of Constructed Wetland Device

In order to study the treatment effect of constructed wetland on heavy metal wastewater and optimize its design and operation parameters, a pilot-scale constructed wetland was constructed. The influent water of the constructed wetland is mainly Cr-containing wastewater. The wetland plants are selected as *Leersia hexandra Swartz*, which is a Cr-super-enriched plant, and the substrates is selected as the mixture of red soil and zeolite (volume ratio 1:1). The main design parameters are shown in Table 1.5. The section view of constructed wetland is shown in Fig. 1.1.

Table 1.5 The main design parameters of constructed wetland

Design parameters	Value
Maximum daily processing capacity	$0.36\ m^3\ d^{-1}$
Hydraulic loading	$0.20\ m^3\ (m^2\ d)^{-1}$
Hydraulic gradient	0.5%
Length	2 m, thereinto, Influent: 0.2 m
Width	1 m
Substrate aggregate	Red soil zeolite: 1–5 mm Gravel: 5–10 mm
Substrate depth	Clay: 0.05 m Red soil zeolite: 0.25 m Gravel: 0.1 m
Over height	0.3 m

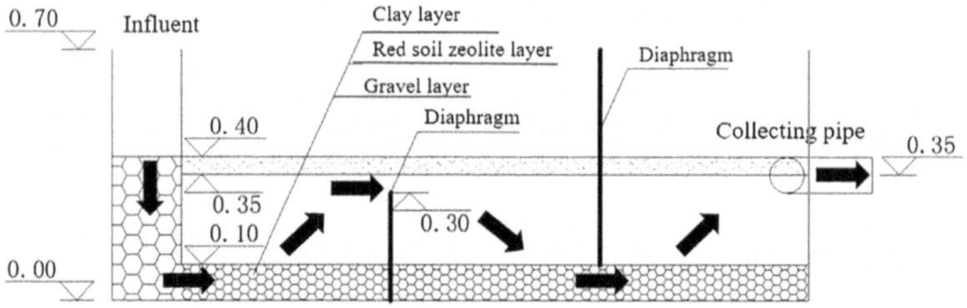

Fig. 1.1 Section view of constructed wetland

1.4.2 Start-Up of Constructed Wetland

The start-up of constructed wetland should ensure the survival of plants and the growth of microorganisms. The planting methods of wetland plants include seedling transplantation, pot transplantation, and harvest plant transplantation, etc. Plants with younger seedling age should not be selected. The planting time of wetland plants should be in spring or early summer and other seasons with suitable temperature, and ensure sufficient water depth. After the plants survive, the water level can be appropriately lowered, which can stimulate the growth of plant roots. The hydraulic loading of wastewater should be increased gradually.

In this constructed wetland, potted plants were used, and the transplanting density was 500–600 plants m^{-2}. After transplantation, the plants were watered with clear water for 15 days, and then the plants were watered with Cr-containing wastewater with a concentration of 10 mg·L-1 at a hydraulic load of 0.04 m^3 $(m^2\ d)^{-1}$ for 15 days, and finally the hydraulic load was gradually increased to 0.10 m^3 $(m2\ d)^{-1}$. The constructed wetland was started up for 40 days. The wetland plants were selected from the Taohua River in Guilin city, and the heavy metal wastewater was collected from an electroplating factory in Guilin City, and then obtained by manual mixing.

1.4.3 Operation of Constructed Wetland

In order to study the reasonable values of the constructed wetland operation parameters, the constructed wetland was operated under different hydraulic loads and different influent concentrations for 16 days respectively, and then the chromium concentration in effluent was tested. When the influent concentration is constant, the average removal efficiency of Cr under different hydraulic loads is shown in Fig. 1.2. When the hydraulic load is constant, the average removal efficiency of Cr under different influent concentrations is shown in Fig. 1.3.

Figure 1.2 shows the removal efficiency of constructed wetland under different hydraulic loads when Cr-concentration of influent is 10 $mg{\cdot}L^{-1}$. The results show that when influent concentration is constant, the lower the hydraulic load is, the higher the removal efficiency will be. The average removal efficiency of Cr under low, medium and high hydraulic loads is 82%, 71.6% and 65.6% respectively. Therefore, in the actual project, in order to improve the removal efficiency of heavy metals, the hydraulic load should be reduced as much as possible, which can be realized by reducing the treated water volume of the constructed wetland or increasing the area of the constructed wetland.

Figure 1.3 shows the removal efficiency of constructed wetland under different influent concentration when the hydraulic load is 0.15 m^3 $(m^2\ d)^{-1}$. The results show that when the hydraulic load is constant, the lower the influent concentration is, the higher the removal efficiency will be. The average removal efficiency of Cr under low, medium and high influent concentration is 82.5%, 77.1%, 71.6% respectively.

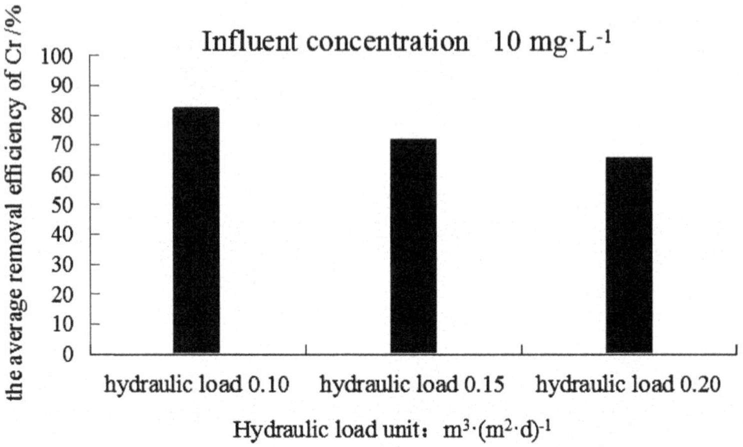

Fig. 1.2 The average removal efficiency of Cr under different hydraulic loads

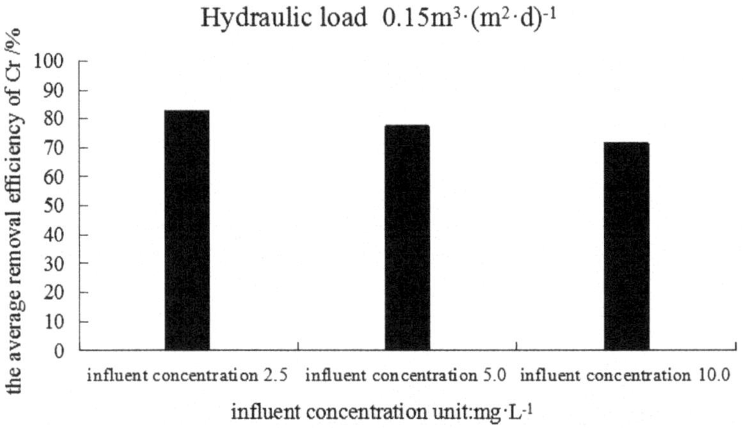

Fig. 1.3 The average removal efficiency of Cr under different influent concentrations

Therefore, in the actual project, in order to improve the removal efficiency of heavy metals, the influent concentration should be reduced as much as possible, which can be realized by increasing the pretreatment process.

The experimental results show the influence of two operating parameters, hydraulic load and influent concentration, on the treatment effect of heavy metal wastewater in constructed wetland. At the same time, the values of hydraulic load and influent concentration in the experiment can also be used as a reference for selecting operating parameters in actual projects. However, it should be noted that the values of operating parameters should be appropriately reduced in order to ensure the long-term vigorous growth of wetland plants and extend the service life of the substrate.

1.5 Conclusion

Based on the research on the mechanism of removing heavy metal pollutants in constructed wetland and the research results of related operation parameters, the constructed wetland system with red soil zeolite as substrate and *Leersia hexandra Swartz* as wetland plant was constructed. On this basis, the effects of the design parameters, hydraulic load, influent concentration and other operating parameters on the removal of heavy metals were further studied. The research results can provide some reference for the development and application of technology for treating heavy metal wastewater in constructed wetland. In the future, it is still necessary to further study and optimize the constructed wetland technology to improve its treatment effect and treatment scale of heavy metal wastewater, and make greater contributions to environmental protection and sustainable development.

1.6 Comment

In order to improve the practicality of this technology, further research can be carried out in the following aspects:

(1) Microorganism. In this paper, we only studied the selection of plants and substrates, but lacked microbial studies. It is necessary to strengthen the research on the microbial species, domestication process, and activity of constructed wetlands, give full play to the role of microorganisms, and improve the effect of constructed wetlands in the treatment of heavy metal wastewater.
(2) Continuous operation effect. We took 140 days to start-up and operate the constructed wetland, while the designed service life of the constructed wetland is generally 10 to 15 years. With the increase of the running time, the treatment effect of constructed wetland will gradually weaken. Therefore, the continuous detection of the operation effect of constructed wetland and the analysis of long-term test data will be more helpful to the promotion and application of this technology.
(3) Wetland plants. In this experiment, only a single plant species, *Leersia hexandra Swartz*, which is a super-enriched plant for a specific heavy metal, was selected as the wetland plant. However, the heavy metal wastewater produced by the chemical industry generally contains a variety of heavy metal pollutants and other toxic substances. Therefore, further research can consider using a combination of differentplants, in this way, the advantages of different plants can be used to achieve targeted treatment of heavy metal wastewater produced by a certain industry.
(4) Combination process. Due to the toxic effects of heavy metal pollutants in wastewater on plants and microorganisms, constructed wetlands are more suitable for treating wastewater with low concentration of heavy metals, so pretreatment is required for wastewater with high concentration of heavy metals, such

as Redox, electrolysis, PH adjustment, precipitation and other pretreatment processes [16]. If the water quality requirements of the treated effluent are high but have not been reached by the constructed wetland process, it is necessary to increase the advanced treatment process after the constructed wetland, such as ion exchange, ultrafiltration, RO, etc.

(5) Operation parameters. Operating parameters such as temperature, hydraulic retention time, pH, Eh also affect the treatment effect of constructed wetland. In order to realize the promotion and application of constructed wetland treatment of heavy metal wastewater, it is necessary to study the reasonable value range of the above operating parameters to ensure the efficient operation of constructed wetland.

Acknowledgements This research is supported by Notice of 2021 Guangxi University Young and middle-aged Teachers Basic Research Ability Improvement Project "Current status and countermeasures of landfill leachate treatment under the background of garbage classification" (No.: 2021KY1165).

References

1. Yang X, Li X, Zhou J et al (2023) Research progress in the treatment of wastewater containing heavy metal ions. Mater Rev 37(9):60–69
2. Fan X, Zhan Z, Gao B et al (2023) Research progress of heavy metal wastewater treatment technology. China Nonferrous Metall 52(4):112–127
3. Mushtaq S, Bareen FE, Tayyeb A et al (2023) Autochthonous strains of Trichoderma isolated from tannery solid waste improve phytoextraction potential of heavy metals by sunflower. Int J Phytorem 25(11):21–20
4. Banerjee A, Roychoudhury A (2022) Assessing the rhizofiltration potential of three aquatic plants exposed to fluoride and multiple heavy metal polluted water. Vegetos 35(4):1158–1164
5. Awa SH, Hadibarata T (2020) Removal of heavy metals in contaminated soil by phytoremediation mechanism: a review. Water, Air, Soil Pollut: Int J Environ Pollut 231(366):2638–2647
6. Zhang X, Luo Y, Huang H et al (2006) Leersia hexandra Swartz, a newly discovered hygrogenic chromium superaccumulator. Acta Ecol Sin 26(3):950–953
7. Liu W, Shu W, Lan C (2003) Viola baoshanensis: a new cadmium superaccumulator. Chin Sci Bull 48(19):2046–2049
8. Chen W, Jiang W, Yang Y et al (2023) Aster subulatus Michx., a newly discovered cadmium-accumulating plant. Acta Ecol Sin 43(13):5592–5599
9. Chen T, Wei Z, Huang Z et al (2002) Arsenic hyperaccumulator plant Centipede grass and its enrichment characteristics for arsenic. Chin Sci Bull 47(3):207–210
10. Wei Z, Chen T, Huang Z et al (2002) A newly discovered plant rich in arsenic. Acta Ecol Sin 22(5):777–778
11. Yang X, Long X, Ni W et al (2002) SedumalfrediiH: a new zinc superaccumulator. Chin Sci Bull 47(13):1003–1006
12. Xue S, Chen Y, Lin Q et al (2003) The first manganese superaccumulator in China: Shanglu. Acta Ecol Sin 23(5):935–937
13. Bu Y, Liu J, You S et al (2017) The difference of the purification effect on Cr(VI) by six kinds of substrate Liwo wetland system. Ind Saf Environ Prot 43(4):64–67+106
14. Dong S (2021) Removal of cadmium from irrigation water and inhibition effect of cadmium in rice by surface flow constructed wetland. Hunan Agricultural University, Changsha

15. Wu M, Lin L, Zhao L et al (2014) Experimental study on arsenic removal in vertical subsurface flow constructed wetland. J Huazhong Univ Sci Technol (Nat Sci Ed) 42(7):38–43
16. Liu K (2023) Treatment of tannery wastewater by combined pretreatment, AO-Fenton oxidation and constructed wetland. Environ Ecol 5(7): 137–141+150

Chapter 2
Water Quality Simulation and Application of Urban Rivers Based on SWMM Model—A Case Study of Xiangjiang River Basin in Zunyi City

Yutong Chen, Tingting Xiao, Li Ma, Ya Zhang, and Qing Zhang

Abstract The ecological environment of the Xiangjiang River region is sensitive, and its water quality directly affects the economic development, environment and people's livelihood of the downstream areas. In this paper, SWMM model is applied to simulate the current situation of water quality in the main and tributaries of Xiangjiang River Basin, which has good accuracy; the water quality changes in the future are predicted. The results show that in 2025 and 2030, Daqiuping section will deteriorate from the current class III water to class IV water, and in 2035 to class V water. The main performance is that NH_3-N exceeds the standard seriously, and pollution control measures need to be taken to improve the water quality.

Keywords SWMM model · Water quality simulation · Water quality prediction

2.1 Introduction

Since the middle and late 1970s, water quality model research has focused on application and gradually improved, forming Mike, wasp, Swat, EFDC, SWMM and other model systems. SWMM (storm water management model, SWMM) is a dynamic hydrological hydraulic model developed by the U.S. Environmental Protection Agency (EPA), which can be used for field or continuous simulation of surface runoff and water quality. Throughout the world, SWMM has been widely used in planning, analysis and design related to rainwater runoff, combined/separated pipe network, and drainage system in urban or non-urban areas [1–6]. In the past 20 years, SWMM has been widely used in China. It has been used in Shenzhen, Guangzhou, Wenzhou, Kunming and other areas to calculate the surface runoff of urban rainstorm flood and estimate the non-point source pollution load. Dong et al. [7] took the Shenzhen River bend area as the research area, used the SWMM model to simulate

Y. Chen (✉) · T. Xiao · L. Ma · Y. Zhang · Q. Zhang
Powerchina Guiyang Engineering Corporation Limited, Guiyang, Guizhou, China
e-mail: chenyutongami@xnmsn.cn

K. Y. Foo et al. (eds.), *Water Nexus Approach for Sustainable Development*, Environmental Science and Engineering, https://doi.org/10.1007/978-3-031-89238-7_2

and analyze its urban drainage system planning scheme, and obtained the impact of the planning scheme on the short-term and long-term water environment quality of the Shenzhen River bend area, indicating that SWMM can effectively and quantitatively analyze the regional sewage discharge. Zhang et al. [8] used the traditional runoff coefficient method to verify SWMM. The research shows that the relative error between the total number of rainfall runoff and rainfall runoff process simulated by SWMM and the results obtained by runoff coefficient method is small, indicating that SWMM can better simulate the total amount and process of urban rainfall runoff. Ma et al. [9] established a non-point source pollution load estimation model for typical residential areas in Wenzhou, Zhejiang Province, using SWMM to simulate the pollution load and cumulative change process of suspended solids under different rainfall scenarios, and the results were satisfactory. Nietiefeng [10] used SWMM to analyze and simulate the non-point source pollution load of rainstorm runoff in the urban area of Guangzhou under different conditions. The peak error of each pollutant content obtained from the simulation was within 20%, and the simulation effect was good. At present, SWMM can reasonably generalize the study area and better simulate the total amount of urban runoff, rainfall runoff process and pollutant load.

In this paper, the SWMM model is used to establish the hydrological hydrodynamic and water quality model of the study area in the Xiangjiang River Basin of Zunyi City. By comparing the existing Zunyi water quality data and simulation data, the SWMM model is calibrated to improve the accuracy of water quality simulation, and then provide support for the follow-up water pollution prevention and control work.

2.2 Overview of the Study Area

Xiangjiang River is a first-class tributary on the left bank of Wujiang River, with a total length of 155 km and a total drainage area of 4913 km^2, The drainage area in Zunyi city is 4878 km^2, Accounting for 92.4% of the total area. The average annual runoff of Xiangjiang River estuary is 2.412 billion m^3, Annual average flow 76.5 m^3/S; The main primary tributaries of Xiangjiang River include Gaoping River, Luojiang River, Yujian River, Renjiang River, Meijiang River, etc., and the main secondary tributaries include Houshui River, Luoan River, Maoguan River, etc. The ecological environment of the Xiangjiang River region is sensitive. The Daqiuping section of the main stream is a state-controlled section. The quality of its water directly affects the economic development, environment and people's livelihood of the downstream areas, especially the continuous improvement of the water quality of the "golden waterway" of the Yangtze River [11]. As the mother river of Zunyi City, Xiangjiang River also has far-reaching influence and symbolic significance on Zunyi's history, economy, culture and many other aspects.

2.3 Model Construction

In order to study the current situation of Zunyi urban area and the characteristics of water environment in the near and long term, combined with the characteristics of pollution sources within the planning scope and the types of receiving water bodies, SWMM model was selected to simulate the hydrology, hydrodynamic and water quality of rivers and lakes. The model construction method and boundary conditions are as follows:

(1) After calibration, the calculation parameters are as follows: (1) depression storage—set the depression storage in impermeable area as 2 mm; The storage capacity of permeable area is set as 5 mm; The ratio of no depression storage is set as 0%. (2) Infiltration—set the capillary head as 290 mm, the hydraulic conductivity as 0.51 mm/h, and the percentage of initial soil moisture status as 0.377. (3) Manning coefficient—the manning coefficient of permeability is 0.3, and the manning coefficient of impermeability is 0.015.
(2) According to the distribution characteristics of Zunyi water system, DEM elevation map and relevant data, the model shown in Fig. 2.1 is built: as shown in the figure, it is divided into 95 sub catchment areas, 2 outfalls and more than 100 rivers.

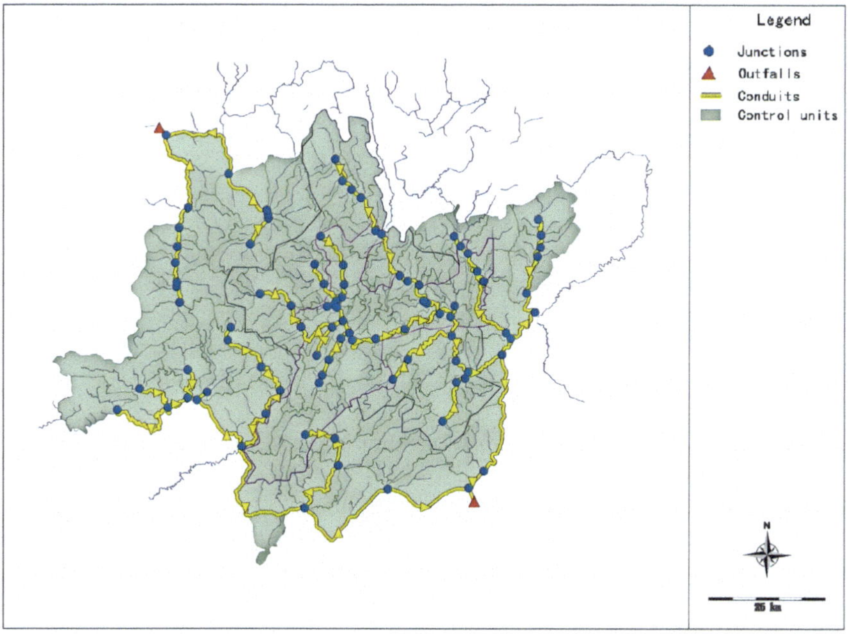

Fig. 2.1 Schematic diagram of model construction

(3) In the planning status assessment stage, the flow in the base year is used as the hydrological boundary condition, and in the water environment capacity calculation and target accessibility analysis stage, the minimum flow with 90% guarantee rate and (for three consecutive months) are used as the design hydrological conditions. At the same time, the rainfall data of Zunyi in 2018 is used as the rainfall boundary condition. According to statistics, the total rainfall in Zunyi City in 2018 was 1086 mm, with the maximum rainfall of the rainfall reaches 46.5 mm/h, mainly from April to October every year, and less in other months.

(4) The water environment pollution load and target water quality grade of the river section refer to the actual monitoring data and the water quality data collected in recent three years as the water quality boundary conditions. COD and NH_3-N are selected to compare the measured values with the simulated values to evaluate the accuracy of the model.

2.4 Model Validation

Through the construction of SWMM model, the matching degree between the simulated water quality of 17 river sections and the measured target water quality grade was compared. Among them, 13 sections, such as Ant River, Ximaqiao, Daqiuping and so on, the simulated value of the target water quality grade was consistent with the measured mean value, and the simulation matching degree reached 76.5%. See Table 2.1 for simulated and measured values of pollutant concentrations and corresponding target water quality grades. Nash efficiency coefficient (NSE) is used as the efficiency evaluation index of the SWMM model to determine the simulation effect. Daqiuping section is further selected as a typical section to evaluate the simulation effect. The comparison curve of COD and NH_3-N water quality in the reference year of Daqiuping section is shown in Fig. 2.2. The simulated average value of COD is 12.85 mg/L, and the simulated average value of NH_3-N is 0.66 mg/L; the measured mean values are 13.28 mg/L and 0.59 mg/L respectively. The simulated and measured values are class III water quality standards. According to the Nash coefficient calculation formula, the Nash coefficients of COD and NH_3-N in Daqiuping section are 0.78 and 0.72 respectively.

The closer NSE is to 1, the better the simulation effect. Generally, $NSE > 0.9$ means that the simulation effect is very good, $0.7 < NSE < 0.9$ means that the simulation effect is good, $0.5 < NSE < 0.7$ means that the simulation effect is reasonable, and $NSE < 0.5$ means that the simulation effect is poor. It can be seen that the model simulation results are reasonable and reliable, which can provide technical support for the simulation and prediction of pollution in the Xiangjiang River Basin in Zunyi City.

Table 2.1 Simulated and measured values of pollutants at each section

Serial number	Section name	Model value (mg/L)		Measured value (mg/L)		Target water quality grade		Match or not
		COD	NH_3-N	COD	NH_3-N	Measured	Model	
1	Ant River	52.24	3.90	26.00	8.47	Inferior V	Inferior V	Yes
2	Xima Bridge	9.88	0.81	14.83	0.55	III	III	Yes
3	Gaoni River	7.57	0.65	19.33	0.95	III	III	Yes
4	Gaoping River	9.07	0.82	15.00	0.09	III	III	Yes
5	Dayuanzi	6.15	0.78	7.00	0.09	II	III	No
6	Xinhongjiang Bridge	2.40	0.29	6.43	0.07	II	II	Yes
7	Sheung Shui	3.21	0.28	9.67	0.07	II	II	Yes
8	Yangzhutou ferry	16.54	2.01	24.14	2.63	Inferior V	Inferior V	Yes
9	Daqiuping	12.85	0.66	13.28	0.59	III	III	Yes
10	Dry base	14.01	2.06	28.75	6.79	Inferior V	Inferior V	Yes
11	Luojiang River	5.75	0.42	11.67	0.37	II	II	Yes
12	Lemin River	21.37	1.94	16.00	0.12	III	V	No
13	Yujian River Bozhou exit	5.70	0.79	15.00	0.15	III	III	Yes
14	Tuanxi River	5.10	0.66	16.00	0.10	III	III	Yes
15	Xingti Bridge	33.98	2.45	28.00	4.13	Inferior V	Inferior V	Yes
16	Tuanze	6.75	0.66	12.50	0.09	II	III	No
17	Yunmen store	3.97	0.51	8.25	0.07	II	III	No

2.5 Model Application

Due to the continuous increase of urban population and pollutant load, the water quality in Zunyi City has a gradual deterioration trend. In order to comprehensively improve the water environment quality in the urban area of Zunyi City and ensure the stability of the state controlled section within the planning scope to meet the requirements of class III water quality assessment, this topic simulates the water quality of daqiuping state controlled section through the model, and analyzes the water quality changes in 2025, 2030 and 2035, as shown in Fig. 2.3. It can be seen from the figure that the daqiuping section will deteriorate from the current class III water to class IV water in 2025 and 2030, and will deteriorate to class V water in 2035, mainly showing that NH_3-N exceeds the standard. The phenomenon of NH_3-N

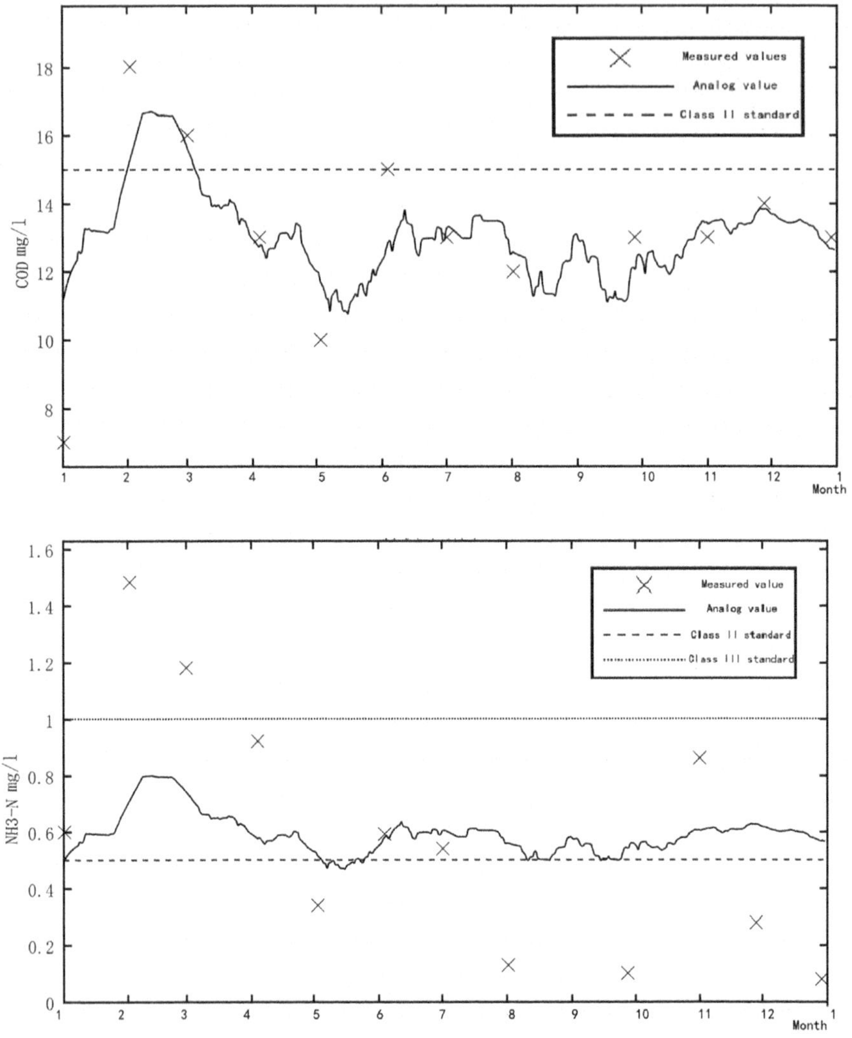

Fig. 2.2 Comparison of simulated and measured water quality values of Daqiuping section in 2018

exceeding the standard is serious from January to March 2035, and pollution control measures should be taken to improve the water quality.

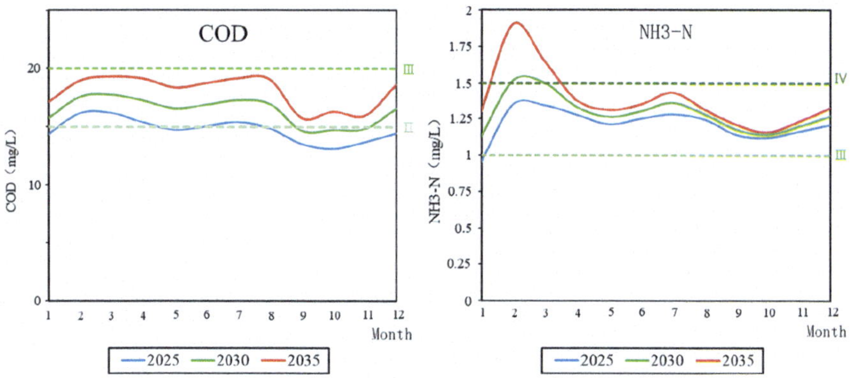

Fig. 2.3 Water quality prediction of Daqiuping section in the planning level year

2.6 Conclusions

Based on the above water quality simulation, the Daqiuping section is gradually experiencing water quality deterioration, and it is recommended to focus on controlling urban domestic pollution. Firstly, measures such as repairing the original main pipes and adding sewage interceptors will be taken to improve the sewage collection rate. Through model simulation, the sewage collection rate has been increased to 75%, and the water quality of the Daqiuping section can meet the Class III water quality standard.

References

1. Yu Y, Zhou Y, Guo Z, van Duin B, Zhang W (2022) A new LID spatial allocation optimization system at neighborhood scale: integrated SWMM with PICEA-g using MATLAB as the platform. Sci Total Environ 831:154843
2. Yang B, Zhang T, Li J, Feng P, Miao Y (2023) Optimal designs of LID based on LID experiments and SWMM for a small-scale community in Tianjin, north China. J Environ Manage 334:117442
3. Kang D, Sun J, Kuang S et al (2019) Research, application and development trend of storm flood management model (SWMM). Purif Water Technol 38(3):45–50
4. Duyingen, Hou S, Mahongli, et al (2021) Simulation study on spatial pattern optimization of lid facilities based on SWMM. China Water Supply and Drainage 37(19):120–125
5. Lisebee WA, Hathaway JM, Winston RJ (2022) Modeling bioretention hydrology: quantifying the performance of drainage urban and the SWMM lid module. J Hydrol 612:128179
6. Huang B, Zhu X, Wang S et al (2012) Simulation of combined drainage system in Chuanfang District of Kunming Based on SWMM. China Water Supply Drain 28(019):15–18
7. Dong X, Chen J, Zhao D (2006) Application of SWMM model in urban drainage system planning. Water Supply Drain 32(05):106–109
8. Zhang Q, Su B, Yuan J (2012) Simulation of SWMM rainfall runoff process in urban residential areas—a case study of Guidu Garden community in Yingkou City. J Beijing Normal University (Nat Sci Ed) 48(3):276–281

9. Ma X, Zhu Y, Mei K et al SWMM model is applied to the simulation calculation of non-point source pollution load in urban residential areas. Res Environ Sci 25(1):12–15
10. Nietiefeng (2012) Research on non-point source pollution load accounting technology of rainstorm runoff in Guangzhou. South China University of Technology, Guangzhou
11. Zhang Z, Zhu X (2018) Division of water environment control units and pollution prevention countermeasures in the Xiangjiang River Basin of Zunyi City. Environ Protect Sci Technol 24(03):43–46

Chapter 3
Research on Sewage System Issue Diagnosis in Districts Based on Water Quality Data Analysis

Tingting Xiao, Can Yu, Yutong Chen, Qin Zhang, and Ya Zhang

Abstract This study focuses on the sewage treatment plant in the central urban area of a specific city. Through the analysis of operational data from the sewage treatment plant and supplementary water quality sampling and testing data, the study systematically investigates the issues within the sewage system serving the plant. The results indicate that the sewage treatment plant faces challenges related to significant fluctuations in influent water flow and instances of extraneous water infiltration. Deficiencies in the construction of the sewage pipeline network within the service area are identified, leading to the entry of river water through discharge outlets, thereby reducing pollutant concentrations in the incoming sewage. The findings of this research hold positive significance for guiding improvements in the quality and efficiency of services within the sewage treatment plant's service area.

Keywords Wastewater treatment plant · Water quality characteristics · Data analysis

3.1 Introduction

Urban wastewater treatment plants play a crucial role in the development of cities, serving as essential infrastructure for urban water pollution control and environmental protection. With the introduction of China's "Dual Carbon" goals, there are new challenges for the sewage treatment industry to enhance influent concentrations at wastewater treatment plants and promote energy-saving and consumption reduction in sewage treatment systems. This study focuses on a wastewater treatment plant in the central urban area of a specific city. By analysing the influent characteristics at the end of the pipeline network and the variation in pollutant concentrations in the

T. Xiao (✉) · C. Yu · Y. Chen · Q. Zhang · Y. Zhang
Power China Guiyang Engineering Corporation Limited, Guiyang, China
e-mail: 2961513706@qq.com

Q. Zhang
Power China Ecological Environment Design Research Corporation Limited, Suzhou, China

K. Y. Foo et al. (eds.), *Water Nexus Approach for Sustainable Development*, Environmental Science and Engineering, https://doi.org/10.1007/978-3-031-89238-7_3

service area's sewage, the research aims to provide data support for the operation and management of the wastewater treatment plant. The findings are significant for improving the quality and efficiency of the sewage system in the designated urban area.

3.2 Project Background

3.2.1 Area Overview

The study area is located between 22°58′N to 23°34′N latitude and 112°46′E to 113°02′E longitude, situated at the northwest end of the Pearl River Delta in the central part of Guangdong Province, China. The research scope is shown in Fig. 3.1. It experiences a subtropical monsoon climate with indistinct seasonal variations. The average annual rainfall is 1696.7 mm, with the majority occurring from April to September, accounting for approximately 80% of the annual precipitation. The average annual relative humidity is 80%, reaching its highest monthly average of 86% in June.

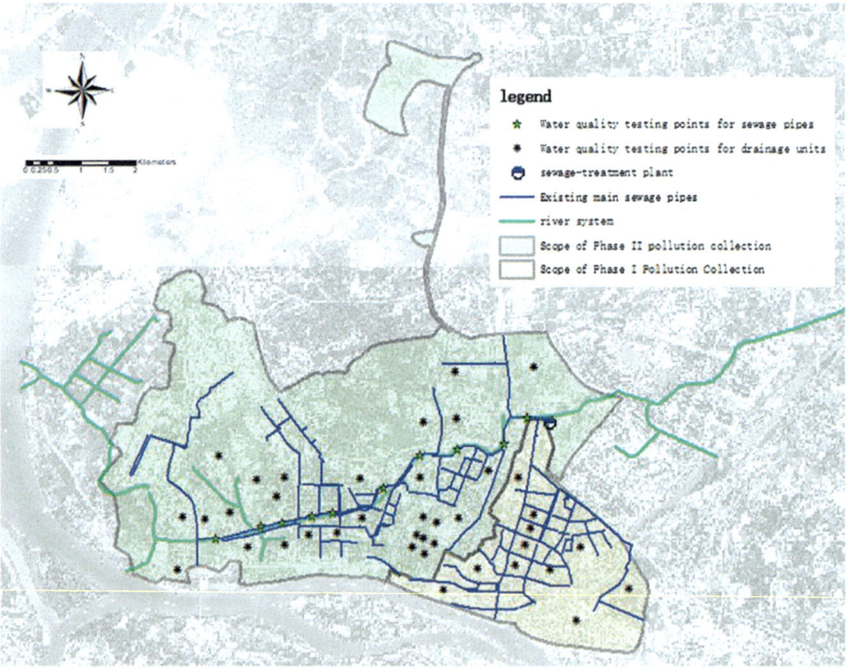

Fig. 3.1 Scope of study

3.2.2 *Introduction to the Wastewater System in the Study Area*

There is a single wastewater treatment plant within the study area, constructed and operationalized in two phases, each with a capacity of 50,000 m^3/day. The sewage treatment process involves a combination of CAST (Chemically Assisted Settling Tank) technology, high-efficiency sedimentation tanks, and denitrifying deep-bed filtration. Sodium hypochlorite is used for disinfection. The residual sludge is subjected to gravity concentration and belt-type mechanical dewatering, and then transported for external disposal with a moisture content of 80%. The treated effluent quality complies with the Class A standard of the "Discharge Standard of Pollutants for Urban Wastewater Treatment Plants" (GB18918-2002) and the more stringent Class A standard of the second phase of urban secondary wastewater treatment plants specified in the local standard "Discharge Limits of Water Pollutants" (DB44/26-2001) in Guangdong Province.

The primary inlet for the first-phase wastewater treatment plant is the Tongfu Road culvert, covering a pollution catchment area of approximately 7.41 km^2, with a total length of municipal pipelines reaching 101.5 km. For the second-phase wastewater treatment plant, the main inlet is the pollution interception pipe along the river Creek, covering a pollution catchment area of about 29.80 km^2, with a total length of municipal pipelines reaching 167.2 km. The existing drainage system follows a combined sewer system, constituting 64% of the total, with a few areas adopting a separate storm water and wastewater system.

3.3 Identification and Diagnosis of Wastewater System Issues

3.3.1 *Research Method*

This study focuses on the service area of a wastewater treatment plant in a specific region. The research utilizes data from the entire year of influent wastewater at the treatment plant (from September 2022 to September 2023) and water quality sampling and testing data from accredited third-party testing agencies at the discharge outlets and municipal trunk sewer points. The analysis aims to provide insights and reference points for improving the quality and efficiency of the wastewater system within the service area of the sewage treatment plant.

3.3.2 Analysis of Influent Conditions in the Wastewater Treatment Plant

Figure 3.2 illustrates the influent water quantity and pollutant concentrations in the wastewater treatment plant. From the graph, it is evident that both the first and second phases of the wastewater treatment plant experience unstable influent water volumes with significant fluctuations. The BOD_5 concentration in the influent remained relatively high from December 2022 to April 2023, though it still fell short of the national policy requirement of 100 mg/L. From April 2023 to September 2023, there is a decrease in influent pollutant concentrations, likely attributed to increased rainfall during this period. Issues in the combined sewer system led to rainwater entering the wastewater treatment plant, resulting in a reduction in pollutant concentrations.

The results of statistical analysis of the annual intake nutrient ratio data of the sewage treatment plant are shown in Table 3.1. From the data in the table, it can be seen that:

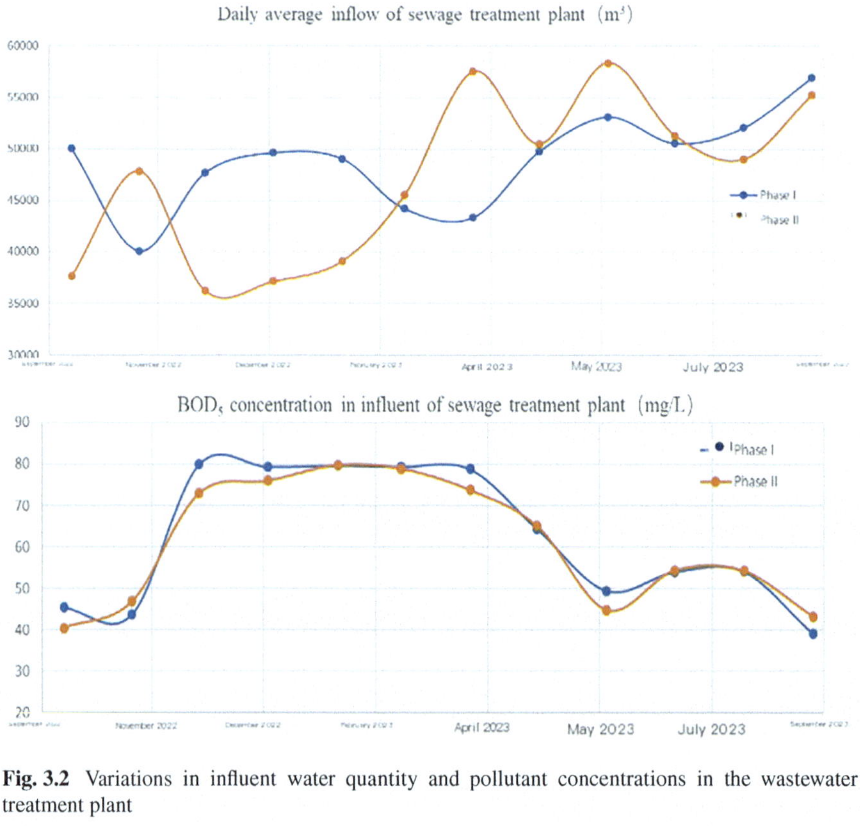

Fig. 3.2 Variations in influent water quantity and pollutant concentrations in the wastewater treatment plant

Table 3.1 Statistics of the proportion of influent pollutants in the sewage treatment plant

Online monitoring data	BOD_5/COD_{cr}		BOD_5/TN		BOD_5/TP		SS/BOD_5	
	Phase I	Phase II	Phase I	Phase II	Phase I	Phase II	Phase I	Phase II
October 2022	0.57	0.52	2.04	1.78	20.36	19.15	1.10	1.38
November 2022	0.48	0.49	1.89	1.90	18.33	20.80	1.67	1.53
December 2022	0.64	0.62	3.12	2.78	37.34	33.64	0.93	1.01
January 2023	0.54	0.58	3.41	3.00	34.21	32.34	1.07	0.98
February 2023	0.53	0.55	3.20	2.86	30.52	29.46	1.34	1.14
March 2023	0.49	0.48	3.20	2.95	25.70	27.33	1.18	1.27
April 2023	0.45	0.50	3.31	3.15	28.87	34.50	1.20	1.25
May 2023	0.56	0.58	2.87	2.83	33.02	33.24	1.12	1.19
June 2023	0.60	0.53	2.72	2.50	27.64	23.86	1.21	1.54
July 2023	0.45	0.58	2.98	2.88	25.59	27.47	0.94	1.08
August 2023	0.52	0.50	2.85	2.66	26.14	26.62	1.11	1.25
September 2023	0.55	0.61	2.80	3.29	26.85	31.22	1.05	0.94
Maximum	0.64	0.62	3.41	3.29	37.34	34.50	1.67	1.54
Mean value	0.53	0.55	2.87	2.71	27.88	28.30	1.16	1.21
Minimum value	0.45	0.48	1.89	1.78	18.33	19.15	0.93	0.94

(1) The BOD_5/COD_{cr} value is between 0.45 and 0.64. Generally, BOD_5/COD_{cr} is used to characterize the biochemical performance of sewage [1]. $BOD_5/COD_{cr} \leq 0.3$ means that sewage is difficult to be treated by biochemical treatment, $0.3 < BOD_5/COD_{cr} \leq 0.5$ means that sewage can be treated by biochemical treatment, and $BOD_5/COD_{cr} > 0.5$ means that sewage is easy to be treated by biochemical treatment [2, 3]. It can be seen that the influent quality of the sewage treatment plant has good biodegradability.

(2) The BOD_5/TN value is between 1.78 and 3.41, and the BOD_5/TP value is between 18.33 and 37.34. According to the denitrification equation, 2.86 g BOD_5 is required for every 1 g of nitrate nitrogen denitrification [4]. In order to ensure the effect of nitrogen and phosphorus removal, the general process requirements for nitrogen and phosphorus removal are $BOD_5/TN > 4$ and $BOD_5/TP > 17$. Through statistical analysis, the problem of low BOD_5/TN exists in the sewage treatment plant, which requires process regulation, and even adding carbon source to ensure the removal effect of TN when necessary.

(3) The value of SS/BOD_5 is between 0.93 and 1.21. Some literatures show that about 65% of CODcr in urban domestic sewage exists in the form of SS [5]. It can be seen that there is a certain proportion between the influent SS and the influent organic matter. SS/BOD_5 can characterize the effect of influent suspended solids on sludge yield and the proportion of active components in sludge. The higher the proportion of influent SS/BOD_5, the lower the proportion of volatile sludge concentration (MLVSS) in activated sludge. When SS/ BOD_5, the denitrification effect is reduced, and the utilization efficiency of particulate

matter in anoxic and anoxic zones is reduced [6–8]. A large number of particulate carbon sources enter the aerobic section and are consumed in the aerobic section through a relatively long residence time, resulting in poor nitrogen and phosphorus removal effects. At the same time, the active components in the activated sludge are low. In order to ensure the full nitrification reaction, it is necessary to increase the sludge concentration, which will further lead to sludge aging and even sludge running from the secondary sedimentation tank, affect the subsequent treatment, and even cause the effluent SS to exceed the standard.

3.3.3 Characteristics of Pollutants in Effluents from Different Typical Residential Areas

Figure 3.3 displays the water quality sampling results from various types of drainage units within the study area. The pollutant concentrations are highest in the wastewater from commercial areas, where pollutant levels are notably elevated. Public utility units, often employing separate storm water and wastewater systems, exhibit relatively higher pollutant concentrations. In shantytown areas and certain old neighbourhoods with outdated sewage systems, issues such as misconnections between storm water and wastewater, as well as suspected instances of "clean water" infiltrating sewage pipes, are observed [9, 10]. As a result, the effluent from these areas shows lower pollutant concentrations at the discharge points.

By comparing Fig. 3.3 with Fig. 3.2, it can be observed that the wastewater from the drainage unit outlets undergoes a significant decrease in pollutant concentrations as it enters the wastewater treatment plant through the sewer network. Considering the potential self-purification effect of pollutants during wastewater transport, there is suspicion that the decrease in pollutant concentrations is attributable to the infiltration of "external water" into the sewage system, leading to the observed decline in pollutant concentrations in the wastewater.

3.3.4 Analysis of the Impact of River Water Levels on Pollutant Concentrations in Wastewater Pipeline Networks

In the region where the main sewer pipeline is closely adjacent to the river, to substantiate the hypothesis of "external water infiltration," several monitoring points were strategically placed along the main pipeline. Simultaneously, technical measures were employed to artificially lower the river water level, ensuring that the river's water level was lower than the depth of the buried sewer pipeline. As shown in Table 3.2, the water quality results from the monitoring points before and after the reduction in river water levels are presented in the table. The results indicate that the concentrations of pollutants in the wastewater within the main sewer pipeline are significantly

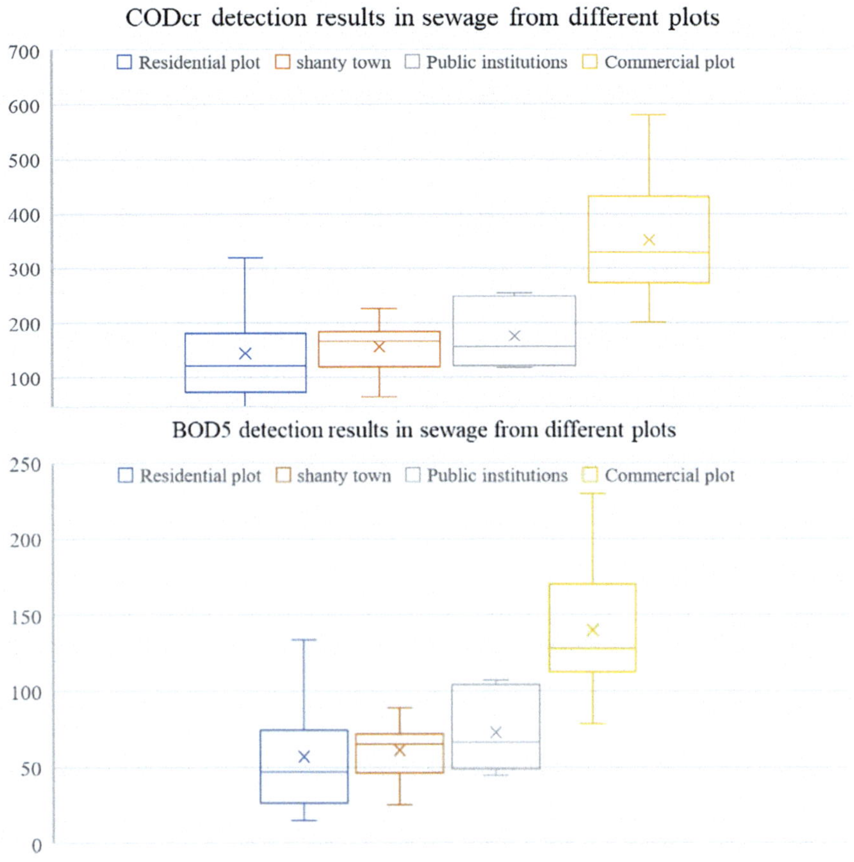

Fig. 3.3 Effluent concentrations of wastewater from different types of drainage units

influenced by variations in the river's water level. After lowering the river water level, there is an observable increase in pollutant concentrations at the monitoring points, confirming the existence of substantial "external water infiltration" in the region.

Table 3.2 Impact analysis of river water level on water quality of sewage main pipe

Serial number	Average concentration of CODcr (mg/L)			Variation
	Before the water level drops	During the process of water level decrease	After the water level drops	
1	60.69	64.56	55.49	– 14.05
2	38.81	51.75	66.96	29.39
3	47.31	59.87	54.26	– 9.37
4	63.11	99.07	78.54	– 20.72
5	65.14	80.27	71.68	– 10.7
6	63.35	82.72	72.54	– 12.31
7	84.37	130.39	155.37	19.16
8	58.25	66.16	57.51	– 13.07
9	48.85	66.97	61.07	– 8.81
10	60.24	83.98	72.97	– 13.12

3.4 Conclusion

This paper conducts a diagnostic analysis of a sewage system in a specific area based on water quality data and presents the following observations:

(1) The sewage treatment plant is capable of meeting the treatment demands for domestic wastewater within the pollution catchment area. The actual influent concentration, lower during periods of higher influent volume, coincides predominantly with the spring and summer seasons. The main inlet channel, along the river for intercepting pollution, is significantly influenced by variations in river water levels, leading to the infiltration of river water into the wastewater treatment plant.

(2) The urban sewage collection network still exhibits combined sewer overflow, with instances of misconnections and cross-connections in the rainwater and sewage drainage pipes. Factors such as local backflow from river water levels and the mixing of rainwater contribute to a significant decrease in pollutant concentrations from the discharge outlets to the entrance of the wastewater treatment plant.

(3) Due to historical reasons, the urban water environment management in the study area lacks systematic top-level planning and design, making it challenging to meet the requirements of sustainable urban development. The governance of the urban water environment lacks scientific principles, particularly in enhancing the quality and efficiency of the urban sewage system and addressing deficiencies in the pipeline network. Strengthening efforts in upgrading the urban sewage system and improving the pipeline network is crucial, and effective comprehensive measures should be implemented to address the identified issues in urban water environment management.

Acknowledgements This article is supported by the research project "Key Technologies for the Digital Application of the Whole Process of Urban Drainage System Existing Pipeline Network Restoration and Reconstruction (YJ2022-10)," funded by Power China Guiyang Engineering Corporation Limited.

References

1. Zou LX, Li HB, Zheng KK (2019) Analysis on the characteristics of influent water quality from wastewater treatment plants in Taihu Basin. Water Wastewater Eng 45(7):39–45
2. Wei YY, Zhang J, Tang JG (2022) Analysis of key points in technical specification for investigation and elimination of illicit discharges of municipal drainage pipes. Water Wastewater Eng 48(8):123–127
3. Sham JFC, Lai WWL, Chan W et al (2019) Imaging and condition diagnosis of underground sewer liners via active and passive infrared thermography: a case study in Singapore. Tunnell Underground Space Technol 84:440–450
4. Dirckx G, Fenu A, Wambecq T et al (2019) Dilution of sewage: is it, after all, really worth the bother. J Hydrol 571:437–447
5. Cory RM, Mcknight DM (2005) Fluorescence spectroscopy reveals ubiquitous presence of oxidized and reduced quinones in dissolved organic matter. Environ Sci Technol 39(21):8142–8149
6. Chen W, Westerhoff P, Leenheer JA (2003) Fluorescence excitation-emission matrix regional integration to quantify spectra for dissolved organic matter. Environ Sci Technol 37(24):5701–5705
7. Maqbool T, Sun MM, Chen L (2022) Exploring the fate of dissolved organic matter at the molecular level in the reactive electrochemical ceramic membrane system using fluorescence spectroscopy and FT-ICR MS. Water Res 210:117979
8. Phanwilai S, Noophan P, Li CW (2020) Effect of COD:N ratio on biological nitrogen removal using full-scale step-feed in municipal wastewater treatment plants. Sustain Environ Res 30(1):24
9. Zhang T, Ma H, Hong ZC (2022) Photo-reactivity and photo-transformation of algal dissolved organic matter unraveled by optical spectroscopy and high-resolution mass spectrometry analysis. Environ Sci Technol 56(18):13439–13448
10. Wen L, Yang F, Li X (2023) Composition of dissolved organic matter (DOM) in wastewater treatment plants influent affects the efficiency of carbon and nitrogen removal. Sci Total Environ 857:159541

Chapter 4
Study on Characteristics of Stormwater Runoff Pollution and Pollution Load Prediction in Residential Areas of Guangzhou

Yue Wu, Xixiang Yue, Jingsong Wang, Sili Chen, Wenli Zheng, and Sha Chang

Abstract Taking a residential area in Guangzhou as the research object, rainwater runoff from roof, main road and green area was monitored and sampled during rainfall to analyze the characteristics of pollutant concentration changes over time and the correlation between different pollutants in the study area. Based on the mean event mean concentration (EMC) of pollutants as the evaluation standard, a three-layer BP neural network model is constructed, which can predict the pollution load in stormwater runoff of residential areas in Guangzhou after training, calculation and testing. The results show that the pollution degree of stormwater runoff in the main road of the residential area is relatively serious. With the increase of rainfall, the peak time of the mass concentration of various pollutants is advanced and the value is increased, and the EMC value of pollutants in stormwater runoff generally presents a trend of first increasing and then decreasing. The chemical oxygen demand (COD) and total phosphorus (TP), total nitrogen (TN) and suspended solids (SS) in different types of subsurface runoff showed strong correlation. After training with the cross-validation method, the average relative error of the BP neural network model for predicting the EMC value of different pollutants in the test set is < 10%. Therefore, it can be considered that the model has certain reference value for predicting the stormwater runoff pollution load in residential areas of Guangzhou.

Y. Wu · X. Yue · J. Wang (✉)
College of Civil Engineering, University of South China, Hengyang, China
e-mail: xhwjs@163.com

S. Chen · W. Zheng · S. Chang
South China Institute of Environmental Science, Ministry of Ecology and Environment, Guangzhou, China
e-mail: chensili@scies.org

W. Zheng
e-mail: zhengwenli@scies.org

S. Chang
e-mail: changsha@scies.org

K. Y. Foo et al. (eds.), *Water Nexus Approach for Sustainable Development*, Environmental Science and Engineering, https://doi.org/10.1007/978-3-031-89238-7_4

Keywords Stormwater runoff · Pollution characteristics · Event mean concentration (EMC) · BP neural network · Cross validation · EMC prediction

4.1 Introduction

With the advancement of urban construction in China, the proportion of impervious surface area of urban underlying surface has increased rapidly [1]. A large amount of rainwater cannot penetrate through the soil, which enhances the erosion effect of pollutants accumulated on the underlying surface of the city, resulting in urban rainwater runoff pollution becoming one of the sources of water pollution [2–5]. In recent years, domestic and foreign scholars have carried out a large number of investigations and studies on the pollution of rainwater runoff on different underlying surfaces in different cities [6–9]. The results show that the water quality of rainwater runoff on different types of underlying surfaces is affected by many different factors, such as local traffic conditions and atmospheric deposition. Therefore, it is very important to clarify the influence mechanism of different factors on rainwater runoff quality, which can provide a practical and reliable theoretical basis for local rainwater runoff pollution control.

4.2 Study Area Profile

Guangzhou is located in the subtropical coast. It is warm and rainy, with sufficient light and heat, long summer and short frost period. It belongs to the subtropical marine monsoon climate. The average annual rainfall in the urban area is about 1720 mm, and the average relative humidity is 77%. The study area selected a residential area in Guangzhou, with a total area of about 37.14 hm^2. The area of roof, main road and green space is about 4.97 hm^2, 5.21 hm^2 and 23.47 hm^2 respectively. By obtaining the elevation data of the region and analyzing the aspect and slope in ArcGIS, it is found that the overall terrain of the region is relatively flat, with an average slope of 0.785%.

4.3 Monitoring and Sampling Program

In this study, a total of 8 rainfalls were monitored in 2022. Due to the interruption or early disappearance of runoff during the rainfall process, 5 sets of valid data were obtained. The rainfall data recorded synchronously by the tipping bucket rain gauge and the environmental monitoring system during the study period are shown in Table 4.1.

Table 4.1 5 groups of effective rainfall data obtained during the monitoring period

Serial number	Rainfall date	Rainfall (mm)	Rainfall duration (min)	Previous dry days (d)
1	2022-07-30	4.5	250	0
2	2022-08-09	11.6	310	3
3	2022-09-09	25.3	360	3
4	2022-09-18	6.3	200	6
5	2022-10-19	9.2	300	4

This paper mainly focuses on the sampling of rainwater runoff on different underlying surfaces in the study area. Before the start of rainfall, the sampling containers were placed in the roof rain pipe mouth, the road rainwater catchment mouth, and the catchment pit at the lower terrain of the lawn. When the runoff began to form, it was collected every 5 min in the first 30 min, every 10 min from 30 to 60 min, and every half an hour after 60 min. In general, 15 water samples (240 min) were collected or the runoff disappeared. The collected water samples were continuous water samples of the same rainfall. After the sample collection is completed, it is immediately loaded into a clean and dry ethylene plastic bottle and sent to the laboratory for testing and analysis in time. The roof rainwater runoff sampling points are located in the single villa area of building 11 and the high-rise residential area of building 16. The main road rainwater runoff sampling points are located in the west branch of the community and the entrance of Maoshan Avenue on the east side of the community. The green space rainwater runoff sampling points are all located in the green area of the community. The specific location distribution of the sampling points is shown in Fig. 4.1.

The underlying surface of the sample is divided into three categories: roof, main road and green space. The analysis indicators include chemical oxygen demand (COD), total phosphorus (TP), total nitrogen (TN) and solid suspended matter (SS). The mass concentration of COD in the sample was determined by dichromate method (HJ828-2017), the mass concentration of TP was determined by ammonium molybdate spectrophotometry (HJ670-2013), the mass concentration of TN was determined by potassium persulfate ultraviolet spectrophotometry (HJ636-2012), and the mass concentration of SS was determined by gravimetric method (GB11901-89) [10].

4.4 Analysis of Pollution Characteristics of Stormwater Runoff on Different Underlying Surfaces

In the five rainfall events monitored in 2022, the peak values of COD, TP, TN and SS mass concentrations all appeared in the rainwater runoff of the main road on September 9 (rainfall of 25.3 mm), and their specific values were 492.71 mg/L, 1.84 mg/L, 28.12 mg/L and 1386.85 mg/L, respectively. The pollution degree of rainwater runoff in the main road is obviously higher than that in the roof and green

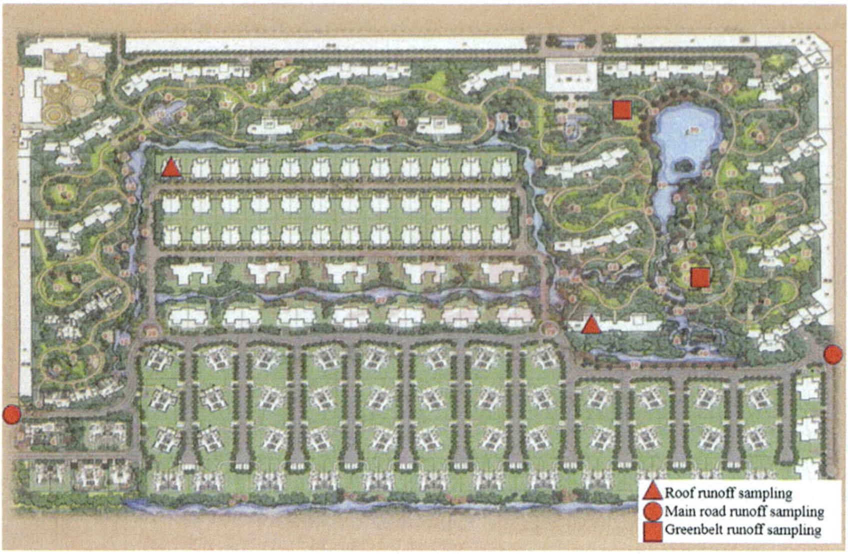

Fig. 4.1 Distribution of sampling points in the study area

space. It is speculated that the possible reason is that the factors such as traffic flow, road construction and municipal greening and fertilization produce a large amount of pollutant accumulation. It may also be that there are more restaurants and other shops near the main road, and leisure and entertainment activities are more concentrated, resulting in a large accumulation of pollutants on the ground. The accumulation of pollutants on the roof will be affected by factors such as atmospheric dry and wet deposition and roof asphalt shedding, and the mass concentration of TN is relatively high, which may also be related to the frequent occurrence of lightning in the rainy season. Compared with the roof, the accumulation of pollutants on the green space is affected by factors such as landscaping and conservation work in addition to the role of atmospheric dry and wet deposition. Sometimes the residual phosphorus-containing compounds in plant fertilizers may lead to an increase in the mass concentration of TP in the rainwater runoff of green space. In addition, it can be seen from the change data of pollutant mass concentration obtained by monitoring 5 rainfall events that with the increase of rainfall, the peak value of mass concentration of various pollutants appears earlier and the value increases. The main reason is that the increase of rainfall intensity leads to the enhancement of erosion ability and scouring effect of rainwater, and the mass concentration of pollutants contained in rainwater will increase accordingly [11].

4.4.1 Correlation Analysis of Pollutant Concentration in Stormwater Runoff from Different Underlying Surfaces

In order to explore the correlation of pollutant concentration in rainwater runoff of different underlying surfaces, taking the typical rainfall event on August 9 as an example, the change of pollutant concentration in rainwater runoff of different underlying surfaces is shown in Fig. 4.2. Pearson correlation analysis was performed on the mass concentrations of COD, TP, TN and SS in rainwater runoff from three underlying surfaces of roof, main road and green space under the same rainfall duration using SPSS software. The correlation coefficient r value and the significant level p value obtained are shown in Table 4.2.

It can be seen from Table 4.2 that the Pearson correlation coefficient r value of COD and TP in roof rainwater runoff is 0.816, and the p value is < 0.001, showing a significant strong correlation, indicating that COD and TP in roof rainwater runoff may have good homology, which is consistent with the research conclusion of Xu et al. [12]. It is speculated that the main source of COD and TP in roof rainwater runoff is atmospheric dry and wet deposition. The Pearson correlation coefficient r value of COD and TP in green rainwater runoff was 0.672, showing a strong correlation (0.6–0.8), and the significant level p value was < 0.01. In addition, in the rainwater runoff of three different underlying surfaces of roof, main road and green space, the Pearson correlation coefficient r values of TN and SS were 0.632, 0.763 and 0.635, respectively, showing a strong correlation, and the significant level p value was $<$ 0.05. The TN and SS showed similar changes in the rainwater runoff of three different underlying surfaces, which indicated that most of the nitrogen-containing substances in the study area may be attached to the presence of solid suspended particles.

4.4.2 Calculation of Mean Runoff Concentration (EMC) for Rainfall Events

Due to the randomness of each rainfall event, the mass concentration of pollutants in rainwater runoff cannot be determined by factors such as underlying surface type and rainfall intensity. Therefore, the event mean concentration (EMC) of rainfall event runoff can be used to more intuitively evaluate the pollution degree of rainwater runoff in a single rainfall event [13, 14]. The calculation formula of EMC (mg/L) is as follows:

$$\mathrm{EMC} = \frac{M}{V} = \frac{\int_0^T C_t Q_t dt}{\int_0^T Q_t dt} \cong \frac{\sum C_t Q_t \Delta t}{\sum Q_t \Delta t} \tag{4.1}$$

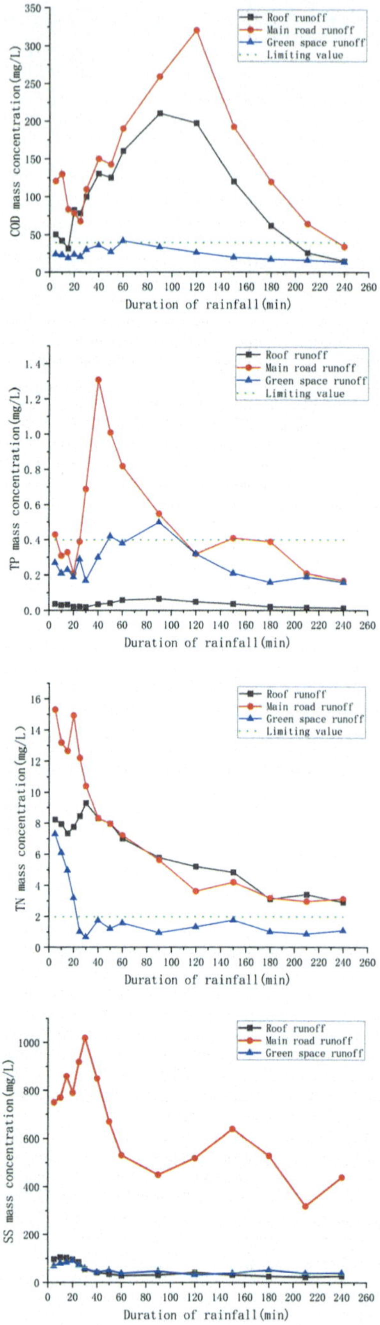

Fig. 4.2 Variation of pollutant mass concentration in stormwater runoff at different underlying surfaces during rainfall event on August 9

Table 4.2 Pearson correlation coefficient and significant level of different pollution in stormwater runoff

			COD	TP	TN	SS
Roof	COD	r	1			
		p				
	TP	r	**0.816**	1		
		p	**0.000**			
	TN	r	0.142	0.152	1	
		p	0.613	0.588		
	SS	r	− 0.381	− 0.179	**0.632**	1
		p	0.161	0.522	**0.012**	
Main road	COD	r	1			
		p				
	TP	r	0.247	1		
		p	0.375			
	TN	r	− 0.323	0.013	1	
		p	0.241	0.962		
	SS	r	− 0.255	0.285	**0.763**	1
		p	0.350	0.303	**0.001**	
Green space	COD	r	1			
		p				
	TP	r	**0.672**	1		
		p	**0.006**			
	TN	r	− 0.116	− 0.154	1	
		p	0.680	0.583		
	SS	r	− 0.199	− 0.245	**0.635**	1
		p	0.476	0.379	**0.011**	

The bold values indicate that the degree of correlation between the two pollutants has reached a certain threshold, suggesting that there is a certain correlation between the two pollutants

In the formula, M is the total mass of pollutants in a single rainfall process, mg; V is the total runoff generated by a single rainfall process, L; t is the total runoff time, min; C_t is the instantaneous mass concentration of pollutants in rainwater runoff at time t, mg/L; Q_t is the runoff at time t, L/min; Δt is the interval time of each sampling, min. Q_t can be estimated by the runoff coefficient of different underlying surfaces and the cumulative rainfall at time t. According to the *Outdoor Drainage Design Code GB50014-2021* [15], the runoff coefficients of roof, road and green space are 0.9,0.85 and 0.15, respectively. The calculation results of EMC values of 3 different types of subsurface stormwater runoff pollutants in 5 rainfall events monitored in 2022 during the study period are shown in Table 4.3.

Table 4.3 EMC values of stormwater runoff of 3 different underlying surfaces for 5 rainfall events (mg/L)

Rainfall event (amount of rainfall)	Underlying surface type	Value of pollution EMC			
		COD	TP	TN	SS
07-30 (4.5 mm)	Roof	17.78	0.023	2.2	35.42
	Main road	31.41	0.078	3.18	72.45
	Green space	10.89	0.212	0.68	52.01
08-09 (11.6 mm)	Roof	85.44	0.028	5.28	36.52
	Main road	180.71	0.488	5.52	565.00
	Green space	24.62	0.250	1.93	49.82
09-09 (25.3 mm)	Roof	70.94	0.020	4.18	24.91
	Main road	122.74	0.333	4.57	316.81
	Green space	15.71	0.170	1.64	36.52
09-18 (6.3 mm)	Roof	54.81	0.026	4.73	36.94
	Main road	101.35	0.536	6.70	478.22
	Green space	20.15	0.195	2.38	58.88
10-19 (9.2 mm)	Roof	56.37	0.027	5.58	37.64
	Main road	150.88	0.562	6.21	544.99
	Green space	22.97	0.279	2.21	47.85
Mean value	Roof	57.07	0.025	4.39	34.29
	Main road	117.42	0.399	5.24	395.49
	Green space	18.89	0.221	1.77	49.02

It can be seen from Table 4.3 that with the increase of rainfall, the EMC value of pollutants in rainwater runoff of each underlying surface generally shows a trend of rising first and then falling. This is because the increase of rainfall will lead to the increase of rainfall intensity. When the rainfall intensity increases to a certain extent, the pollutants stored on the surface are close to the completion of transportation, and the mass concentration of pollutants in rainwater runoff will decrease with the increase of runoff [16].

According to the EMC mean value, among the four pollutants in the rainwater runoff of three different underlying surfaces in the residential area of Guangzhou, the mass concentration of COD and TN showed the law of main road > roof > green space, while the mass concentration of TP and SS showed the law of main road > green space > roof, and the mass concentration of pollutants in the rainwater runoff of main road was much higher than that of roof and green space. The reason why the mass concentration of TP in green rainwater runoff is higher than that in roof rainwater runoff may be that the residual phosphorus-containing compounds of plant fertilizer in the process of greening maintenance lead to the increase of TP mass concentration in runoff after rainwater scouring. The reason why the mass concentration of SS in rainwater runoff of green space is higher than that of roof rainwater runoff may be that during the process of rainwater erosion, the green space has a strong permeability to rainwater, and a large amount of rainwater contains soluble pollutants to penetrate into the soil, while SS is difficult to penetrate, resulting in an increase in mass concentration.

On the whole, the average EMC of SS in the rainwater runoff of the three underlying surfaces in the residential area of Guangzhou is higher than the secondary standard (30 mg/L) in the *Discharge Standard of Pollutants for Municipal Wastewater Treatment Plant* (GB18918-2002). The EMC mean values of COD and TN in roof and main road rainwater runoff were higher than Class V standard limits of *Surface Water Environmental Quality Standard* (GB3838-2002) except TP and COD and TN in green space rainwater runoff. It can be seen that the pollution of rainwater runoff in the residential area is serious, and it is urgent to take corresponding measures to control it.

4.5 Evaluation and Prediction of Pollution Load by BP Neural Network Model

4.5.1 BP Neural Network Hierarchical Structure Construction

The six influencing factors of runoff coefficient, confluence area (m^2), rainfall (mm), rainfall duration (min), early dry period (d) and cumulative rate constant of the underlying surface were used as the input units of the neural network [17], that is, the number of neurons in the input layer was 6. The EMC values of COD, TP, TN and SS in rainwater runoff are taken as the output units of the neural network, that is, the number of neurons in the output layer is 4. The initial number of nodes of the hidden layer neurons is 2/3 of the number of neurons in the input layer plus 2/3 of the number of neurons in the output layer, and the initial number of hidden layer neurons is determined to be 6 by the trial algorithm [18].

The neural network training function selects the trainlm function, which is based on the L-M algorithm and has a good performance in the training process of medium-sized neural networks. The activation function selects the nonlinear continuously differentiable sigmoid function, which includes the bipolar S function tansig (hyperbolic tangent function) and the unipolar S function logsig (S-type logarithmic function) as the transfer functions of the hidden layer and the output layer, respectively. The topology of the neural network model is shown in Fig. 4.3.

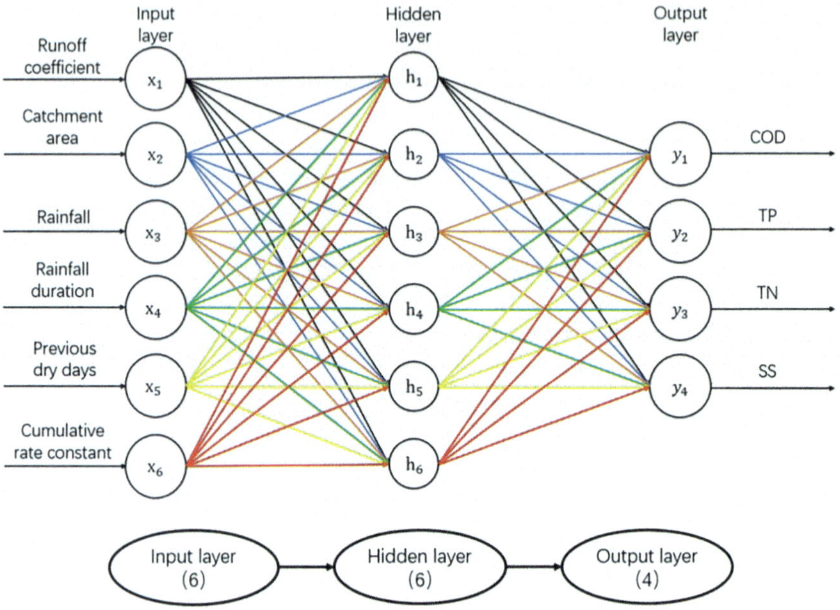

Fig. 4.3 BP neural network topology structure diagram

4.5.2 Model Training Calculation and Test Evaluation

In order to improve the accuracy of the training results, in addition to the measured data, a total of 100 sets of sample data were collected through the data disclosure platform of Guangzhou Ecological Environment Bureau and related research literature [9, 19–23] in the vicinity. Among them, 80 groups of sample data are input into the neural network as a training set for training, and the other 20 groups are used as a test set to evaluate the generalization ability of the neural network model. In the training process, combined with the K-folder Cross Validation (K-CV) method, the training set was randomly divided into 4 parts (20 sets of sample data per part). One of the 4 parts was taken as the verification set, and the remaining 3 parts were used as the training set to obtain 4 models respectively. Since 3.1 demonstrates that there may be a correlation between different pollutants, the error results of each model can only be averaged for analysis, and the model with the smallest average error is selected and tested on the test set. In this way, the model is verified for each sample data, and the prediction results are more objective. The schematic diagram of the fourfold cross-validation method used in this paper is shown in Fig. 4.4; the cross-validation results of the neural network model are shown in Table 4.4.

The cross-validation results of the neural network model show that the overall average relative error of Model 2 is 11.50%, which is the lowest among the four models. Therefore, Model 2 is selected for testing on the test set. Taking COD as an example, according to the research results of the cumulative rate of pollutants on

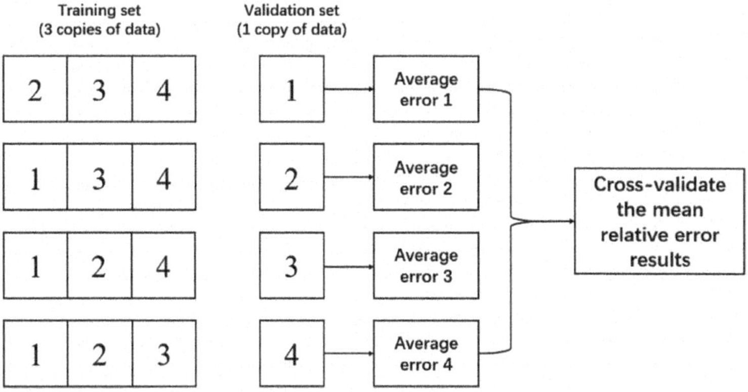

Fig. 4.4. 4 fold cross validation diagram

Table 4.4 Results of cross-validation of neural network models

Model number	COD average error (%)	TP average error (%)	TN average error (%)	SS average error (%)	The overall average relative error model (%)
1	13.58	15.58	10.14	10.74	12.51
2	8.52	10.26	13.69	13.51	11.50
3	12.46	16.61	9.93	14.28	13.32
4	11.83	11.47	15.94	10.06	12.33

different underlying surfaces in the relevant literature [16, 24], the cumulative rate constants of roofs, main roads and green spaces in the study area were set to 0.5, 0.8 and 0.2, respectively.

According to the test results of the neural network model, after training, the average relative errors of the neural network model for predicting the EMC values of COD, TP, TN and SS in residential areas of Guangzhou are 7.47%, 5.91%, 6.29% and 9.75%, respectively, which are all < 10%. The calculation results show that it is feasible to use the BP neural network model to calculate and predict the pollution load in residential stormwater runoff in Guangzhou.

4.5.3 Prediction of Pollution Load by BP Neural Network Model Under Large-Scale Rainfall Scenarios

Due to the low frequency of large-scale rainfall scenarios, usually once every two years, once every five years or even once every ten years, data on the concentration of pollutants in rainwater runoff during large-scale rainfall is scarce. In order to explore the characteristics of pollution load in rainwater runoff of different underlying surfaces under large-scale rainfall scenarios, BP neural network model can be used to predict the EMC value of pollutants under different scenarios.

According to the rainfall standard determined by the rainstorm yellow warning signal and the rainstorm orange warning signal, the rainfall was designed to be 25 mm and 50 mm respectively, the rainfall duration was 180 min, and the early dry period was 3 days. The three impact factors contained in different underlying surfaces and the three impact factors contained in rainfall scenarios are input into the neural network model. The predicted results and errors are shown in Fig. 4.5.

According to the prediction results of the neural network model, under the designed two large-scale rainfall scenarios, the rainfall duration decreases, the rainfall increases, and the rainfall intensity increases, resulting in an increase in the initial erosion effect of the rainfall. The transferable amount of pollutants stored on the surface is close to the limit and the transfer rate of rainwater runoff is accelerated. The time for pollutants in rainwater runoff to maintain high concentrations is reduced. The EMC values of pollutants in rainwater runoff on different underlying surfaces are significantly lower than before, and the range is larger. It can be seen that the pollution degree of rainwater runoff in the whole rainfall event under the heavy rainfall scenario is lower than that under the moderate rainfall scenario, but the pollution degree of rainwater runoff in the main road is still the most serious among the three underlying surfaces.

4.6 Conclusion

In this study, a residential area in Guangzhou was taken as the research object. Through the actual monitoring and sampling of rainwater runoff on different underlying surfaces in the study area, the pollution characteristics of rainwater runoff on different underlying surfaces were analyzed. Taking the EMC value of pollutants in rainwater runoff as the evaluation standard, a BP neural network model that can predict the pollution load of rainwater runoff in the study area was constructed. The following conclusions are drawn:

(1) The pollution degree of the main road runoff in this residential area in Guangzhou is obviously higher than that of the roof runoff and green runoff. TN and SS have strong correlation in the runoff of each underlying surface in the study area. There is a strong correlation between COD and TP in roof and green runoff.
(2) With the increase of rainfall, the EMC value of pollutants in rainwater runoff of each underlying surface in the residential area of Guangzhou showed a trend of rising first and then falling. According to the EMC mean value, more than half of the rainwater runoff pollution indexes of different underlying surfaces in the residential area of Guangzhou are at a high level compared with the water quality standards stipulated in the relevant documents, and it is urgent to take corresponding measures to control them.

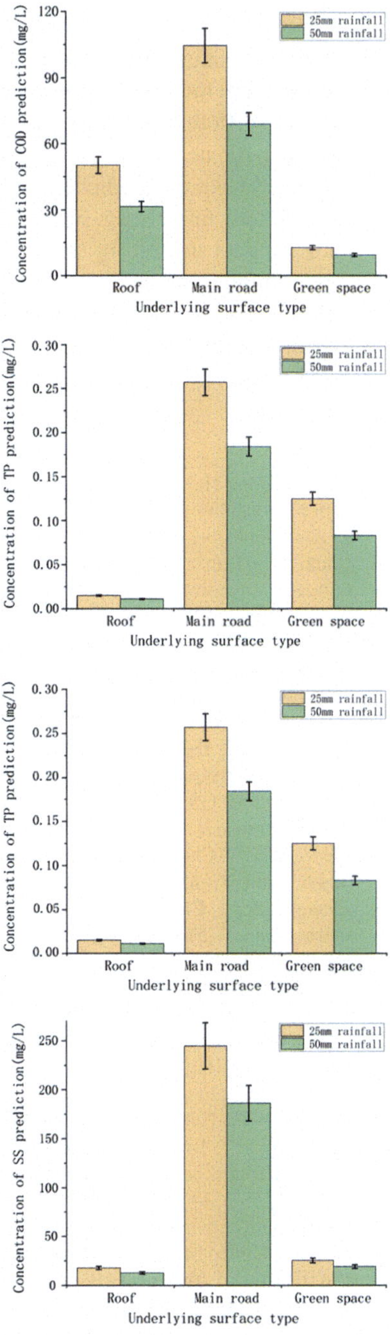

Fig. 4.5 Prediction results and errors of EMC values of various pollutants in different underlying surfaces by neural network model

(3) The results of training calculation and test evaluation of BP neural network model are satisfactory. The overall average relative error of the cross-verified model is the lowest 11.50%, and the average relative error of EMC value prediction for different pollutants in the test set is also < 10%. It can be concluded that the BP neural network model has a certain reference value for the data obtained from the prediction of EMC values of pollutants in stormwater runoff at different underlying surfaces of residential areas in Guangzhou, and can provide data support for stormwater runoff pollution load control in residential areas in the future.

References

1. Zhao Q, Cong H, Wang Z, Tang Z, Jiang X, Wang Y (2018) Study on the reduction effect of high permeability concave green space on urban rainfall runoff. Water Conserv Hydropower Technol 49(09):41–48
2. Booth DB, Hartley D, Jackson R (2010) Forest cover, impervious-surface area, and the mitigation of stormwater impacts. Jawra J Am Water Resour Assoc 38(3):835–845
3. Haq M, Akhtar M, Muhummad S et al (2012) Techniques of remote sensing and GIS for flood monitoring and damage assessment: a case study of Sindh province, Pakistan. Egypt J Remote Sens Space Sci 15(2):135–141
4. Zhang W, Che W, Liu DK et al (2012) Characterization of runoff from various urban catchments at different spatial scales in Beijing, China. Water Sci Technol 66(1):21–27
5. Li YX, Ma JH, Yang ZF et al (2011) Influence of non-point source pollution on water quality of Wetland Baiyangdian, China. Desalin Water Treat 32(1/2/3):291–296
6. Yang YY, Toor GS (2017) Sources and mechanisms of nitrate and orthophosphate transport in urban stormwater runoff from residential catchments. Water Res 112:176–184
7. Liu W, Wei W, Chen WP et al (2019) The impacts of substrate and vegetation on stormwater runoff quality from extensive green roofs. J Hydrol 576:575–582
8. Zhang K, Sui T, Chen S (2008) Analysis of urban rainwater quality characteristics in Jinan. J Irrigat Drain 27(5):119–121
9. Huang G, Nie T (2012) Non-point source pollution characteristics and pollution load of stormwater runoff in Guangzhou city. J South China Univ Technol Nat Sci Ed 40(2):142–148
10. General Administration of Environmental Protection of the People, ', s Republic of China (2002) Water and wastewater monitoring and analysis methods, 4th edn. China Environmental Science Press, Beijing
11. Gu Y, Wang X, Fang Z (2020) Investigation and analysis of pollution characteristics of rainfall runoff in typical cities. China Rural Water Conserv Hydropower 2020(06):46–50+57
12. Xu Y, Gong Y, Bi J, Wang Y, Cheng J, Wang K (2020) Analysis of the characteristics of rainwater runoff pollution on the underlying surface of typical cities in Ningbo. Environ Sci 41(07):3275–3284
13. United States Environmental Protection Agency (1983) Results of the nationwide urban runoff program. United States Environmental Protection Agency, Washington DC
14. Qin HP, Khu ST, Yu XY (2010) Spatial variations of storm runoff pollution and their correlation with land-use in a rapidly urbanizing catchment in China. Sci Total Environ 408(20):4613–4623
15. Zhang C, Li C, Chen Y (2022) Outdoor Drainage Design Standard (GB 50014–2021) ten questions and ten answers. Water Supply Drain 58(01):13–18
16. Wang C, Chen R, Wu L et al (2018) Accumulation and erosion of rainfall and runoff pollutants on different underlying surfaces of typical villages. Water Purif Technol 37(04):21–27

17. Li Y, Tian H, Zhang M, Wang W (2017) Evaluation model of rainwater runoff pollution load based on BP neural network. China Rural Water Conserv Hydropower 03:68–74
18. Yi Z, Li J, Qian J, Lu B, Zhang Y, Li F (2023) Based on BP and PSO, the inverse method of estuary double boundary conditions. Ren Min Changjiang 54(02):141–146
19. Lu J, Fan Z, Luo Q, Jiang Y, Fan S, Lin S, Liu A (2022) Analysis of pollution characteristics and algal toxicity of rainwater runoff in small towns in Guangdong. Environ Sci Technol 45(04):69–78
20. Gan H, Zhuo M, Li D, Zhou Y (2006) Water quality characteristics of urban road rainwater runoff in Guangzhou. Ecol Environ 05:969–973
21. Huang G, Zhao X, Mai Y (2021) Control effect of low impact development measures on rainwater runoff. Water Resour Protect 37(04):29–36
22. Tu J, Ai N (2018) Characteristics of urban stormwater runoff pollution: A case study of Foshan New City, Guangdong. Water Resour Hydropower 09:33–39
23. Peng X, Zheng H, Zhao D et al (2015) Preliminary study on non-point source pollution of early rainwater in the central city of Guangzhou. Guangdong Water Resour Hydropow 2015(09):40–43+46
24. Wang P, Dong G, Yin Z et al (2023) Sensitivity analysis of SWMM water quality parameters for complex scenarios. J Hohai Univ (Nat Sci) 51(05):31–37

Chapter 5
Soft Measurement of Key Effluent Parameters in Municipal Wastewater Treatment Processes Based on Deep Online Transfer Learning

Hao Zhuang, Guanghui Xu, and Yuhua Weng

Abstract Municipal wastewater treatment processes are non-stationary, leading to degradation of the performance of trained soft measurement models. To address this problem, a deep online transfer learning (OTL) framework is developed in this article. First, offline modeling is carried out by long short-term memory networks (LSTM), and the temporal information in the data is fully extracted. Second, first determine whether the model needs to be reconstructed in the online phase through multiple kernel variants of maximum mean discrepancy (MK-MMD), then determine whether the model needs to be fine-tuned through prediction error, and complete the online update of the model. Then, the final prediction results are obtained by combining the models with different knowledge through ensemble learning. Utilizing data from BSM1, simulation outcomes demonstrate the superiority of the introduced methodology over prevailing approaches, evidencing its capability to achieve elevated precision in forecasting within the context of the erratic wastewater treatment process.

Keywords Wastewater treatment process · Key effluent parameters · Online transfer learning · Soft measurement

5.1 Introduction

The process of urban wastewater treatment is a complex and highly delayed biochemical reaction process, and the detection and control of key effluent water quality parameters are crucial prerequisites for the consistent and effective functioning of facilities for treating wastewater. However, challenges such as harsh working environments, expensive hardware costs, and high maintenance expenses have made it difficult to directly measure important effluent parameters such as ammonia nitrogen

H. Zhuang (✉) · G. Xu · Y. Weng
Faculty of Information Technology, Beijing University of Technology, Beijing, China
e-mail: Arthurlbj@163.com

K. Y. Foo et al. (eds.), *Water Nexus Approach for Sustainable Development*,
Environmental Science and Engineering, https://doi.org/10.1007/978-3-031-89238-7_5

(NH_4-N) and total nitrogen (TN) in the wastewater treatment process. Soft measurement can address the issue of difficult direct measurement of critical variables. It provides an effective auxiliary means for process monitoring, optimization, and control, and is thus widely applied in process control.

Due to their excellent learning ability and non-linear approximation capability, artificial neural networks have been successfully validated in many nonlinear system modeling problems. Consequently, many researchers have begun applying them to wastewater treatment process modeling. Han et al. [1] introduced a methodology for soft measurement, employing interval type-2 fuzzy neural networks, and developed a model for the soft measurement of effluent NH_4-N, enabling real-time detection of NH_4-N in the effluent. Despite the significant success of the aforementioned soft measurement models in practical applications, their performance may not be ideal in highly complex tasks with multiple external disturbances. Specifically, these soft measurement models are based on shallow neural networks, which fail to extract and mine effective features from the complex data of wastewater treatment plants, thereby reducing the predictive accuracy of the models. Based on deep learning, neural networks can improve data mining and feature extraction. They can overcome the limitations of shallow neural networks in effectively extracting features from complex real-world data, such as in wastewater treatment plants. Therefore, they have attracted considerable attention. Qiao et al. [2] proposed an adaptive lasso echo state network (ALESN) for predicting the concentration of effluent NH_4-N, achieving good results. Wang et al. [3] introduced a deep belief network with event-triggered learning (DBN-EL) to enhance the efficiency and accuracy of soft measurement models in wastewater treatment plants. However, the aforementioned methods primarily rely on static deep networks and cannot adapt to urban wastewater treatment processes' complex nonlinear dynamic behavior. To address this issue, Ref. [4] used a recurrent neural network (RNN), a widely utilized dynamic neural network is employed to grasp the temporal dynamics inherent in time-series datasets. However, standard RNN suffers from the vanishing gradient problem [5], making it difficult to model long sequences. To tackle this problem, Hochreiter and Schmidhuber introduced LSTM networks, which use memory cells and three non-linear gates to replace the basic activation units in RNN. LSTM can forget irrelevant past information and selectively store useful information in memory cells. Consequently, LSTM neural networks exhibit superior efficacy in capturing dependencies over long temporal sequences. Yaqub et al. [6] employed LSTM neural networks to construct a predictive model for effluent TN, using monitoring data from wastewater treatment plants such as TOC, COD, TN, NH_4-N, SS, TP, DO, MLSS, and ORP as inputs, achieving TN prediction in the effluent.

The aforementioned soft measurement methods assume that the training and testing data follow the same distribution. Even if the soft measurement model is reasonably constructed, the time-varying nature, uncertainty, and external environmental changes in urban wastewater processes can lead to non-stationary data distributions, which may still degrade the model's predictive performance. In response to this challenge, recent studies have concentrated on the application of transfer learning, with the goal of adapting models trained within a source domain for use in a target domain [7]. It is often utilized for handling scenarios with similarities between different domains, by extracting features and integrating relevant knowledge for predictions. However, in the context of wastewater treatment processes, in the target domain, data usually appears sequentially. Online Transfer Learning (OTL), a variant of transfer learning, stands out not only as an efficient learning model but also as adept at managing streams of sequential data. Lin et al. [8] integrated deep learning with OTL and applied it to the diesel hydrofining process in petrochemical workshops, effectively improving the accuracy and stability of online predictions. Mao et al. [9] proposed a deep OTL based method for early fault detection of rolling bearings, enhancing the reliability of online detection results. Although OTL has been applied in various contexts and has shown satisfactory results, there is currently no research on the nitrogen content in the effluent of urban wastewater treatment processes. Therefore, further exploration of OTL is still needed in future research in this domain.

In this study, a deep OTL framework is designed, consisting of two phases: offline and online. In the offline phase, an LSTM neural network is employed for offline modeling to exploit the temporal information in the data fully. Then, in the online phase, the model is evaluated for the need of reconstruction using the MK-MMD. Additionally, the model is fine-tuned based on prediction errors. Based on the evaluation results, the model is updated online, and an ensemble learning approach is used to integrate models with different knowledge to obtain the final prediction results. Ultimately, simulation experiments with BSM1 data serve to confirm the efficacy and superiority of the method proposed. The developed soft measurement model based on deep online transfer learning can be effectively applied in wastewater treatment processes and is of significant importance for achieving real-time optimization control of these processes.

5.2 Materials and Methods

This section elaborates on the outlined OTL framework, presenting the EMFTNN model structure depicted in Fig. 5.1. Detailed descriptions are provided for the deep neural network leveraging LSTM, alongside its MK-MMD-based update strategy and fine-tuning process. Furthermore, to attain the model's currently optimal prediction, an ensemble method has been developed.

5.2.1 LSTM

LSTM networks, introduced by Hochreiter and Schmidhuber, address a well-known limitation of traditional RNNs: the challenge of learning long-term dependencies due to the vanishing gradient problem. By incorporating three distinct gates—input,

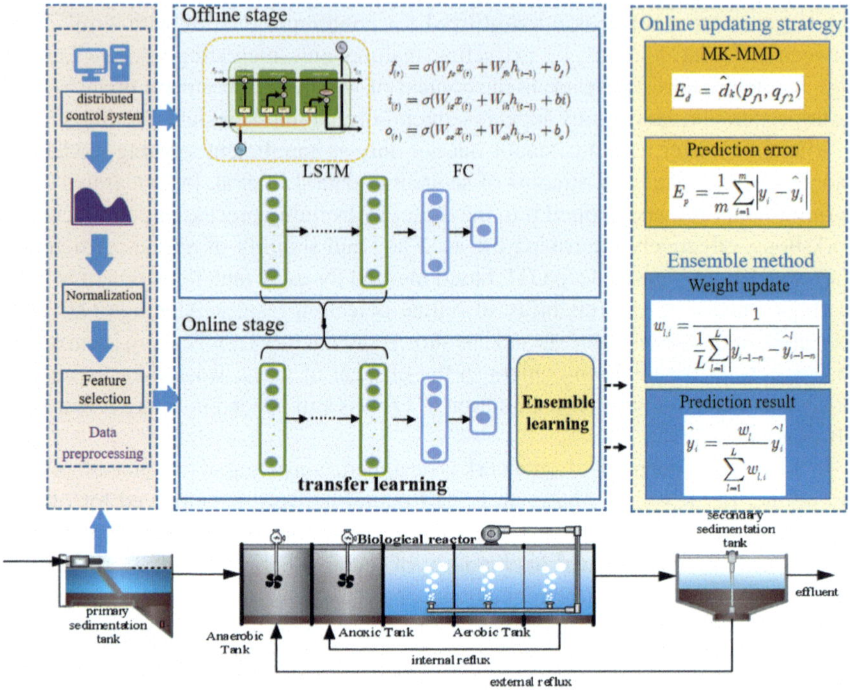

Fig. 5.1 Structure of the EMFTNN

output, and forget gates—alongside four interconnected layers within its architecture, LSTM enhances the ability to regulate and safeguard the cell state, facilitating the preservation of temporal memory. This advanced structure enables the LSTM to effectively manage long-term dependencies by modulating the gate operations, thereby mitigating the issue of gradient disappearance and making LSTMs suitable for a broad range of applications. Indicate the input vector $x_{(t)}$ and the antecedent hidden state $h_{(t-1)}$, with the preceding cell state $c_{(t-1)}$serving as the external inputs. Subsequently, the activation of the forget gate $f_{(t)}$ ensues as follows:

$$f_{(t)} = \sigma\left(W_f \cdot \left[h_{(t-1)}, l_{(t)}\right] + b_f\right) \tag{5.1}$$

And, the input gate $i_{(t)}$ and new candidate vectors $\hat{c}_{(t)}$ are computed as follows:

$$i_{(t)} = \sigma\left(W_i \cdot \left[h_{(t-1)}, l_{(t)}\right] + b_i\right) \tag{5.2}$$

$$o_{(t)} = \sigma\left(W_o \cdot \left[h_{(t-1)}, l_{(t)}\right] + b_o\right) \tag{5.3}$$

$$\hat{c}_{(t)} = \tau\left(W_c \cdot \left[h_{(t-1)}, l_{(t)}\right] + b_c\right) \tag{5.4}$$

The new cell state $c_{(t)}$ in LSTM is updated by the following equation.

$$c_{(t)} = c_{(t-1)} * f_{(t)} + \hat{c}_{(t)} * i_{(t)} \tag{5.5}$$

The output gate vector $h_{(t)}$ can be calculated as follows:

$$h_{(t)} = \tau(c_{(t)}) * o_{(t)} \tag{5.6}$$

5.2.2 *Online Updating Strategy*

In the offline stage, a deep LSTM network with some predictive performance has been trained. In this context, the historical data is designated as the source domain D_s, while the online data is characterized as the target domain D_t. Using MK-MMD, the discrepancy in distribution distance between extracted features is assessed. This method maps feature representations from both source and target domains into the regenerative kernel Hilbert space (RKHS) before measuring the mean distance between these two types of data. The MK-MMD technique is characterized as follows:

$$d_k^2(p, q) \triangleq \left\| E_p\left[\phi(x^s)\right] - E_q\left[\phi(x^t)\right] \right\|_{\mathcal{H}_k}^2 \tag{5.7}$$

where ϕ is the nonlinear transformation, $\mathcal{H}_k$ represents the RKHS, which is furnished with a distinctive kernel k.

When there is a significant variance in online data distribution, the model's feature extractor might struggle to adapt and extract relevant features for the current context. To address this, the present study first calculates the MK-MMD for n randomly selected samples from the source domain D_s and compares them with the latest n instances. If the difference surpasses a pre-established threshold, the entire model undergoes reconstruction. If not, the decision is made based on the model's prediction error. It's worth noting that the predictive accuracy is assessed every n ($n > 1$) new instances are received, reducing the influence of random variability. A surpassing prediction error threshold indicates the need for LSTM fine-tuning. However, when updating the model, there may be sample forgetting issues. This refers to instances where a model fails to make accurate predictions for those very samples despite being trained on specific samples. More successful deep-learning models tend to embrace repeated training to mitigate this issue. In contrast, traditional online learning algorithms often learn from each incoming example just once. Consequently, the model can still be more accurate with previously known samples. To counteract this, our approach focuses on the n currently arriving pieces during each model update and revisits a subset of b once received samples. This strategy promotes multiple learning passes over the incoming samples and ultimately aims to mitigate the model's issue of sample forgetting.

5.2.3 Ensemble Approach for Online Prediction

Multiple models M_l are trained, where each model M_l focuses on learning a portion of historical knowledge K_b in the dataset. The ensemble weights $w_{l,i}$ are updated based on the prediction error Δy_{i-n} at the previous moment $i - 1$ to give more weight to better-performing models:

$$w_{l,i} = \frac{1}{\frac{1}{L}\sum_{l=1}^{L} \left| y_{i-1-n} - \hat{y}_{i-1-n}^{l} \right|} \quad (l = 1, 2, \ldots, L) \tag{5.8}$$

Then, the Model M_l generates predictive results at time instance after undergoing online updates. These results are then integrated $\hat{y}_i^l$ into the final prediction $\hat{y}_i$ of multiple models using a weighted average approach:

$$\hat{y}_i = \frac{w_l}{\sum_{l=1}^{L} w_{l,i}} \hat{y}_i^l \tag{5.9}$$

Based on the methods described above, the final prediction result $\hat{y}_i$ can be determined. A concise overview of the entire framework for online prediction is presented in Fig. 5.2.

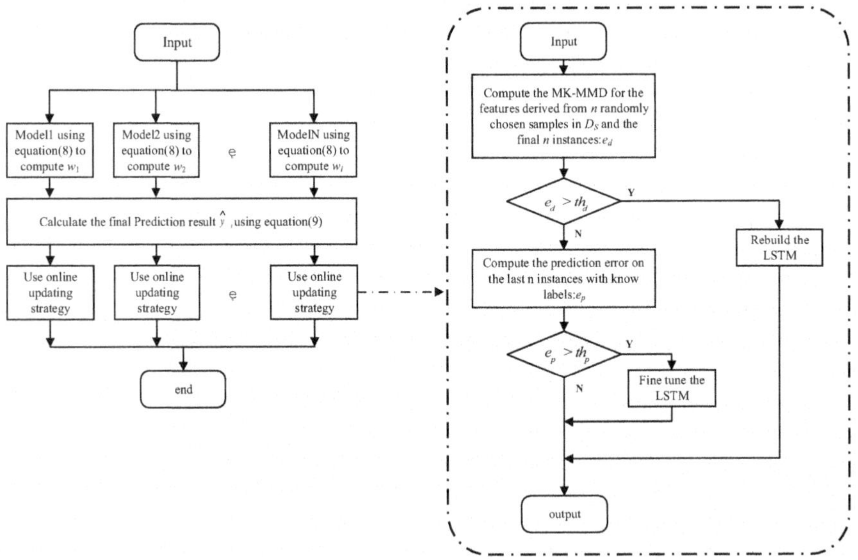

Fig. 5.2 Proposed framework for online prediction

5.3 Results and Discussion

This section utilizes the BSM1 simulation dataset to showcase the efficacy of the method proposed. Every experiment conducted in this study operates within an identical environment. The programs are coded in the TensorFlow 2.6.0 platform. The computer is configured as Intel(R) Core(TM) i9-12900H CPU @ 2.50 GHz, NVIDIA GeForce RTX3070, and Microsoft Windows 11 operating system. In this study, data obtained during rainy weather conditions was analyzed. There is a significant fluctuation in the influent flow between the 8th and 11th days, attributable to the rainfall. This weather event concurrently caused substantial variability in the concentration of pollutants in the influent stream. Such fluctuations compromise the efficacy of the soft-measurement model. The underlying reason is that the offline model isn't designed to adjust rapidly to these new environmental conditions. Consequently, updating the LSTM during the online phase is imperative to maintain accuracy.

To accurately gauge the predictive capability of the model, the evaluation of its accuracy was conducted using root mean square error (RMSE) and mean absolute percentage error (MAPE). A comparative analysis between the test output of the EMFTNN and the desired output is presented in Fig. 5.3. This comparison indicates a commendable alignment between the predicted and observed trajectories. Furthermore, the modeled measurements of effluent NH_4-N, as derived from the proposed model, closely mirror the actual sampled values.

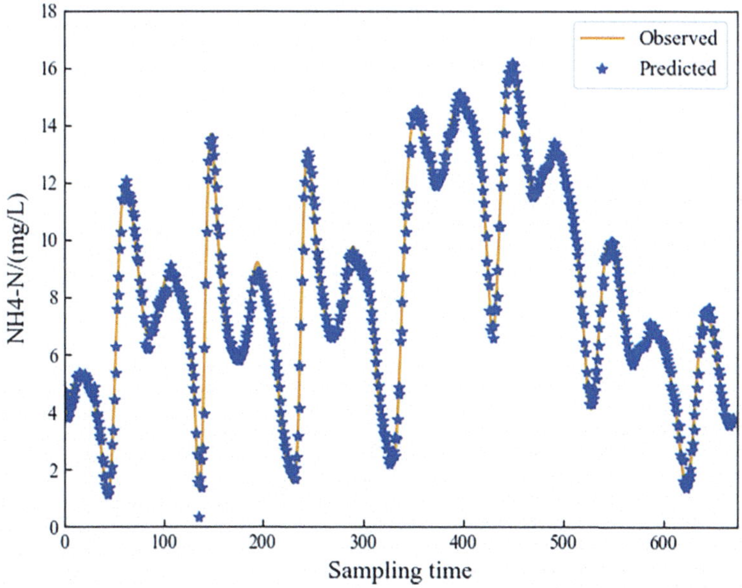

Fig. 5.3 Prediction results of effluent NH_4-N with EMFTNN

To rigorously assess the efficacy of the online detection algorithm introduced in this study, a comparative analysis was conducted between the EMFTNN algorithm and alternative methods. Comparison methods include SW-LSTM, FTNN, MFTNN, and EMFTNN, are depicted in Fig. 5.4. The prediction error associated with EMFTNN is notably lesser than those of the other three methods, making it adept for real-time detection of NH_4-N in effluent water. Furthermore, statistical results from 20 repetitions for each method are presented in Table 5.1 for a quantitative evaluation and comparison of different methods. Upon examination, the algorithms introduced in this study manifest a distinct superiority in prediction accuracy compared to the competing algorithms. While the FTNN method boasts the swiftest testing time amongst all the algorithms, it underperforms in terms of accuracy when measuring NH_4-N in effluent water. Utilizing the established EMFTNN soft measurement model for detecting effluent NH_4-N yields results in a mere 0.41 s, substantiating the claim that the proposed methodology is capable of real-time detection of pivotal effluent parameters.

To underscore the effectiveness of the methodologies proposed in this study, Fig. 5.5 presents a box plot illustrating the absolute prediction errors of the four methods concerning effluent NH_4-N. The median of the data set, represented by the orange-red line centered within the box, signifies the average value of the absolute prediction error. The upper and lower boundaries of the box demarcate the third and first quartiles of the absolute error, respectively. This box encompasses half of the data, and its width reflects the data's variability. Notably, the error associated with the EMFTNN model consistently spans a narrower range at a comparatively lower mean

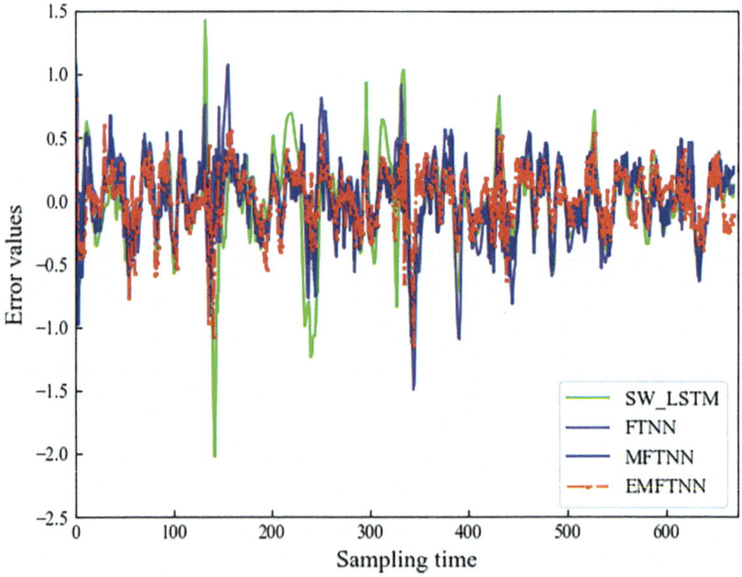

Fig. 5.4 Comparison of prediction errors of different models for effluent NH_4-N

Table 5.1 Comparison of the performance of different models for predicting effluent NH_4-N

Methods	Testing RMSE	Testing MAPE (%)	Testing time (s)
SW-LSTM	0.46	5.10	0.40
FTNN	0.34	4.42	**0.23**
MFTNN	0.27	3.52	0.27
EMFTNN	**0.24**	**3.21**	0.41

The bold values indicate the best-performing model for each evaluation metric

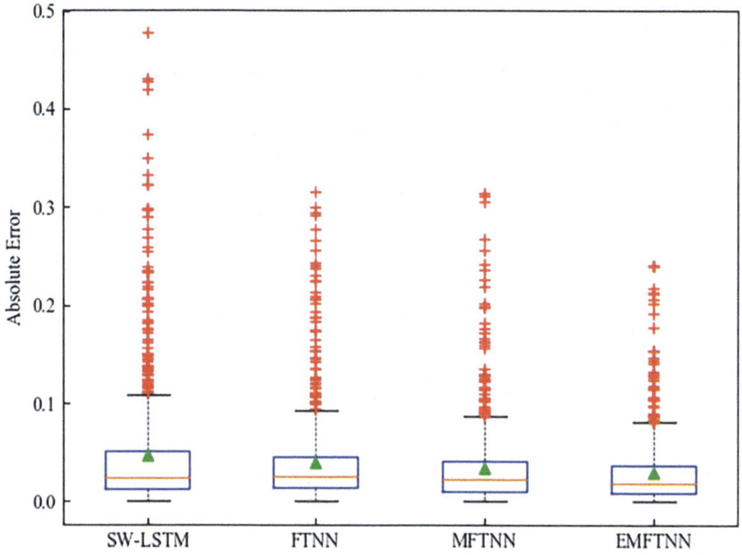

Fig. 5.5 Box plots of the absolute prediction errors of effluent NH_4-N with different models

level than its counterparts. In summation, the EMFTNN effectively tailors its model in accordance with the specific task at hand, facilitating real-time and high-precision detection of critical effluent parameters in the wastewater treatment process.

5.4 Conclusions

This paper proposes a method combining deep learning and OTL for soft measurement of key effluent parameters in the municipal wastewater treatment process, in which an offline-to-online framework is developed to handle online data with time-distributed variations. In the offline training phase, LSTM is used to extract the temporal information in the data to realize the complete mining of data knowledge. In the online phase, an update discrimination mechanism combining MK-MMD and prediction error is proposed to reduce the negative impact of online data due to distributional differences. In addition, ensemble learning combines models containing different knowledge to improve the accuracy and robustness of the model. Experimental results demonstrate the effectiveness of the proposed method.

Acknowledgements This work was financially supported by the National Science Foundation of China under Grants 61903012.

References

1. Han HG, Chen ZY, Qiao JF (2017) Soft-sensor method for effluent ammonia nitrogen based on interval type-2 fuzzy neural networks. CIESC J 68(3):1032–1040
2. Qiao JF, Wang L, Yang CL (2019) Adaptive lasso echo state network based on modified Bayesian information criterion for nonlinear system modelling. J Neural Comput Appl 31:6163–6177
3. Wang GM, Jia QS, Zhou MC, Bi J, Qiao JF (2021) Soft-sensing of wastewater treatment process via deep belief network with event-triggered learning. J Neurocomput 436:103–113
4. Wang JY, Li XL, Li JZ, Sun QH, Wang HY (2022) NGCU: a new RNN model for time-series data prediction. J Big Data Res 27:100296
5. Yadav SP, Zaidi S, Mishra A, Yadav V (2022) Survey on machine learning in speech emotion recognition and vision systems using a recurrent neural network (RNN). J Arch Comput Methods Eng 29(3):1753–1770
6. Yaqub M, Asif H, Kim S, Lee W (2020) Modeling of a full-scale sewage treatment plant to predict the nutrient removal efficiency using a long short-term memory (LSTM) neural network. J Water Process Eng 37:101388
7. Zhu ZD, Lin KX, Jain AK, Zhou JY (2023) Transfer learning in deep reinforcement learning: a survey. J IEEE Trans Pattern Anal Mach Intell 45(11):13344–13362
8. Lin YH, Chang L (2021) An online transfer learning framework for time-varying distribution data prediction. J IEEE Trans Indus Electron 69(6):6278–6287
9. Mao WT, Chen JX, Liu J, Liang XH (2022) Self-supervised deep domain-adversarial regression adaptation for online remaining useful life prediction of rolling bearing under unknown working condition. J IEEE Trans Indus Inf 19(2):1227–1237

Chapter 6
Research Progress on Chemical Phosphorus Removal

Hao Cai

Abstract In the field of sewage treatment, the use of aluminum salts and iron salts as chemical phosphorus removal agents is of great significance for improving the efficiency of biological phosphorus removal. This study comprehensively discusses the impact of chemical phosphorus removal agents on the biological phosphorus removal process, focusing on the analysis of the action mechanisms of aluminum salts and iron salts and their impact on biological phosphorus removal performance. By reviewing the existing research results, the current problems in understanding the impact mechanism of chemical agents on biological phosphorus removal are pointed out, and the directions and suggestions for future research are put forward. This aims to provide theoretical support and practical guidance for the optimization of the synergy between biological and chemical phosphorus removal in sewage treatment plants.

Keywords Sewage treatment · Chemical phosphorus removal · Biological phosphorus removal

6.1 Introduction

In order to deal with the problem of eutrophication of water bodies, especially when phosphorus is the dominant factor, controlling the phosphorus content in water has become an important environmental protection task. The U.S. Environmental Protection Agency (EPA) stipulates that when the phosphorus content in water exceeds 0.02 mg/L, the water body will enter a state of eutrophication. In this case, ecological problems such as algal blooms are prone to occur. After the blue-green algae outbreak in Taihu Lake, China established more stringent emission standards, requiring urban sewage treatment plants to meet or exceed the Class A standard specified in the

H. Cai (✉)
School of Civil Engineering and Architecture, Wuhan University of Technology, Wuhan City, Hubei Province, China
e-mail: 903777228@qq.com

K. Y. Foo et al. (eds.), *Water Nexus Approach for Sustainable Development*, Environmental Science and Engineering, https://doi.org/10.1007/978-3-031-89238-7_6

"Pollutant Discharge Standards for Urban Sewage Treatment Plants" (GB 18918-2002). This requirement has promoted the advancement of sewage treatment technology, especially in improving the phosphorus removal effect. Due to the competition between biological denitrification and phosphorus removal processes in operation, pure biological phosphorus removal methods are often difficult to stably meet strict emission requirements [1]. Therefore, many sewage treatment facilities have begun to use chemical phosphorus removal agents supplemented by biological treatment strategies to ensure that the total phosphorus content in the effluent reaches stable standards.

In the process of urban sewage treatment, chemical phosphorus removal technology can be subdivided into three main technical routes according to the specific location of adding chemicals: pre-precipitation, simultaneous precipitation and post-precipitation. Synchronous sedimentation technology requires less chemical dosage than pre-precipitation and does not require the construction of additional flocculation and sedimentation facilities compared to post-precipitation, thereby saving infrastructure costs and becoming the preferred process for many urban sewage treatment plants. Way. In the actual sewage treatment process, in order to ensure that the effluent phosphorus standards are met, the dosage of chemical phosphorus removal chemicals often exceeds actual needs. These excess chemicals will re-enter the biochemical treatment process along with the return sludge, which will have a negative impact on the efficiency of biological phosphorus removal [2]. In addition, the continued use and accumulation of phosphorus removal agents in the system may, in some extreme cases, lead to the complete failure of the biological phosphorus removal function of the sewage treatment system. This situation forces wastewater treatment facilities to rely entirely on chemical phosphorus removal to meet effluent phosphorus discharge standards, significantly increasing operating costs and reducing overall phosphorus removal effectiveness. This article comprehensively discusses the impact of chemical phosphorus removal agents on the biological phosphorus removal process and the related mechanism of action, and provides suggestions and prospects on how to effectively unleash the potential of biological and chemical synergistic phosphorus removal in the sewage treatment process. It aims to provide practical reference information for urban sewage treatment facilities to optimize operating costs while ensuring that the phosphorus content of effluent meets standards [3].

6.2 Selection and Dosage of Chemical Phosphorus Removal Agents

At present, in the field of sewage treatment, chemical phosphorus removal agents are mainly divided into four categories: calcium salts, iron salts, aluminum salts and magnesium salts. These categories cover various chemical substances such as lime, ferrous sulfate, ferric chloride, polyaluminum chloride (PAC), aluminum sulfate, magnesium ammonium phosphate, etc., as well as a series of advanced compound phosphorus removal agents. In urban sewage treatment plants, the process of selecting phosphorus removal agents should comprehensively consider a variety of factors, including the specific conditions of the water quality to be treated, the process flow used, the configuration of mechanical equipment, cost considerations during the treatment process, and possible ponds [4]. Problems such as wall scaling and pipe blockage. Such comprehensive considerations are designed to ensure that the phosphorus removal process is efficient and economical while reducing maintenance challenges that may be encountered during operations.

In order to ensure that the total phosphorus content of the effluent stably meets the standards, many sewage treatment facilities adopt the strategy of continuous and excessive dosing of phosphorus removal agents. The dosage of this agent is usually set based on the results of laboratory tests or medium-scale trials. In addition, based on experimental data and metal salt precipitation theoretical models, some researchers have further optimized the dosage of phosphorus removal agents, aiming to improve the efficiency and economy of chemical phosphorus removal, reduce the amount of agents used, and ensure that processing effect. For example, by introducing the concept of dosage coefficient β, the calculation process of pharmaceutical dosage can be simplified. This method can accurately calculate the required dosage of chemical phosphorus removal agents based only on the total phosphorus concentration in the sewage and the predetermined phosphorus removal efficiency. This approach makes the determination of pharmaceutical dosage more flexible and diverse, and at the same time improves the rationality and reliability of the calculation results [5].

To improve the empirical analysis of this study, we propose incorporating mathematical models and simulation programs designed to predict the outcomes of different dosages and combinations of chemical phosphorus removal agents. For instance:

$$\text{Efficiency} = \frac{\text{Initial P} - \text{Final P}}{\text{Initial P}} \times 100\%$$

where InitialP and FinalP represent the initial and final phosphorus concentrations, respectively. Moreover, a simulation program can be developed to model the dynamics of phosphorus removal under varying conditions, allowing for the optimization of chemical dosages based on specific objectives, such as minimizing costs while achieving regulatory compliance. This approach not only enhances the scientific rigor of the study but also provides practical tools for wastewater treatment plant operators [6].

6.3 Effect of Chemical Phosphorus Removal Agents on Biological Phosphorus Removal Effect

With the popularization of simultaneous chemical-assisted phosphorus removal technology in sewage treatment, the impact of chemical phosphorus removal agents on biological phosphorus removal has attracted increasing attention. This is because the large amount of chemical sludge produced by the chemical phosphorus removal process will directly affect the biological phosphorus removal process when it enters the biological treatment tank along with the return sludge (the tank often adopts anaerobic/aerobic alternating operation mode)., affecting the anaerobic phosphorus release and aerobic phosphorus uptake activities of microorganisms. This effect becomes a key factor in judging whether chemical and biological phosphorus removal can effectively synergize. The study by Liu et al. explored the effects of three metal ions, aluminum (Al^{3+}), ferrous iron (Fe^{2+}) and ferric iron (Fe^{3+}), on the biological phosphorus removal effect. Research results show that aluminum ions (Al^{3+}) have a significant inhibitory effect on the biological phosphorus removal process, including the anaerobic phosphorus release and aerobic phosphorus uptake stages. However, ferrous iron (Fe^{2+}) and ferric iron (Fe^{3+}) will only have adverse effects on biological phosphorus removal at higher dosages. This finding provides important guidance for the selection of phosphorus removal agents in wastewater treatment. The research by Lu Xiubin and others focused on the impact of high dosage of polyaluminum chloride (PAC) on the biological phosphorus removal process. Compared with the control group without PAC, when the dosage of PAC reached 50 mg/L and 100 mg/L, the amount of phosphorus released was reduced by 25% and 77% respectively, and the amount of phosphorus absorbed was reduced by 34% and 34% respectively [7]. 74%. This study clearly pointed out that PAC has a significant negative impact on the biological anaerobic phosphorus release and aerobic phosphorus uptake processes, and this impact intensifies as the PAC dosage increases. This result further emphasizes the need to carefully select chemical phosphorus removal agents and their dosage in practical applications to avoid excessive inhibition of the biological phosphorus removal process [8].

Research shows that when chemical phosphorus removal agents are used at lower concentrations, they may have a positive effect on the biological phosphorus removal process. Especially when the added concentrations of ferric iron (Fe^{3+}) and aluminum ions (Al^{3+}) do not exceed 8 mg/L and 6 mg/L respectively, such additives can significantly enhance the phosphorus release and absorption of phosphorus-accumulating bacteria ability, thus improving the efficiency of biological phosphorus removal to a certain extent. In addition, Fan Jie et al. used chemically synthesized iron phosphate to simulate the impact of chemical sludge on the phosphorus removal process of phosphorus-accumulating bacteria. They found that when the iron phosphate concentration was lower than 0.075 mmol/L, the phosphorus release and phosphorus release of phosphorus-accumulating bacteria were reduced. The phosphorus absorption capacity will be increased by more than 20% compared with that without addition. These findings reveal that under specific conditions, the appropriate use of chemical phosphorus removal agents can promote the biological phosphorus removal process, rather than just inhibit it. According to existing research results, the impact of chemical phosphorus removal agents on the biological phosphorus removal process is two-sided: it may exhibit either an inhibitory effect or a promoting effect. The specific manifestation of this impact depends on factors such as the treatment process, the type of chemicals used, and the specific dosage of the chemicals. Therefore, in practical applications, it is necessary to conduct detailed analysis and make reasonable judgments and adjustments based on the specific conditions of these variables.

To address the complex interactions in the wastewater treatment process, it's essential to consider a broader range of factors that influence the efficacy of aluminum and iron salts in biological phosphorus removal. This includes water quality characteristics such as pH, conductivity, the presence of competing ions, and organic matter concentration; sludge age, which affects the biomass adaptation and the accumulation of intracellular storage compounds; and the microbial population structure, emphasizing the balance between phosphorus-accumulating organisms (PAOs) and glycogen-accumulating organisms (GAOs). These factors contribute to the dynamic environment in which chemical phosphorus removal agents operate and interact with biological systems. Further research should integrate these variables to develop a more holistic understanding of their combined impact on phosphorus removal efficiency [9].

6.3.1 The Mechanism of Chemical Phosphorus Removal Agents Affecting Biological Phosphorus Removal

Oxygen consumption rate is a key indicator for evaluating oxygen consumption and matrix degradation efficiency during the aerobic process of activated sludge. It is directly related to the metabolic activity of aerobic microorganisms and is an important parameter for judging the activity of sludge. While exploring the impact of chemical phosphorus removal agents on activated sludge efficiency, researchers such as Hou Yanling found that adding ferric iron (Fe^{3+}) and aluminum ions (Al^{3+}) would lead to a decrease in the oxygen consumption rate of activated sludge, thereby inhibiting microorganisms activity. At the same concentration, the inhibitory effect of aluminum ions (Al^{3+}) is more significant than that of ferric iron (Fe^{3+}). Qiu et al. pointed out in the study that when the dosage of polyaluminum chloride (PAC) increased from 0 to 50 mg/L, the oxygen consumption rate of activated sludge was observed to decrease by nearly 60%. In addition, Chen Yasong studied the effects of three different chemicals, PAC, aluminum chloride ($AlCl_3$) and ferrous sulfate ($Fe_2(SO_4)_3$), on microbial activity through on-site measurements of activated sludge oxygen consumption rate. The research results show that these three chemicals all have an inhibitory effect on microbial activity, and the order from large to small is: PAC > $AlCl_3$ > $Fe_2(SO_4)_3$. This shows that as a chemical phosphorus removal agent, ferrous iron (Fe^{2+}) will also reduce the activity of microorganisms [10].

Microbial dehydrogenases are key enzymes that are essential for microbial decomposition of organic pollutants and energy acquisition. The activity of dehydrogenase not only shows the number of active microorganisms, but also reflects the ability of these microorganisms to metabolize organic matter. Therefore, dehydrogenase activity has become an important criterion for evaluating sludge activity and microbial treatment capacity. Research by Zhao Chunlu and others revealed the inhibitory effect of two aluminum salts, polyaluminum chloride (PAC) and aluminum chloride ($AlCl_3$), on dehydrogenase activity in activated sludge. The study found that this inhibitory effect intensified with the prolongation of the contact time between the agent and activated sludge, and the inhibitory effect of $AlCl_3$ on dehydrogenase was significantly stronger than that of PAC. In contrast, when ferrous sulfate ($Fe_2(SO_4)_3$) is used as a chemical phosphorus removal agent, ferric iron (Fe^{3+}) is crucial for microbial metabolism, and only when the concentration of $Fe_2(SO_4)_3$ exceeds 60 mg/L, the dehydrogenase activity will be significantly inhibited. This finding highlights the differences in the effects of different chemical phosphorus removal agents on microbial activity.

In the process of treating urban sewage, iron salts and aluminum salts, as commonly used chemical phosphorus removal agents, generally inhibit the activity of microorganisms, and this inhibitory effect is closely related to the type of agent and its usage. In particular, aluminum salts usually have a more significant inhibitory effect on microbial activity than iron salts. This phenomenon may be related to the biotoxicity properties of aluminum. The interaction mechanism between chemical phosphorus removal agents and biological phosphorus removal microorganisms remains insufficiently explored. It is crucial to delve deeper into how aluminum and iron salts, beyond their physical removal of phosphorus, potentially affect the metabolic pathways and signaling mechanisms within microbial communities. This includes investigating the impact of these agents on microbial cell membrane integrity, energy transfer processes, and the expression of genes related to phosphorus uptake and storage. Advanced analytical techniques, such as metaproteomics and metatranscriptomics, should be employed to uncover these mechanisms at a molecular level, providing insights into optimizing the balance between chemical and biological phosphorus removal methods.

6.3.2 Microbial Population Structure

In biological phosphorus removal systems, phosphorus-accumulating bacteria (PAOs) and polysaccharide-containing bacteria (GAOs) play key roles. PAOs are the main players in the phosphorus removal process, releasing phosphorus under anaerobic conditions and absorbing excess phosphorus under aerobic conditions. In contrast, although GAOs do not have the function of removing phosphorus, they can compete with PAOs for limited carbon source resources. This competition directly affects the efficiency of biological phosphorus removal. Therefore, the impact of chemical phosphorus removal agents on the community structure of PAOs and GAOs in activated sludge is closely related to the phosphorus removal effect of the entire system. Jin Hu and other research teams used fluorescence in situ hybridization technology to measure the community changes of PAOs and GAOs before and after adding polyaluminum chloride (PAC) in the AAO process. Research results show that after adding PAC, the proportion of PAOs dropped from 11.72 to 9.77%, while the proportion of GAOs increased from 1.14 to 3.41%. Different from this, Pei Hao and others also used fluorescence in situ hybridization technology to study the impact of adding iron salts on the community structure of PAOs and GAOs. Studies have found that long-term addition of chemical phosphorus removers will cause the proportion of phosphorus-accumulating bacteria in the total bacteria to gradually decrease [11].

Jin Hu and other research teams used fluorescence in situ hybridization technology to measure the community changes of PAOs and GAOs before and after adding polyaluminum chloride (PAC) in the AAO process. Research results show that after adding PAC, the proportion of PAOs dropped from 11.72 to 9.77%, while the proportion of GAOs increased from 1.14 to 3.41%. Different from this, Pei Hao and others also used fluorescence in situ hybridization technology to study the impact of adding iron salts on the community structure of PAOs and GAOs. Studies have found that long-term addition of chemical phosphorus removers will cause the proportion of phosphorus-accumulating bacteria in the total bacteria to gradually decrease.

6.3.3 PHA Synthesis

Biological organisms play an important role in the phosphorus removal process. In an anaerobic environment, phosphate-polymer plant microorganisms (PAOs) release energy by decomposing polyphosphate particles in the body and produce adenosine triphosphate (ATP) to absorb volatile fatty acids in the liquid phase. These volatile fatty acids are stored within cells in the form of poly-beta-hydroxyalkanoic acid (PHA). In an aerobic environment, PAOs will break down PHA and obtain energy from it to absorb phosphorus and replenish glycogen. Research shows that the metabolism of PHA plays a leading role in the biological phosphorus removal process. Chemical phosphorus removal agents will affect the synthesis of PHA after entering the anaerobic section, which actually illustrates the impact of chemical phosphorus removal agents on the intrinsic phosphorus removal ability of phosphorus-accumulating bacteria.

Studies have pointed out that as the Al/P ratio increases from 1:1 to 4:1, the amount of PHA produced by activated sludge in the anaerobic stage decreases significantly. A study conducted by Wen Qinxue and others found that when polymerized aluminum ferro was added to the AAO reactor to examine the effect of the agent on biological phosphorus removal, when the dosage was 3 mg/L, the PHA content in the anaerobic and anoxic zones The biggest difference. As the dosage increases, the difference in PHA content between the anaerobic and anoxic zones gradually decreases, indicating that increasing the dosage of polymerized aluminum iron will have a negative impact on the synthesis of PHA. In addition, Gao Ying and others used ferric chloride as a simultaneous phosphorus removal agent to conduct research. The results showed that after 40 days of administration, the PHA produced in the anaerobic stage was only about 50% of that before the administration. This shows that the addition of ferric chloride will have a certain inhibitory effect on the synthesis of PHA by phosphorus-accumulating bacteria.

6.4 Conclusion and Outlook

Chemical phosphorus removal process has been widely used in urban sewage treatment plants because of its simple operation, easy control and relatively stable phosphorus removal effect. By adding chemical phosphorus removal agents, the total phosphorus concentration in the effluent of the sewage treatment system can be effectively reduced to ensure that the effluent reaches stable standards. However, continuous excessive addition of chemical phosphorus removal agents often has adverse effects on biological phosphorus removal, which directly affects whether chemical phosphorus removal and biological phosphorus removal can play a synergistic role in phosphorus removal. At present, there is basically a consensus that the addition of phosphorus removal agents can effectively improve the phosphorus removal effect of the system. However, the specific mechanism by which phosphorus removal agents affect biological phosphorus removal is still unclear.

It is recommended that future research on the impact of chemical phosphorus removal agents on biological phosphorus removal can be carried out from the following aspects:

- Laboratory research can appropriately reduce the dosage of phosphorus removal agents and be as close as possible to actual production conditions. For different chemical phosphorus removal agents, the impact mechanism of their addition on biological phosphorus removal was studied in depth to provide theoretical support for practical engineering applications.
- Based on the incoming water quality, treatment processes and types of phosphorus removal agents used in different sewage treatment plants, establish a model of the impact of phosphorus removal agent dosage on biological phosphorus removal. On the basis of ensuring that the total phosphorus in the effluent reaches the standard, the optimal dosage range of phosphorus removal agents is determined to reduce the operating cost of the sewage treatment plant.
- Explore indicators that are easy to monitor and can reflect the impact of chemical phosphorus removal agents on biological phosphorus removal. This is of great significance for guiding actual production, helps to better understand the impact of chemical phosphorus removal agents on biological phosphorus removal, and provides a reference for optimizing the sewage treatment process.

References

1. Zhang M, Chi Y, Li S (2022) Chemical phosphorus removal optimization from coating wastewater using iron–calcium salt. Desalin Water Treat 264:164–171
2. Group CE (2022) Extending the operating temperature of biological phosphorus removal. Chem Eng 2022(8):129
3. Ge G, Zhao J, Lu CY (2022) Nitrogen and phosphorus removal in sequencing biofilm batch reactors: impact of chemical oxygen demand and aerobic operational modes. Environ Eng Sci 39(8):731
4. Hu Y, Zhang Y, Wang K et al (2023) Variations in the morphological phosphorus in a full-scale A_2O+MBR process with heavy chemical P removal addition dosing. Water Environ J 37(4):782–791
5. Bakhmutov VI, Contreras-Ramirez A, Drake H et al (2023) On phosphorus chemical shift anisotropy in metal (IV) phosphates and phosphonates: a P-31 NMR experimental study. Magn Reson Chem 61:16–21. https://doi.org/10.1002/mrc.5313
6. Bastounopoulou M, Gasparatos D, Haidouti C, Massas I (2022) Chemical fractionation and sorption of phosphorus in Greek inceptisols
7. Zhao W, Bi X, Bai M et al (2022) Research advances of the phosphorus-accumulating organisms of *Candidatus accumulibacter*, *Dechloromonas* and *Tetrasphaera*: metabolic mechanisms, applications and influencing factors. Chemosphere 307:135675
8. Fang X, Zhang D, Chang Z et al (2023) Phosphorus removal from water by the metal-organic frameworks (MOFs)-based adsorbents: a review for structure, mechanism, and current progress. Environ Res 243:117816
9. Zhang P, He M, Huo S et al (2022) Recent progress in metal-based composites toward adsorptive removal of phosphate: mechanisms, behaviors, and prospects. Chem Eng J 446:137081
10. Huang X, Yao K, Yu J et al (2022) Nitrogen removal performance and microbial characteristics during simultaneous chemical phosphorus removal process using Fe^{3+}. Biores Technol 363:127972
11. Hu H, Tong YW, He Y (2023) Current insight into enhanced strategies and interaction mechanisms of hydrogel materials for phosphate removal and recovery from wastewater. Sci Total Environ 892:164514

Chapter 7
Comparative Analysis of Water Quality Prediction Using BPNN and GA-BP Algorithms

Xinyue Tao

Abstract With the ongoing urbanization and industrialization, the development and application of smart water quality prediction models are of paramount importance for ensuring water quality safety and advancing environmental sustainability. Therefore, this study introduces two types of machine learning methods, namely the Back Propagation Neural Network (BPNN) and the Genetic Algorithm optimized BPNN (GA-BP). The performances of the GA-BP model and the BPNN are then compared in predicting water quality indexes. Here, five different water quality indexes, including dissolved oxygen (DO), ammonium nitrogen (AN), chemical oxygen demand (COD), total phosphorus (TP), and potassium permanganate index (PPI), were selected from the Zhuting section of the Xiang River Basin between July 2017 and September 2022. The findings revealed that the GA-BP model significantly surpassed the BPNN in terms of predictive performance, exhibiting both superior accuracy and robustness. Specifically, in forecasting the PPI within the testing dataset, the GA-BP model achieved an exceptional accuracy rate of 96.4%. Additionally, its strong global search capability prevents the BPNN model from being trapped in local extremes. Therefore, the GA-BP water quality forecast model adeptly integrates the advantages of GA, offering enhanced predictive capability and stability for anticipating future water quality conditions. This method provides robust technical support for water environment management and early warning systems.

Keywords BPNN · GA-BP · Sliding window · Water quality forecast

X. Tao (✉)
Department of Statistics and Mathematics, Zhejiang Gongshang University, Hangzhou, China
e-mail: txy567815@163.com

K. Y. Foo et al. (eds.), *Water Nexus Approach for Sustainable Development*,
Environmental Science and Engineering, https://doi.org/10.1007/978-3-031-89238-7_7

7.1 Introduction

The water environment system is facing many urgent problems due to the rapid development of society and economy. Water quality prediction and assessment can accurately reflect the pollution status and quality of the water environment. This provides a scientific basis for decision-making and management of the water environment.

Various methods are commonly used to predict water quality, including chaos theory model (CTM) [1, 2], gray system Model (GSM) [3], and machine learning networks [4, 5]. The CTMs describe system evolution through dynamical equations that are typically nonlinear and sensitive to initial conditions. Simulating system evolution allows observing behavior over time and predicting long-term system behavior [6]. However, there are limitations to the CTM [7], including the difficulty of accurately predicting behavior for small changes and the high demand for computational resources. Second, modeling requires knowledge of the system dynamics equations, which are difficult to access for complex systems. The GSM is a mathematical tool for prediction and analysis designed to deal with sparseness and incompleteness of data. It excels in small sample and nonlinear systems [8]. By partitioning data into known and unknown parts, the model uses gray differential equations or models to describe known data patterns and predict unknown data trends. However, it has high data requirements, is sensitive to outliers and noise, and may be less effective for complex and nonlinear systems than alternative models. Instead, machine learning methods offer the advantage of learning from data without requiring explicit knowledge of system dynamics equations [9]. This ability makes them particularly suitable for modeling complex and nonlinear systems, where traditional approaches may struggle [10]. This paper has introduced two machine learning methods, including Backpropagation Neural Network (BPNN) and an optimized Genetic Algorithm (GA) applied to BPNN [11–13] to investigate the potential of predicting water quality indexes. The accuracy of these models was thoroughly assessed using metrics including the coefficient of determination (R^2), root mean square error (RMSE), and mean absolute percentage error (MAPE). The results obtained can significantly inform decision-making processes in water quality management, thereby facilitating the development of more efficient and reliable prediction models that ensure the safety and sustainability of water resources.

7.2 Machine Learning Methods

7.2.1 Back Propagation Neural Network

The Back Propagation Neural Network (BPNN) is a type of multilayer feedforward neural network that employs the error backpropagation. It has demonstrated considerable efficacy in capturing and modeling complex nonlinear relationships with a desired degree of accuracy [14].

The BPNN operates by propagating signals forward and adjusting the weights and thresholds of the connections during the backward error propagation process [15]. The modifications to the weight and threshold values of the network are intended to be carried out in the direction of the negative gradient, which reflects the fastest decline of the loss function. This adjustment is performed when the output layer fails to achieve the desired prediction. The objective is to minimize the sum of the mean square error in the output layer. Through this iterative process, the BPNN gradually refines its predictions, bringing them closer to the desired results.

7.2.2 BPNN with Genetic Algorithm Optimization

The Genetic Algorithm (GA) is a heuristic, globally-searching evolutionary method inspired by the biological evolution process found in nature [16]. It operates based on the principle of "survival of the fittest" proposed by Darwin. When dealing with complex nonlinear problems that have numerous sub-optimal solutions, BPNN often encounters the drawback of getting trapped in local minima. To overcome this limitation, the BPNN with genetic Algorithm optimization (GA-BP) is adopted.

During the optimization process, GA is utilized to optimize the initial weights and thresholds for training the neural network [17]. In contrast to BPNN, the integration of BPNN with GA exhibits superior capabilities in expanding search spaces and converging towards the global optimum [18]. Moreover, GA-BP leverages the advantage of parallel processing offered by GA, which enhances the efficiency of the algorithm. The optimization process in GA-BP resembles genetic evolution, involving selection, crossover, and mutation [19]. These operations mimic the natural selection and genetic variation observed in biological systems. The implementation details of these steps are illustrated in Fig. 7.1. In GA-BP, the selection operator employs the roulette-wheel algorithm to choose individuals based on their fitness [19, 20]. Higher fitness values increase the probability of selection, favoring individuals with superior fitness as parents for the next generation. This method enhances the genetic information passed on to subsequent generations. Crossover, a pivotal step, exchanges genetic information between individuals to create new population members. In this paper, a single-point crossover method [21] is used, randomly selecting a point in the genetic information of parents and combining fragments to produce offspring with a blend of both information of parents. This process increases population size and search space while maintaining constraint boundaries to prevent solutions from straying outside. Mutation introduces random changes to individual genes [22], enhancing genetic diversity and broadening the search scope to avoid local optima. These steps iterate until terminal conditions are met, guiding the GA towards a robust and optimal solution.

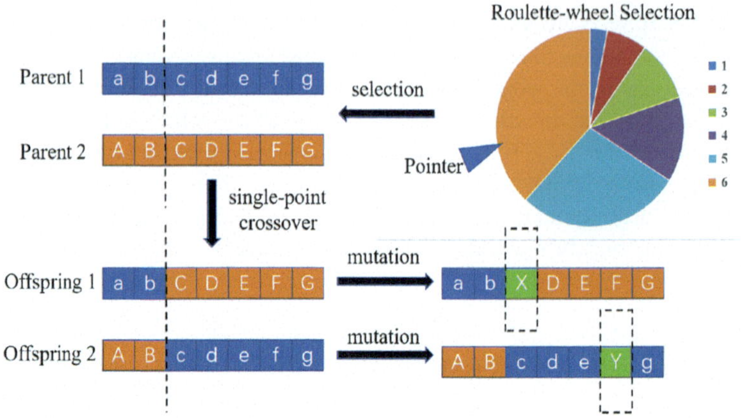

Fig. 7.1 The implements of selection, crossover and mutation

7.3 Performance Indicators

This paper utilizes three different norms to assess the performance of the proposed models. The coefficient of determination (R^2) is used to measure the goodness of fit, indicating the extent to which the model aligns with the data. Root mean square error (RMSE) calculates the square root of the average squared differences between predicted and actual values, providing a measure of overall model accuracy. Mean absolute percentage error (MAPE) represents the average percentage difference between predicted and actual values. Equations (7.1)–(7.3) provide the formulations for R^2, RMSE, and MAPE.

$$R^2 = \frac{\left[\sum_{i=1}^{N}(Y_i - \overline{Y})\left(\widehat{Y}_i - \overline{\widehat{Y}}\right)\right]^2}{\sum_{i=1}^{N}(Y_i - \overline{Y})^2 \sum_{i=1}^{N}\left(\widehat{Y}_i - \overline{\widehat{Y}}\right)^2} \tag{7.1}$$

$$RMSE = \sqrt{\frac{1}{N}\sum_{i=1}^{N}\left(Y_i - \widehat{Y}_i\right)^2} \tag{7.2}$$

$$MAPE = \frac{1}{N}\sum_{i=1}^{N}\left|\frac{Y_i - \widehat{Y}_i}{Y_i}\right| \times 100\% \tag{7.3}$$

where N is the total number of data, Y_i and $\widehat{Y}_i$ are the target value and predicted value of ith sample, and $\overline{Y}$ and $\overline{\widehat{Y}}$ are the average of target values and predicted values.

7.4 Database Acquisition

7.4.1 Data Description

This study utilized water quality automatic monitoring data sourced from the Zhuting section of the Xiang River in Zhuzhou City, Hunan Province, spanning from July 2017 to September 2022. The datasets was acquired from the Ecological Environment Bureau of Zhuzhou City, encompassing measurements of five key water quality parameters (dissolved oxygen (DO), ammonia nitrogen (AN), chemical oxygen demand (COD), total phosphorus (TP), and potassium permanganate index (PPI)). A comprehensive total of 315 valid data points were gathered, with data sets collected on a monthly basis.

Table 7.1 presents statistical characteristics of the data, encompassing measures of central tendency (mean, quartiles, mode), data dispersion (min, max, standard deviation), and data shape and symmetry (skewness, kurtosis). Skewness between -3 and 3 and kurtosis between -10 and 10 suggest a distribution close to normal [23]. In Table 7.1, all variables exhibit near-normal distributions, with skewness and kurtosis values well within acceptable ranges. Additionally, Fig. 7.2 illustrates histograms depicting the distribution and concentration levels of each water quality indicator, providing a visual aid for data analysis and water quality evaluation.

Table 7.2 presents the classes and divisions of national surface water quality. By combining Table 7.2 and Fig. 7.2, it is clear that all samples belong to Class I for both TP and PPI, with a range of less than 0.1 mg/L and less than 5 mg/L, respectively. Additionally, more than 85% of AN indicators in the samples were less than 0.2 mg/L and also belong to Class I. Furthermore, it can be clearly seen that all samples fall within the range of Class I and Class II standards for DO and COD. Therefore, the water quality in the Zhuting section of the Xiang River appears to be good, with relatively low levels of contaminants in the water bodies.

Table 7.1 Statistical summary of the water quality datasets

	DO	PPI	COD	AN	TP
Min	5.6	1.1	5	0.03	0.01
Max	11	3.4	15	0.47	0.07
Mean	8.1805	1.6857	9.746	0.1279	0.0416
Quantile (25%)	7.3	1.5	8	0.06	0.03
Quantile (50%)	8.1	1.6	10	0.094	0.04
Quantile (75%)	8.7	1.9	12	0.172	0.05
Mode	7.8	1.5	8	0.06	0.04
Std	1.22826	0.39138	2.48199	0.09821	0.01221
Skewness	0.453	1.526	− 0.059	1.923	0.4
Kurtosis	− 0.244	4.751	− 0.834	3.738	0.626

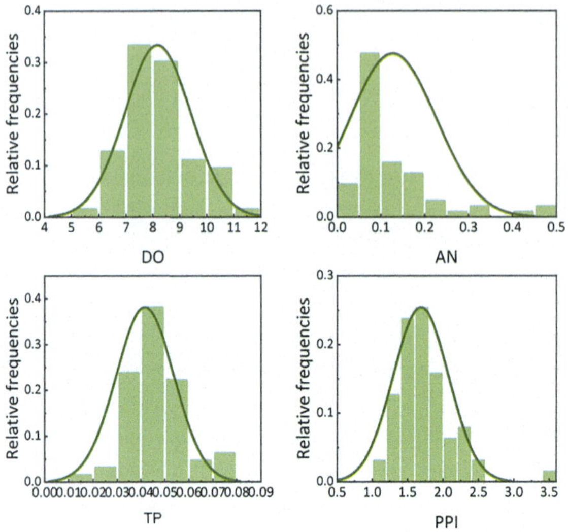

Fig. 7.2 Histogram showing the distribution of water quality indexes

Table 7.2 Water quality grade

	Class I	Class II	Class III	Class IV	Class V
DO (mg/L)	7.5–6.0	6.0–5.0	5.4–4.0	3.0–2.0	2.0–0
AN (mg/L)	0–0.15	0.15–0.5	0.5–1.0	1.0–1.5	1.5–2.0
COD (mg/L)	0–10	10–15	15–20	20–30	30–40
TP (mg/L)	0–0.02	0.02–0.10	0.10–0.20	0.20–0.30	0.30–0.40
PPI (mg/L)	0–2.0	2.0–4.0	4.0–6.0	6.0–10	10–15

7.4.2 Data Pre-processing

Data pre-processing is essential for ensuring the quality and reliability of analysis results in any data-driven research. By addressing inconsistencies, errors, and outliers in the data, pre-processing enhances the accuracy and effectiveness of the subsequent analysis processes. The dataset utilized in this research was devoid of any missing values, obviating the necessity for corrective measures. To standardize the datasets effectively, the study employed the z-score normalization (ZSN) method. It is a valuable technique for standardizing features across varying scales without altering their relative differences. By transforming features to have a mean of 0 and a standard deviation of 1, ZSN prevents any single feature from dominating the analysis due to its magnitude. Standardized data enables fair contributions from all features, enhancing the performance of algorithms such as clustering, regression, and classification [24]. Moreover, it simplifies the interpretation of model coefficients and feature importance, facilitating a more effective and accurate analysis. The formula for ZSN is given as follows:

$$X_{i,n}^{*} = \frac{X_{i,n} - \mu_i}{\sigma_i} \tag{7.4}$$

where $X_{i,n}^{*}$ represents the normalized value of the ith variable for the nth sample. The ith original variable has a mean value of μ_i and a standard deviation of σ_i.

7.5 The Sliding Window Method

The sliding window method is used to process time series data by moving through a window of fixed length from the beginning to the end of the series, updating the data while maintaining a constant number of data points within the window [24]. Assuming that a water quality indicator X comprises a total of n time series data sets, its time series is structured as $X(n) = [x(1), \ldots, x(t), \ldots, x(n)]$, where $x(t)$ represents the monitoring value of the water quality indicator X at the tth time point. If moment t is the current moment, the time series of the previous k moments of water quality indicator X can be represented as $X(k) = [x(t-k+1), x(t-k+2), \ldots, x(t)]$, where k denotes the size of the sliding window. In the prediction of a single water quality indicator, based on the value of k, the historical monitoring data of water quality indicator X is entered into the sliding window as the input of the BPNN model, and the predicted value $x(t+1)$ at the time of $(t+1)$ can be obtained. The corresponding sliding window structure is shown in Fig. 7.3.

In this paper, sliding windows are utilized to construct time series and process the indicator datasets with a fixed window size. The optimal sliding window size parameters can be obtained by comparing the mean absolute error of prediction using BPNN for each indicator at different window sizes through the sliding window method. The optimal sliding window sizes for DO, AN, COD, TP and PPI are identified as 8, 8, 6, 7, and 6, respectively, since they correspond to the minimum mean absolute error values.

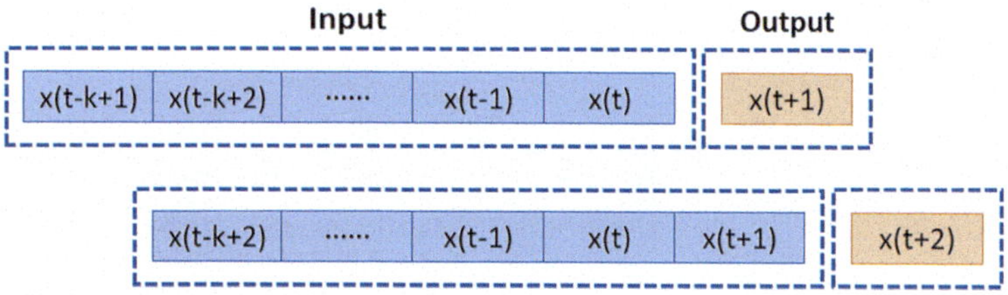

Fig. 7.3 A sliding window for time series

7.6 Analysis of Prediction Results

7.6.1 DO Prediction Results

DO means the amount of molecular oxygen dissolved in water, expressed in milligrams of oxygen per liter of water. The prediction of DO is essential for the protection of aquatic ecosystems and the identification of the effects of pollution. The values of performance metrics R^2, RMSE and MAPE of each model for the DO parameters are listed in Table 7.3. Using the traditional BPNN model, the RMSE and MAPE in the training set are 0.337 and 3.54%, respectively, and the R^2 of the BPNN is 0.924. Compared to the training set, the performance of the BPNN on the testing set is slightly worse, with an R^2 of only 0.852. However, since the RMSE and MAPE of the BPNN model are 0.363 and 5.15%, respectively, it still has a high fitting accuracy. Moreover, the BPNN model optimized by the GA algorithm has a better agreement between the testing set R^2 (0.946) and the R^2 (0.969) in the training set samples. This indicates that the BPNN model improved by the GA algorithm has a more stable prediction effect in water quality prediction than the traditional BPNN model.

The reason is that GA-BP differs significantly from traditional BPNN in parameter selection and search methods. While traditional BPNN adjusts initial parameters and sets learning rates, maximum training iterations, and error thresholds, it is susceptible to local minima. In contrast, GA-BP employs a genetic algorithm for global search capabilities. This genetic algorithm extensively explores parameter spaces using multiple populations and iterations to discover the optimal parameter combination. This global search feature enables GA-BP to thoroughly investigate parameter space possibilities, enhancing model generalization and prediction accuracy. In our experiment, GA optimization enhanced the R^2 prediction evaluation index of traditional BPNN on the testing set from 0.852 to 0.946. Furthermore, RMSE and MAPE decreased by 39.7% and 68.5% respectively. These enhancements minimized the parameter gap between the training and test sets, ensuring algorithm prediction robustness.

Table 7.3 Evaluation parameters for the BPNN and GA-BP in the training and testing set

		Training set					Testing set				
		DO	PPI	COD	AN	TP	DO	PPI	COD	AN	TP
R^2	BP	0.924	0.975	0.943	0.942	0.925	0.852	0.941	0.926	0.829	0.917
	GA-BP	0.969	0.979	0.970	0.967	0.976	0.946	0.964	0.959	0.917	0.958
RMSE	BP	0.337	0.063	0.599	0.021	0.003	0.363	0.075	0.623	0.011	0.005
	GA-BP	0.218	0.058	0.433	0.016	0.002	0.219	0.059	0.463	0.008	0.004
MAPE	BP (%)	3.54	2.40	2.71	8.79	14.35	5.15	3.39	3.89	2.49	11.88
	GA-BP (%)	0.90	1.47	1.33	6.56	12.17	1.62	1.56	2.19	2.27	10.01

7.6.2 AN Prediction Results

The concentration of AN is a widely used indicator for assessing the level of eutrophication in water bodies and the effectiveness of wastewater treatment. High AN concentrations can harm aquatic organisms and cause eutrophication, which leads to algal blooms and disrupts ecological balance. Therefore, it is important to predict AN concentrations in water. Table 7.3 shows that the BPNN model has a larger decrease in predicting AN water quality indicators on the testing set compared to the training set, with R^2 of only 0.829. The experimental results in the table indicate that the RMSE and MAPE of BP and GA-BP on the testing set are smaller than those on the training set. This phenomenon may be due to over-fitting of the models on the training set. The model over-fit the features and noise of the data, resulting in a complex model that learned the noise or specific sample features in the training set and did not generalize well in the testing set.

7.6.3 COD Prediction Results

COD is the total amount of organic and inorganic matter in a water sample that can be oxidized by chemical oxidants. As shown in Table 7.3, the GA-BP model outperforms the BPNN model in various quality parameters. Specifically, the GA-BP model demonstrates higher accuracy in predicting COD values in both the training and testing sets. In the training set, the GA-BP model achieved an R^2 of 0.970, an RMSE of 0.433, and a MAPE of 1.33%. The GA-BP model demonstrated superior predictive capabilities for COD values, with R^2 and RMSE metrics of 0.959 and 0.463, respectively, and a MAPE of 2.19% in the testing set. The optimization process, which involved the Genetic Algorithm (GA), significantly improved the performance of the BPNN model in predicting COD values. This optimization resulted in a 3.56% increase in fitness for the testing set, a substantial 25.68% decrease in RMSE, and a remarkable 43.70% reduction in MAPE. The effectiveness of leveraging GA to refine the BPNN model and enhance its predictive accuracy for COD values is highlighted by these improvements. This ultimately contributes to more reliable water quality predictions.

7.6.4 TP Prediction Results

TP is the sum of all forms of phosphorus present in a water body, including dissolved and suspended phosphorus. Phosphorus is an essential nutrient for the growth and survival of living organisms. However, excessive amounts of phosphorus can lead to eutrophication of water bodies, which can trigger algal blooms and the formation of red tides or cyanobacterial blooms. Therefore, predicting and controlling the total phosphorus content in water bodies is very important for preventing eutrophication.

Focusing on TP prediction in the training set, the optimized GA-BP model significantly improved prediction accuracy. The GA-BP model (R^2 = 0.976, RMSE = 0.002, MAPE = 12.17%) demonstrated slightly better predictive abilities than the unoptimized model (R^2 = 0.925, RMSE = 0.003, MAPE = 14.35%). The optimized model showed a notable 5.51% increase in R^2, a substantial 33.3% decrease in RMSE, and a significant 15.19% decrease in MAPE compared to the original model. These enhancements underscore the impact of the optimization on boosting the accuracy of BPNN in predicting TP concentrations in the training data. Beyond the training set, the comparison between the GA-BP model and the unoptimized model extends to the testing set. The optimized GA-BP model further improved in the testing set, with a significant 4.47% increase in R^2, a noteworthy 20% decrease in RMSE, and a substantial 15.74% decrease in MAPE compared to the BPNN model. These results confirm the effectiveness of the optimization process in enhancing predictive performance and highlight the optimized model's robustness and generalizability across diverse datasets.

7.6.5 PPI Prediction Results

The PPI reflects the amount and oxidizability of organic matter in a water body. The prediction results of PPI are shown in Table 7.3. Based on the evaluation parameters in Table 7.3, the GA-BP model is optimal in terms of R^2, RMSE, and MAPE for both the training and testing sets. It maintains high values on both sets, with R^2 value of 0.979 on the training set and 0.964 on the testing set. The GA optimization reduced the RMSE to 0.058 on the training set, with an MAPE of only 1.47%. On the testing set, the RMSE decreased to 0.059, with an MAPE of 1.56%. These results suggest that the GA-BP model has better fitting accuracy than the traditional BPNN model. Specially, among the five types of water quality index predictions, the GA-BP model is found to be more suitable for modeling the PPI.

7.7 Conclusions

(1) The GA-BP model is a more stable and accurate predictor of changes in water quality than the BPNN model. Furthermore, GA-BP prevents the BPNN model from becoming trapped in local optimal solutions and enhances its global search capability, enabling the model to more effectively adapt to changes in data.
(2) The GA-BP model proved to be reliable in predicting water quality of PPI (R^2 = 0.964, RMSE = 0.059 and MAPE = 1.56%). Among the five water quality indicators examined in this study, PPI was the most effective and was suitable for modeling the GA-BP with sliding window treatment.

References

1. Sivakumar B (2004) Chaos theory in geophysics: past, present and future. Chaos, Solitons Fractals 19(2):441–462
2. Qiao J, Hu Z, Li W (2016) Soft measurement modeling based on chaos theory for biochemical oxygen demand (BOD). Water 8(12):581
3. Baohui M, Rishang L, Yawei Z et al (2015) Correlation analysis and forecasting changes in Yongding River water quality based on information entropy and gray system theory. Nat Environ Pollut Technol 14(3)
4. Tian Y, Liu Q, Ji Y et al (2024) Prediction of sulfate concentrations in groundwater in areas with complex hydrogeological conditions based on machine learning. Sci Total Environ 171312
5. Zhu M, Wang J, Yang X et al (2022) A review of the application of machine learning in water quality evaluation. Eco-Environ Health 1(2):107–116
6. Li N, Zhang H, Feng L et al (2024) Analyzing and identifying predictable time range for stress prediction based on chaos theory and deep learning. Health Inf Sci Syst 12(1):16
7. Biswas HR, Hasan MM, Bala SK (2018) Chaos theory and its applications in our real life. Barishal Univ J Part 1(5):123–140
8. Chen X (2023) Overview of gray system theory. In: Application of gray system theory in fishery science. Springer Nature Singapore, Singapore, pp 1–20
9. Jadhav AR, Pathak PD, Raut RY (2023) Water and wastewater quality prediction: current trends and challenges in the implementation of artificial neural network. Environ Monit Assess 195(2):321
10. Kaur J, Khehra BS, Singh A (2023) Back propagation artificial neural network for diagnose of the heart disease. J Reliable Intell Environ 9(1):57–85
11. Li X, Wang J, Yang C (2023) Risk prediction in financial management of listed companies based on optimized BP neural network under digital economy. Neural Comput Appl 35(3):2045–2058
12. Sivakumar A, Singh NB, Arulkirubakaran D et al (2023) Prediction of production facility priorities using back propagation neural network for bus body building industries: a post pandemic research article. Qual Quant 57(1):561–585
13. Zhang T, Shuai J, Shuai Y et al (2023) Efficient prediction method of triple failure pressure for corroded pipelines under complex loads based on a backpropagation neural network. Reliab Eng Syst Saf 231:108990
14. Zhang S, Liu L, Yang Q et al (2023) FBG micro-strain sensor demodulation system based on back propagation neural network. In: 2023 IEEE 6th International conference on pattern recognition and artificial intelligence (PRAI). IEEE, 2023, pp 1167–1171
15. Hou J, Peng J, Liu K et al (2023) A multi-channel optical sensing system based on back propagation neural network for mixed ion solution. Optik 291:171391
16. Mirjalili S, Mirjalili S (2019) Genetic algorithm. In: Evolutionary algorithms and neural networks: theory and applications, 2019, pp 43–55
17. Sun J, Kainz J (2024) State of health estimation for lithium-ion batteries based on current interrupt method and genetic algorithm optimized back propagation neural network. J Power Sour 591:233842
18. Jin L, Duan J, Fan T et al (2023) Using GA-BP coupling algorithm to predict the high-performance concrete mechanical property. KSCE J Civ Eng 27(2):684–697
19. Alhijawi B, Awajan A (2023) Genetic algorithms: theory, genetic operators, solutions, and applications. Evol Intell, 1–12
20. Saini N (2017) Review of selection methods in genetic algorithms. Int J Eng Comput Sci 6(12):22261–22263
21. Zainuddin FA, Abd Samad MF (2021) Comparison of crossover in genetic algorithm for discrete-time system identification. Int Rev Mech Eng (IREME) 15(2):59–66
22. Gülmez B (2023) A novel deep neural network model based Xception and genetic algorithm for detection of COVID-19 from X-ray images. Ann Oper Res 328(1):617–641
23. Singh D, Singh B (2020) Investigating the impact of data normalization on classification performance. Appl Soft Comput 97:105524

24. Hota HS, Handa R, Shrivas AK (2017) Time series data prediction using sliding window based RBF neural network. Int J Comput Intell Res 13(5):1145–1156

Chapter 8
Strengthing Concrete-Ultrafiltration Membrance Pre-treatment Rural Sewage Test Research

Wenxing Lu and Zheng Wang

Abstract In order to develop a pretreatment process suitable for rural sewage in my country, the enhanced coagulation-ultrafiltration membrane process is used to pretreat rural sewage. Examine the removal effect of enhanced coagulation-ultrafiltration membrane on COD, TP. Explore the filtration characteristics of the stable operation of the ultrafiltration membrane; analyze the recovery effect of different cleaning methods on membrane flux. The results show that the removal rates of COD, TP by the enhanced coagulation-ultrafiltration membrane process are 94.5%, 94.1% respectively. SEM and EDS results show that flocs on the surface of the ultrafiltration membrane gradually accumulate, forming a dense filter cake layer and causing membrane pollution. The organic components of membrane pollution are oxygen-containing organic matter, and the inorganic components are Si, Al, and Na. The online chemical cleaning method of NaClO + citric acid can effectively remove membrane fouling. This study provides technical support for the application of ultrafiltration membrane pretreatment of rural domestic sewage.

Keywords Rural sewage · Enhanced coagulation · Ultrafiltration membrane · Membrane fouling control

8.1 Introduction

Since the amount of sewage in rural China is small and scattered, it is difficult to collect, and there is a lack of drainage facilities, which makes it impossible to use centralized water treatment systems for sewage treatment[1]. Direct discharge will not only cause the deterioration of water bodies in rural areas, but even produce black and smelly water bodies[2]. Therefore, solving the problem of rural sewage is of great significance to realizing rural revitalization and building a beautiful China.

W. Lu · Z. Wang (✉)
Xi'an University of Science and Technology, Xi'an, Shanxi, China
e-mail: 2232676908@qq.com

K. Y. Foo et al. (eds.), *Water Nexus Approach for Sustainable Development*,
Environmental Science and Engineering, https://doi.org/10.1007/978-3-031-89238-7_8

Compared with traditional processes, ultrafiltration membranes have become the third-generation water treatment technology due to their good ability to retain pollutants and adapt to water quality. The enhanced coagulation membrane technology adopts a composite process of coagulation/adsorption membrane filtration to separate pollutants such as carbon and phosphorus in wastewater, ensuring stable effluent quality; Greatly reduces equipment size and footprint[3, 4]. Therefore, the prospect of using enhanced coagulation membrane technology to treat rural sewage is broad. Due to membrane fouling[5] being a limiting factor in the application of ultrafiltration membrane technology, pollutants clog membrane pores and accumulate on the surface of the ultrafiltration membrane, thereby reducing membrane flux and causing a rapid increase in transmembrane pressure difference (TMP) of the ultrafiltration membrane[6], leading to a significant increase in operating and maintenance costs.

This study starts with improving operating conditions and membrane cleaning to slow down membrane fouling, reduce operating costs, and better promote the application of ultrafiltration membrane technology in rural areas.

8.2 Material and Methods

8.2.1 Instrument Overview

The enhanced concrete-ultra-filter membrane device process adopted in this study. As shown in Fig. 8.1, the entire set of integrated equipment is highly integrated, including film pools, membrane components, anti-washing systems, intermediate water tanks, pump valve systems, PLC control systems, etc. 1 set of membrane components is installed in the membrane, which is made of PVDF. The effective volume of the membrane pool is 3.6 m^3, the membrane area is 192.6 m^2, the average diameter of the membrane is 0.01 μm, HRT is 1.4 h, and the design processing scale is 50 m^3/d. The volume of the middle water tank is 1 m^3. During the operation of the entire device, the cross -membrane pressure difference (TMP) was recorded in real time during the operation of the device, and instantaneous flow was recorded with a flow meter.

8.2.2 Water Quality and Ultrafiltration Membrane System Run

Experimental device is located at a rural sewage treatment station in Langfang, Hebei. The quality of rural sehwage and water filtered after fine grids is: NH4 + -N 19.6 ~ 49.6 mg/L TN 23.9 ~ 56.0 mg/L COD 128 ~ 240 mg/L,TP 3.0 ~ 6.4 mg/L, SS 160 ~ 210 mg/L.The basic operation of the enhanced concrete-ultrafiltration membrane reactor includes the filtering stage and pulse aeration stage. 10 min is used as a cycle

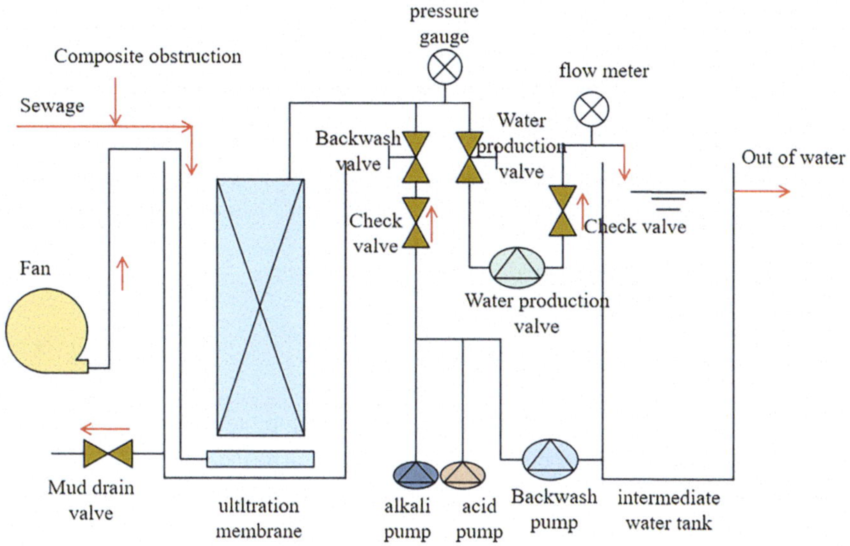

Fig. 8.1 Strengthen the concrete super filter process device diagram

during the trial. Each cycle is divided into 8 min of filtering water in order and 2 min pulse aeration. Online chemical cleaning: refers to soaking and cleaning with a lower concentration chemical agent. After a certain period of filtration cycle, the operation of lower concentration agents (NaCIO, citric acid) soaking, circulating anti -washing, aeration and other operations through anti -washing pumps (NaCIO, citric acid), and finally reached the chemical agent around the membrane and the entire membrane pool. Online chemical cleaning is performed every 2 to 3 months.

8.2.3 Analysis Method

The use of COD, TP, DOC and other indicators are determined in accordance with national water and wastewater detection and analysis standards[7]. The formation of the membrane section structure is determined by scanning electron microscope (Merlin model of the German Zeiss Company); the types and content of membrane section elements are detected by energy color scattered spectrometer (EDS).The molecular weight of dissolved organic compounds was analyzed by ultrafiltration membrane separation method, and ultrafiltration membranes with molecular weight cut-off of 100 kDa, 10 kDa, and lk Da (Millipore, USA) were selected, and the water samples were filtered through ultrafiltration membranes of different specifications by parallel filtration method. The amount of filtrate from different ultrafiltration membranes is differently reduced.Zeta potential adoption Model ZEN3690 Zeta Potentiometer (Malvern Instruments, UK) Mensuration.

(1) Filter resistance

$$R=3.6 \times 10^9 \times \frac{TMP}{\mu J} \tag{8.1}$$

In the formula, R is the filter resistance, m^{-1}; TMP is cross-membrane pressure difference, kPa; μ is the viscosity of the solution, Pa·s; J is a membrane flux, L/(m^2·h).

(2)Cross -membrane pressure difference recovery rate

$$FR = \frac{P_{fw} - P_w}{P_{fw} - P_{iw}} \times 100\% \tag{8.2}$$

In the formula, the FR is the TMP recovery rate. The P_w is the TMP of the TMP when filtering the sewage after the membrane is filtered. The P_{fw} is the TMP when the pollution membrane is filtered with sewage; the P_{iw} is the TMP of the new membrane when filtering the sewage.

8.3 Results and Discussion

8.3.1 Filtration Performance of the Reactor

The start-up of the reactor took 5 days, and the addition of composite blocking agent ensured the rapid start-up of the reactor. Compared with MBR, enhanced coagulation ultrafiltration membranes have the advantage of rapid start-up. During the stable operation stage of the reactor, the daily water production flow rate is 50 m^3/d , The dosage of the compound blocking agent was 60 mg/L. Figure 8.2 shows the variation of flow rate and filtration resistance during a typical membrane filtration cycle. There is no significant increase in filtration resistance within a single filtration cycle, indicating that the addition of composite blocking agents can effectively alleviate membrane fouling. This is consistent with the research results of Hui Gong[8]. This is because the addition of composite blocking agents can remove dissolved organic matter, especially proteins and some humic acid organic matter, thereby reducing the entry of organic pollutants into the membrane structure; Secondly, the addition of composite blocking agents improves the structure and properties of the filter cake layer, increasing its ability to adsorb pollutants. The filter cake layer with adsorption capacity deposits on the membrane surface, playing a role in pre filtration and reducing the entry of pollutants into the membrane pores, thereby reducing irreversible membrane fouling.

Figure 8.3 shows the principle of membrane fouling control,The addition of composite blocking agent to the raw water, the destabilization and coagulation of

Fig. 8.2 Changes in flow rate and filtration resistance during a typical filtration cycle

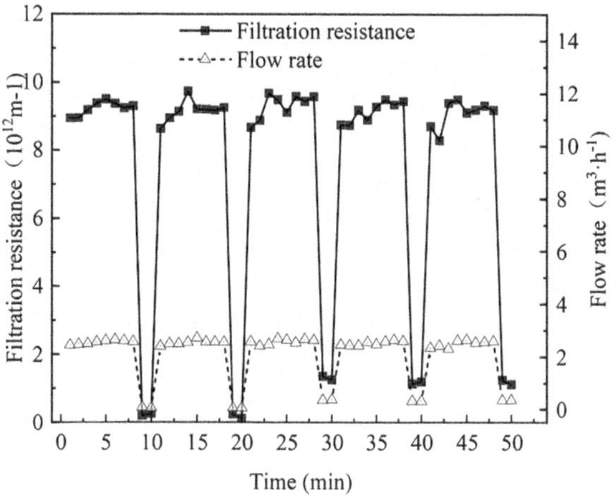

colloidal and dissolved organic matter in the membrane pool make the floc hydrophobicity in the slurry-water mixture larger, and the adsorption filter cake layer formed on the surface of the membrane prevents the organic matter from being tightly bound to the membrane, which improves the separation efficiency of the membrane. The blocking agent can participate in the coagulation process as a coagulation core, forming a larger particle size and strength, increasing the particle size of suspended solids, having more floc branches, better water permeability, and a larger gel mass being accumulated on the surface of the membrane, and the porosity of the filter cake layer on the membrane surface is higher. The dissolved small molecule organic matter is adsorbed by the hydrolysate of the composite blocker and adheres to the gel mass through net capture and sweeping. Therefore, the internal and external resistance of the membrane is reduced, and the internal pollution of the membrane filament is not increased, and the occurrence of membrane pollution is delayed. With the continuous input of raw water and chemicals into the membrane pool, the thickness of the filter cake layer on the surface of the membrane layer increases, and the degree of membrane pollution increases. The concentration of suspended solids in the membrane pool is getting higher and higher, the filter cake layer is thickened during the filtration process, and the degree of membrane pollution is aggravated. Pulse aeration scrubbing is used to shake off the cake layer from the membrane surface to reduce membrane fouling. When the sludge concentration in the membrane tank increases to a certain extent, the sludge in the membrane tank is discharged. With the increase of the amount of filtered water, the concentration of organic matter and inorganic salts in the membrane pool is reduced by sludge discharge, so as to reduce the degree of membrane pollution and ensure the membrane flux. When the TMP reaches the set limit, the membrane is restoratively cleaned by air–water backwash and chemical cleaning to move on to the next cycle.

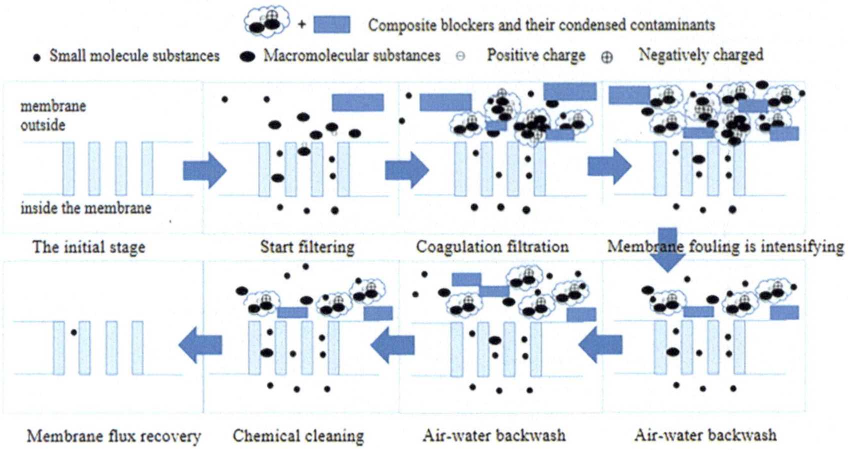

Fig. 8.3 Schematic diagram of membrane pollution control

8.3.2 The Removal Effect of the Reactor on Pollutants

8.4 The Removal Effect of COD.

Figure 8.4 shows the removal effect of enhanced coagulation ultrafiltration membrane on COD. The influent COD fluctuates greatly, with influent COD ranging from 128 to 240 mg/L. After enhanced coagulation membrane filtration, the effluent COD is 10–39 mg/L. The average COD removal rate is 94.5%. The average COD removal rate was 94.5%. The enhanced coagulation-ultrafiltration membrane has a very good removal effect on COD, which is due to the fact that in this experiment, the residence time is only 1.4 h, the aeration time accounts for only 20% of the running time, the aerobic microbial action is weak, and the concentration of EPS and SMP in the membrane pool is low. The colloidal COD in the sewage is negatively charged (-11.8 mV), and after the addition of the composite blocker, the composite blocker will hydrolyze to produce a high-valence positive ion charge, and the high-valence positive ions of the hydrolyzed water produced by the composite blocker will be replaced, making the electric double layer compressed, and the colloidal particles are easier to adsorb the counterions, and the counterions will be neutralized with the charge carried by the colloidal COD, as shown in Fig. 8.5, the Zeta potential will gradually decrease to 0.5 mV, which will reduce the electrostatic repulsion between them and promote the aggregation and precipitation of organic matter. On the other hand, the composite blocking agent will hydrolyze to form a large number of hydrated metal oxides, and due to the three-dimensional structure of the hydrated metal oxides, when it shrinks, the suspended COD mesh roll in the water can be swept and settled, so as to achieve the removal of COD in the water.

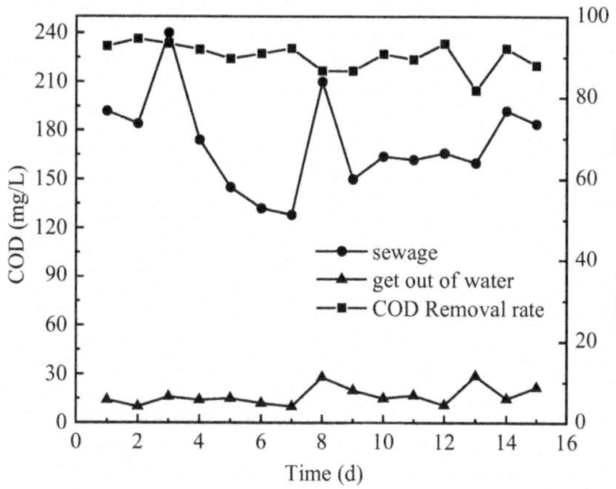

Fig. 8.4 Device to remove COD

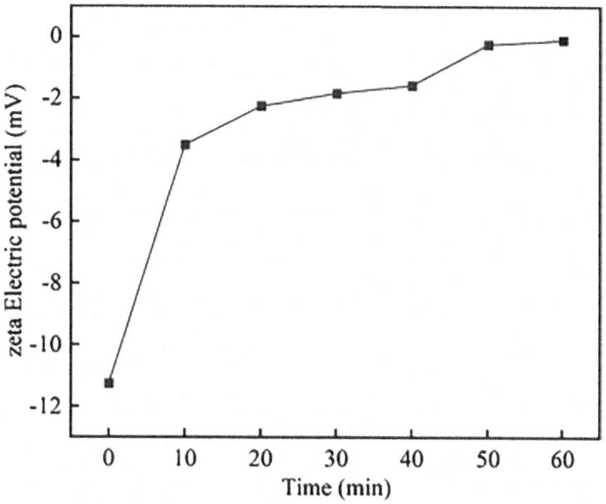

Fig. 8.5 Zeta potential change situation

As shown in Fig. 8.6, due to the coagulation effect of the composite blocker, some organic matter with a molecular weight of < 100 kDa (about 0.01 μm) will flocculate into large and dense alum flowers, and the floc with a large particle size (>0.01 μm) will be generated by net capture and sweeping, adsorption and bridging, and then the COD will be separated and removed by precipitation and membrane filtration (membrane pore size 0.01 μm).

Fig. 8.6 Molecular weight distribution of organic matter

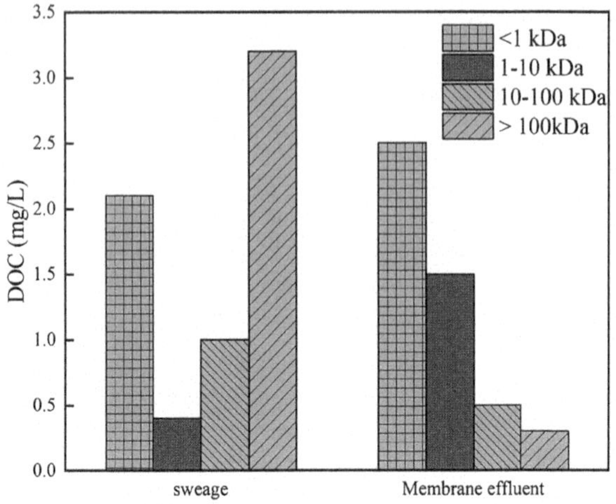

8.5 TP Removal Effect.

Figure 8.7 shows the TP removal effect of the enhanced coagulation-ultrafiltration membrane. The inlet water TP is 31.3 ~ 48.5 mg/L and the effluent TP after enhanced coagulation-membrane filtration is 0.1 ~ 0.5 mg/L. The average total phosphorus removal rate was 94.1%. The enhanced coagulation-ultrafiltration membrane has a higher removal rate of TP. This is because after adding the composite blocking agent, various components in the sewage begin to react, including first the hydrolysis and polymerization reaction of the aluminum salt itself; in addition, There are also complex interactions between Al^{3+} and organic matter in sewage, including adsorption, complexation, electrical neutralization, collision, shearing, etc., which cause the aluminum morphology and structure to continuously change. Phosphorus is eventually separated from domestic sewage in the form of chemical deposition. In this experiment, ultrafiltration membrane filtration ensured the removal of particulate phosphorus in rural sewage. In addition, the addition of composite blocking agent captured dissolved phosphorus through chemical effects such as adsorption, complexation, and electrical neutralization.

When the aluminum salt in the composite blocker is dispersed in the sewage, on the one hand, Al^{3+} reacts with PO_4^{2-} to form $AlPO_4$ precipitate, on the other hand, Al^{3+} is first hydrolyzed to form single-core complexes $Al(OH)_2$, $Al(OH)_2^+$ and AlO_2^-, etc. and the single-core complexes are further condensed through collision to form a series of multi-nuclear complexes Aln(OH)m(3n-m) + ($n > 1, m \leq 3n$), and these aluminum multinuclear complexes often have high positive charge and specific surface area. It can quickly adsorb negatively charged impurities in the water body, neutralize the colloidal charge, compress the electric double layer and reduce the colloidal ξ potential, promote the rapid destabilization, coagulation and precipitation of colloids and suspended solids, and show a good phosphorus removal effect.

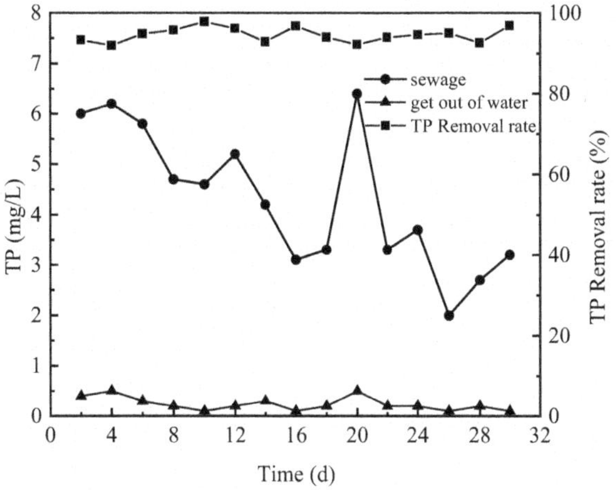

Fig. 8.7 Device effect on TP removal

8.5.1 Impact of Different Cleaning Methods on Membrane Fouling

Reverse membrane pollution can be removed by reflux of gas and water, and irreversible membrane pollution should be removed through online chemical cleaning. There are two ways: separate NaClO (10%) cleaning and NaClO (10%) + citric acid (5%).

It can be seen from Fig. 8.8 that compared with gas and water, chemical cleaning can effectively remove irreversible membrane pollution. Before using NaClO alone, the TMP of the ultra -filter membrane was 49.8 kPa. After cleaning, the TMP was reduced to 3.9 kPa. After being cleaned by NaClO + citric acid, the TMP was reduced from 53.4 kPa before washing to 1.3 kPa. Compared with, the recovery rate of NaClO + citric acid combined with the cross -membrane pressure difference was significantly higher. And NaClO + citric acid combined cleaning is better than that of individual NaClO cleaning. This is because the removal of NaClO is mainly organic pollutants, and citric acid can remove NA, Al, Si and other inorganic pollutants in ultra -filter membrane pollution. NaClO + citric acid chemical cleaning method can make the recovery rate of cross -membrane pressure difference reaching 88% Therefore, the online chemical cleaning method of NaClO + citric acid is better. The purpose of slowing membrane pollution is conducive to the promotion and application of ultra -filter membrane in the deep treatment of rural sewage.

8.5.2 Membrane Surface Appearance and Element Analysis

During the stable operation of the ultrafiltration membrane, cut the ultra -filter membrane wire, and then perform the SEM (zooming 10,000 times) of the membrane

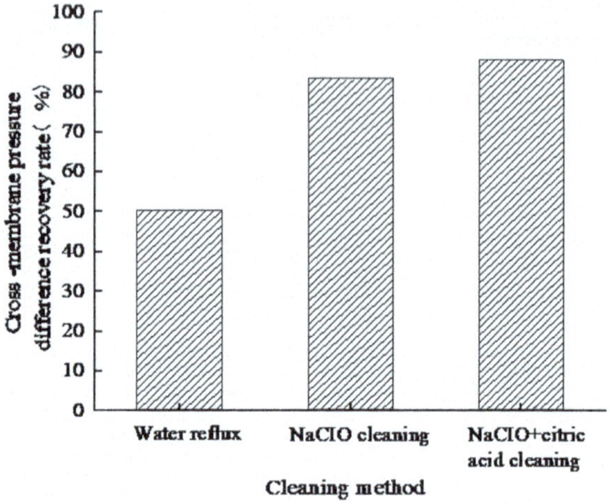

Fig. 8.8 Different cleaning methods under cross -membrane pressure difference recovery conditions

after air drying. As shown in Fig. 8.9 b, the surface of the membrane after cleaning is smooth, the surface of the membrane wire is evenly surface The pollutants that are distributed with membrane holes on the surface of the membrane and membrane holes are cleared. As shown in Fig. 8.9 a, the surface of the contaminated membrane was rough and uneven, the entire membrane hole was blocked by pollutants, the surface of the membrane was severely polluted, and a dense mud cake layer was formed.

Figure 8.10 shows the EDS results before and after membrane cleaning,The EDS analysis is performed on the surface of the membrane before cleaning. As shown in Fig. 8.10 a, in addition to the C, N, O, F elements contained in the ultrafiltration

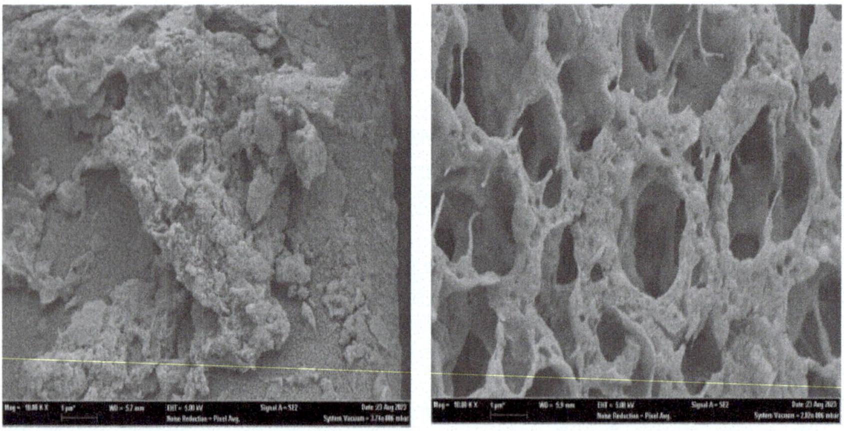

a Before chemical b After chenical cleaning

Fig. 8.9 Clean the surface of the membrane surface SEM

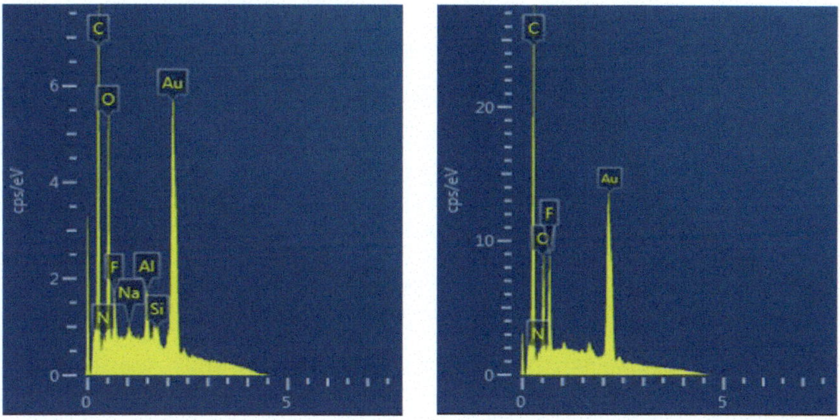

a Before chemical b After chenical cleaning

Fig. 8.10 Clean the surface of the membrane surface EDS

membrane itself, it also contains a small amount of Na, Al, Si elements Later, a large number of floccope was formed in the film pool. These floccope bodies will be intercepted by ultrafiltration membrane or adsorbed on the surface of the membrane to form a clay layer. Compact, eventually cannot be removed by gas and water, which causes a dense filter cake layer to cause membrane pollution [9].

8.6 Conclusions

(1) During the operation of the enhanced coagulation ultrafiltration membrane equipment, the COD and TP removal rates were 94.5% and 94.1%,
(2) The main component causing membrane fouling is oxygen-containing organic compounds, combined with trace inorganic elements Na, Al, Si. The use of NaCIO + citric acid online chemical cleaning can effectively alleviate membrane fouling.
(3) The enhanced coagulation ultrafiltration membrane process is an efficient, environmentally friendly, and economical sewage pretreatment method, which is very suitable for promotion and application in rural sewage treatment.

References

1. Manxue Z, Yu D, Fuquan N. Progress in rural domestic sewage treatment technology. (2017) Water treatment technology. J. 43 (06): 5-10

2. Fuxing L, Xiangfu S, Guoyan Z. (2013) The "4R" theory and engineering practice of rural non-point source pollution control Journal of Agricultural and Environmental Sciences. 32(11):2105–2111.
3. Jin Z, Meng F, Gong H, Improved low-carbon-consuming fouling contral in long-term membrance-based sewage pre-concerntration, (2017) The role of enhanced coagulation process and air backflushing in sustainable sewage treatment. Journal of Membrance Science. 529:252–262
4. S. Nazari,A. Asadi,F. Gholami & M. Dolatshah. (2024). Coupling of augmented PAN membranes by surface immobilization of pincer ligands with an activated sludge to treat a real wastewater.Journal of Industrial and Engineering Chemistry. 410–423.
5. Lin H, Zhang M, Wang F (2012) A critical review of extracellular polymeric substances (EPS) in membrane bioreactors: Characteristics, roles in membrane fouling and control strategies. J Membr Sci 460:110–125
6. The State Environmental Protection Administration (2002) Monitoring and Analysis Methods for Water and Wastewater, 4th edn. China Environmental Science Press, Beijing
7. Lin H, Gao W, Meng F (2012) Membrane bioreactors for industrial wastewater treatment: A critical review [J]. Crit Rev Environ Sci Technol 42(7):677–740
8. Gong H, Jin Z, Xu H (2018) Redesdigning C and N mass flows for energy-neutral wastewater treatment by coagulation adsorption enhanced membrance(CAEM)-based pre-concentration process. Chem Eng J 342:304–309
9. Nicolas Derlon, Alexander Grütter, Fabienne Brandenberger, Anja Sutter, Ute Kuhlicke,Thomas R, Neu &Eberhard Morgenroth. (2016) The composition and compression of biofilms developed on ultrafiltration membranes determine hydraulic biofilm resistance. Water Research. 63–72.

Chapter 9
Hg Stable Isotopes as the Tracer to the Historical Record and Metal Pollution Sources in the Lake Sediments: A Review

Chunliang Gao, Jinxin Wen, Xiuyun Min, Aiying Cheng, and Leiming Li

Abstract Mercury isotope compositions in unpolluted system as well as the anthropogenic Hg sources generally have different isotopic signatures. As a novel tracer, Hg stable isotopes can be useful to interpret the anthropogenic Hg sources. The review shows that: (a) Hg stable isotopic tracer mechanism; (b) Hg isotopic signatures originated from nature and human activities; (c) as a tracer to the historical record and metal pollution sources in the lake sediments. It is proposed that the focus of future study is to use a combination of theoretical and experimental methods to better understand mercury isotope fractionation mechanism, and to effectually recognize the Hg isotope compositions and metal pollution sources in the lake sediments.

Keywords Mercury isotopes · Lake sediment · Hg pollution sources · Environmental tracer

C. Gao (✉) · J. Wen · X. Min · A. Cheng · L. Li (✉)
Key Laboratory of Green and High-End Utilization of Salt Lake Resources, Qinghai Institute of Salt Lakes, Chinese Academy of Sciences, Xining, Qinghai, China
e-mail: gaochunliang@isl.ac.cn

L. Li
e-mail: lileiming@isl.ac.cn

C. Gao · J. Wen · A. Cheng
Qinghai Provincial Key Laboratory of Geology and Environment of Salt Lake, Xining, Qinghai, China

J. Wen
University of Chinese Academy of Sciences, Beijing, China

K. Y. Foo et al. (eds.), *Water Nexus Approach for Sustainable Development*, Environmental Science and Engineering, https://doi.org/10.1007/978-3-031-89238-7_9

9.1 Introduction

As the only toxic metal element that can stably exist in the form of a gaseous elemental mercury (Hg) at room temperature and pressure, Hg can migrate over long distances in the atmosphere and settle in different forms into water and terrestrial environments, transforming into highly toxic methylmercury (MeHg) that accumulates in organisms, causing ecological hazards [1]. Therefore, Hg pollution as a global environmental issue, causing the only heavy metal with Minamata Convention on Mercury. In the past century, human activities had greatly exacerbated Hg emissions and accumulation in the environment. A large amount of evidences indicated that industrial activities had increased atmospheric Hg concentrations by 300% to 500%, and surface marine Hg concentrations by 200% since the 20th century [1].

As an important host for the migration and transformation of various environmental substances in the basin, lake sediments record the information of regional environmental changes, which can reflect the impact of natural changes and human activities on the lake environment. Especially in the lake water-sediment system, more than 90% of Hg is adsorbed by the sediment, so the sediment can be used as a sink and a second source of Hg. Therefore, Hg pollution can be reflected in the lake sediments, and the source and process of Hg pollution can be determined by Hg isotopic characteristics. With the advancements of high-precision detection technology, the metal stable isotopes (e.g. Hg, Pb, Zn, Cd, etc.) has also been widely used in the study on pollution source apportionment [2–4]. Hg isotope tracer technique is a powerful source apportionment tool developed in recent 10 years, compared with the traditional estimation of atmospheric Hg emissions by simple mass balance model method, Hg isotope tracer technique provides more direct and accurate evidences [5]. At present, the focus of Hg stable isotopes as follows: (a) the mechanisms and fractionation [6]; (b) the method development on Hg isotope analysis [7]; (c) the tracer application in the ecological environment system [8, 9].

Finally, Hg stable isotopic tracer mechanism as well as the isotopic signatures originated from nature are briefly summarized, and followed by the tracing of Hg pollution sources in the lake sediments. This paper aims to summarize the progress on the Hg stable isotopes as a tracer to the historical record and metal pollution sources in the Lake Sediments. Particularly, human activities during the pre-industrial to present day period are highlighted.

9.2 Mercury Stable Isotopic Tracer Mechanism

Hg stable isotopes encompass δ^{196}Hg (0.15%), δ^{198}Hg (10.04%), δ^{199}Hg (16.94%), δ^{200}Hg (23.14%), δ^{201}Hg (13.17%), δ^{202}Hg (29.73%) and δ^{204}Hg (6.83%) [10]. In addition, Hg stable isotopes includes mass-dependent fractionation (MDF), odd-numbered isotope (^{199}Hg, ^{201}Hg) mass independent fractionation (odd-numbered

MIF) and even-numbered isotope (^{200}Hg, ^{204}Hg) mass independent fractionation (even-numbered MIF).

MDF and MIF is defined as $\delta^{xxx}Hg$ and $\Delta^{xxx}Hg$, respectively, as follows:

$$\delta^{xxx}Hg = (^{xxx/198}Hg_{sample}/^{xxx/198}Hg_{NIST3133} - 1) \times 10^3 \tag{9.1}$$

$$\Delta^{xxx}Hg = \delta^{xxx}Hg - \beta_{xxx} \times \delta^{202}Hg \tag{9.2}$$

Where ^{xxx}Hg is the measured ^{199}Hg, ^{200}Hg, ^{201}Hg, ^{202}Hg and ^{204}Hg, respectively; β_{xxx} is the scaling factor, generally 0.252 for ^{199}Hg, 0.502 for ^{200}Hg, 0.752 for ^{201}Hg and 1.493 for ^{204}Hg [10].

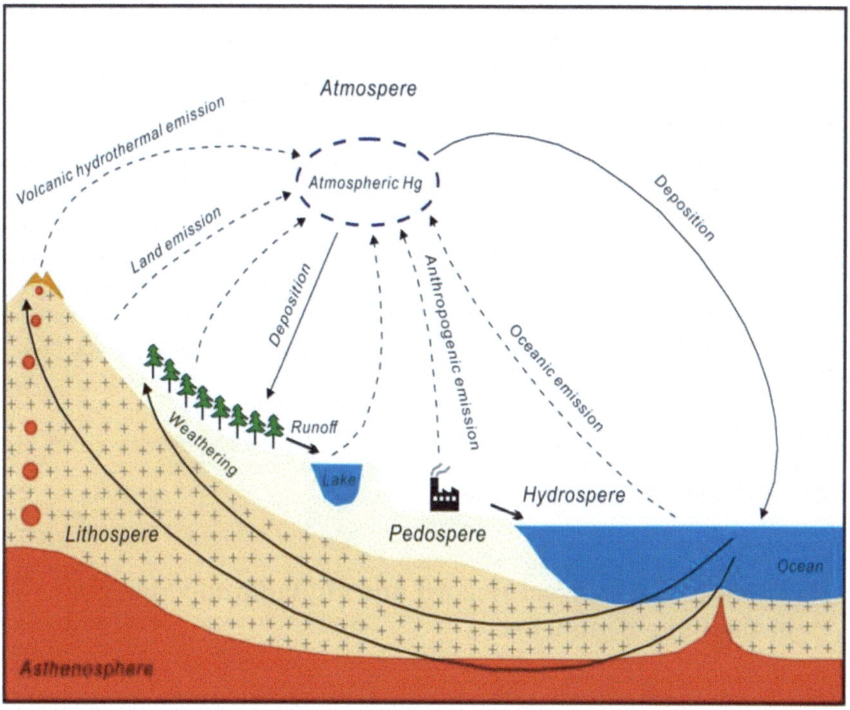

Fig. 9.1 The schematic diagram of Hg biogeochemical cycle

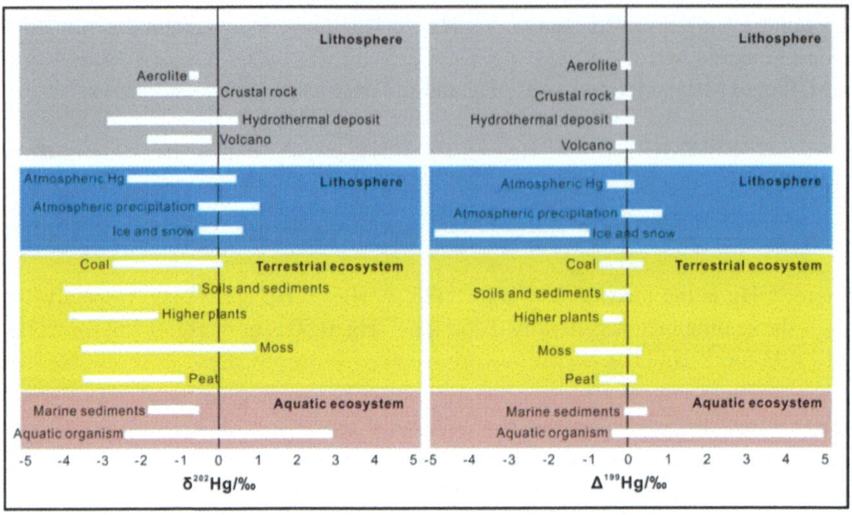

Fig. 9.2 Displaying the Hg isotope compositions in various geological samples [11]

9.3 Mercury Isotopic Characterizations in Nature

As shown in Fig. 9.2, the different Hg isotopic characterizations exist in the different reservoirs including lithosphere, terrestrial ecosystem and aquatic ecosystem on Earth. The changes of Hg stable isotopes in nature are caused by the fractionation effect in the processes of Hg biogeochemical cycle (Fig. 9.1). Therefore, the scientific construction of relationship between Hg isotope variation in nature and Hg isotope fractionation effect is very significant for understanding the global Hg biogeochemical cycle.

9.4 Tracing of Historical Record and Hg Pollution Sources in the Lake Sediments

Since the Industrial Revolution, the atmospheric Hg deposition in the global cryosphere has shown a rapid upward trend (Fig. 9.3), closely related to the large-scale Hg production and the rapid releasing Hg pollutant by global human activities. With the increasing awareness of environmental protection, developed countries in the North America and Europe have taken strong emission reduction measures in recent decades, and the Hg pollutant released by human activities in these regions have also shown a certain downward trend (Fig. 9.3). However, the Hg concentrations in ice core from Mt. Altai and Mt. Geladaindong in Asia have still shown a significant increase trend in recent decades, which is closely related to the rapid economic and industrial development in Asia.

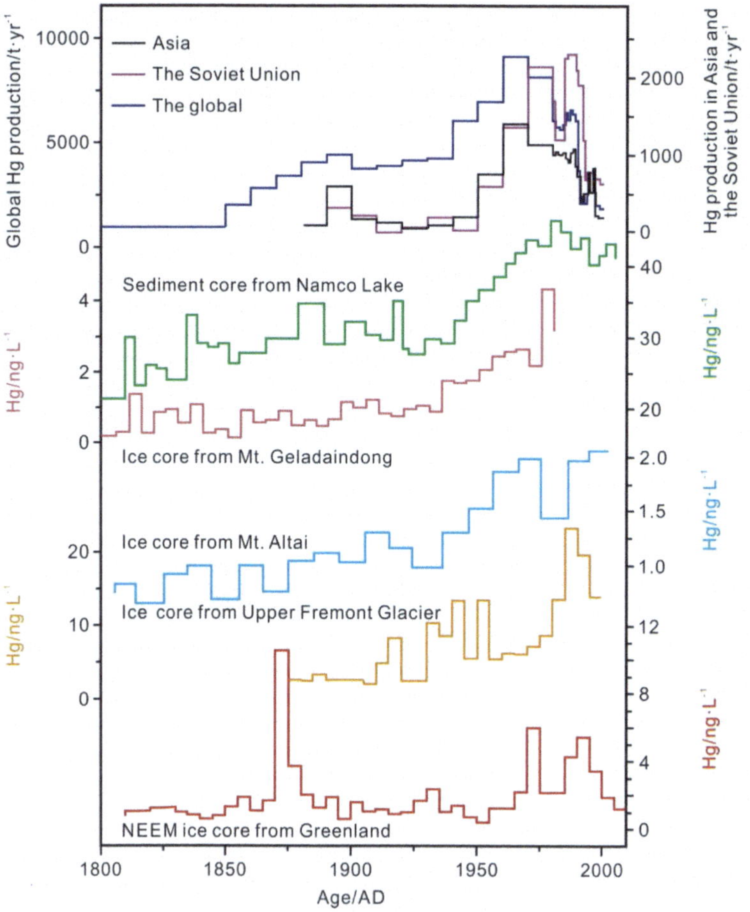

Fig. 9.3 Historical mercury records from the lake sediment core and ice core in different regions of the global cryosphere as well as the changes in Hg production [12]

On the Tibetan Plateau and Himalayas, the records from lake sediment core and ice core showed that, especially after 1950s, the rapidly increasing of atmospheric Hg deposition flux corresponded to the increasing of anthropogenic Hg emissions in the South Asia, indicating that anthropogenic pollutant emissions in the South Asia are the main reasons affecting the background and atmospheric Hg deposition flux on the Tibetan Plateau [3]. In the North America, Hg deposition in the lake sediment cores from northeastern United States was reconstructed, due to the complex redistribution of atmospheric Hg and other metals after deposition, the response was decoupage from the emission reduction [13]. In addition, the $\Delta^{199}Hg$ and $\Delta^{201}Hg$ values in the lake sediments from United States and Mexico ranged from + 0.52‰ to − 0.48‰, indicating that mercury in Florida lake sediments is primarily from the litterfall, anthropogenic Hg source exists in the Erie Lake, while the possible Hg source in

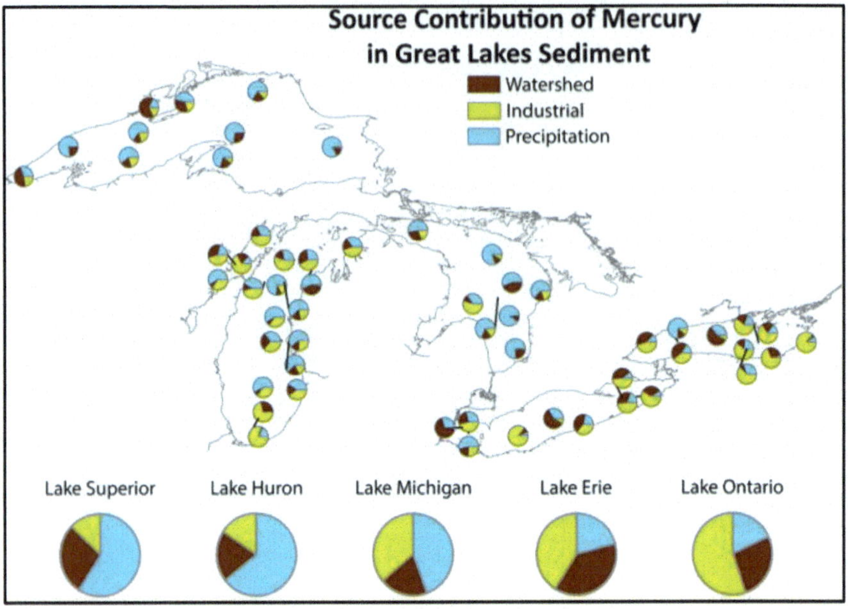

Fig. 9.4 Source contribution of great lakes in United States determined by Hg stable isotopic signatures [15]

Yucatan Peninsula is from the water column [14]. As shown in Fig. 9.4, Hg source contribution in the sediments from Great Lakes in United States were evaluated, the MDF and MIF (Δ^{199}Hg and δ^{202}Hg) results showed that the main Hg sources in the lake sediments are industrial activities, precipitation and watershed [15]. In Canada, the Hg isotope results showed that negative Δ^{199}Hg and negligible Δ^{200}Hg values were alike to the plants, terrestrial soils and atmospheric elemental Hg (0), suggesting that the main source of Hg was primarily originated from the terrestrial particle inputs in watershed [16].

Finally, to evaluate the factors caused by spatiotemporal variation, THg and Hg isotope ratios in the global lake sediment cores were compiled (Fig. 9.5). The results showed that THg together with the positive δ^{202}Hg shifts increased during pre-industrial to present day period was a common phenomenon, which was controlled by the increasing mercury contributions from local and regional anthropogenic sources [17]. It can be seen that, the Hg isotope values in the lake sediments show lower values before being contaminated by human activities or point sources, while higher values after being contaminated. There are differences in the mercury isotope compositions affected by different pollution sources.

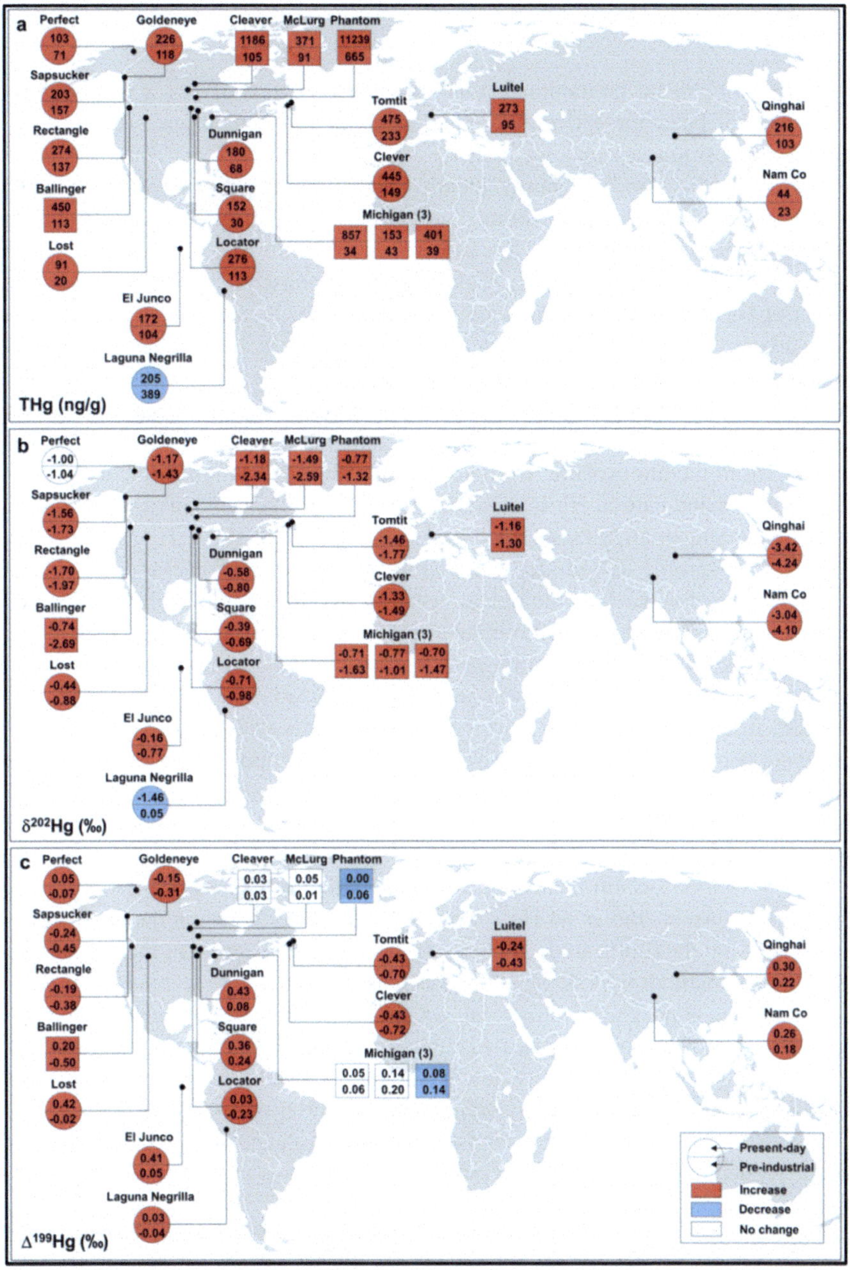

Fig. 9.5 Spatiotemporal variations in THg, δ^{202}Hg and Δ^{199}Hg in the global lake sediment cores during the pre-industrial to present day period [17]

9.5 Discussion and Conclusions

Hg is the only metal contaminant that can transport globally as a gas, thus its environmental behavior and impact have received wide attentions. The mercury isotope compositions are playing an increasingly important role in tracing Hg pollution sources. Most studies mainly identify the different pollution sources based on different Hg isotope compositions, combined with Hg concentrations and the chronology of core sediments. However, there are some issues in the process of tracing pollution sources with Hg isotope compositions: (a) the accurate isotope tracing of Hg source areas is challenging, not only needs to consider the diversity and overlap of Hg isotope compositions in the source areas, but also needs to recognize the fractionation effect of emission sources and atmospheric physicochemical processes on the initial isotope labeling in the source areas, especially the photochemical reduction mechanism and isotope fractionation mechanism of particulate divalent Hg in the atmosphere; (b) many studies clearly distinguish the sources of anthropogenic and natural Hg from sediments in the background area, nevertheless the significant uncertainty exists, such as the negative value of Δ^{199}Hg at the bottom in the vertical profile of sediments is due to the influence of background values, however, the factors such as biological disturbance, migration of pollution sources towards the bottom and physical mixing also caused the this variation; (c) when it is believed that the main sources of Hg pollution originate from the industrial sources, however, with the continuous development of urbanization, a large amount of Hg pollution from cities deposited into the lake sediments with atmospheric circulation, there is no clear distinction between Hg isotope composition of urban sources and industrial sources; (d) the understanding of mechanisms on Hg stable isotopic fractionation in biochemical processes in nature is not sufficient, especially in the atmospheric Hg cycle.

The Hg isotope system is unique as it exhibits multiple types of MIF in nature, whereas the current studies on Hg isotope system mostly focus on the application. We suggest that the future study should emphasize a combination of theoretical and experimental methods to understand the mechanisms of mercury stable isotope fractionation in the atmospheric, terrestrial, oceanic and polar environment, etc. Therefore, more study work should be carried out to investigate Hg stable isotopic fractionation in the different industrial activities, to recognize Hg stable isotopic compositions in sediments and aerosols, and to comprehend the variations of isotopic signatures in the ecosystem, especially the complicated system influenced by multiple sources and the related biogeochemical processes.

Acknowledgements This work was supported by the Science and Technology Plan Project of Qinghai Province (No. 2022–ZJ–982Q), "Kunlun Talents · High-end Innovative and Entrepreneurial Talents" Plan Project (Grant to CL Gao) and Chinese Academy of Sciences "Western Young Scholars" Talent Project (Grant to CL Gao).

References

1. UNEP (2019) Global mercury assessment 2018. https://web.unep.org/globalmercurypartnership/global-mercury-assessment-2018
2. Nitzsche KN, Yoshimura T, Ishikawa NF, Ogawa NO, Suzuki K, Ohkouchi N (2021) Trace metal geochemical and Zn stable isotope data as tracers for anthropogenic metal contributions in a sediment core from lake Biwa. Japan Appl Geochem 134:105107
3. Kang SC, Huang J, Wang FY, Zhang QG, Zhang YL, Li CL, Wang L, Chen PF, Sharma CM, Li Q, Sillanpää M, Hou JZ, Xu BQ, Guo JM (2016) Atmospheric mercury depositional chronology reconstructed from lake sediments and ice core in the Himalayas and Tibetan Plateau. Environ Sci Technol 50:2859–2869
4. Xia YF, Liu YH, Chen T, Xu YD, Qi M, Sun GY, Wu X, Chen MJ, Xu WP, Liu CS (2024) Combining Cd and Pb isotope analyses for heavy metal source apportionment in facility agricultural soils around typical urban and industrial areas. J Hazard Mater 466:133568
5. Yang SC, Li P, Sun KF, Wei N, Liu JL, Feng XB (2024) Mercury isotope compositions in seawater and marine fish revealed the sources and processes of mercury in the food web within differing marine compartments. Water Res 241:120150
6. Zheng W, Demers JD, Lu X, Bergquist BA, Anbar AD, Blum JD, Gu B (2019) Mercury stable isotope fractionation during abiotic dark oxidation in the presence of thiols and natural organic matter. Environ Sci Technol 53:1853–1862
7. Zhang W, Sun GY, Yin RS, Feng XB, Yao ZX, Fu XW, Shang LH (2021) Separation of methylmercury from biological samples for stable isotopic analysis. J Anal At Spectrom 36:2415–2422
8. Jung S, Kwon SY, Hong Y, Yin RS, Motta LC (2021) Isotope investigation of mercury sources in a creek impacted by multiple anthropogenic activities. Chemosphere 282:130947
9. Cai N, Wang XP, Zhu HX, Hu Y, Zhang XY, Wang LQ (2024) Isotopic insights and integrated analysis for heavy metal levels, ecological risks, and source apportionment in river sediments of the Qinghai-Tibet Plateau. Environ Res 251:118626
10. Blum JD, Bergquist BA (2007) Reporting of variations in the natural isotopic composition of mercury. Anal Bioanal Chem 388:353–359
11. Yin RS, Feng XB, Li XD, Yu B, Du BY (2014) Trends and advances in mercury stable isotopes as a geochemical tracer. Trends Environ Anal 2:1–10
12. Kang SC, Huang J, Mu CC, Zhang YL, Xu JZ, Dong ZW, Du WT (2020) Cryospheric chemistry: fingerprint to decipher climate/environmental changes and anthropogenic activities. Proc Chin Aca Sci 35:456–465
13. Taylor VF, Landis JD, Janssen SE (2022) Tracing the sources and depositional history of mercury to coastal northeastern US lakes. Environ Sci Proc Imp 10:1805–1820
14. Das R, Landing W, Bizimis M, Odom L, Caffrey J (2015) Mass independent fractionation of mercury isotopes as source tracers in sediments. Proc Earth Planet Sci 13:151–157
15. Lepak RF, Yin RS, Krabbenhoft DP, Ogorek JM, DeWild JF, Holsen TM, Hurley JP (2015) Use of stable isotope signatures to determine mercury sources in the great lakes. Environ Sci Technol 2:335–341
16. Chen JB, Hintelmann H, Zheng W, Feng XB, Cai HM, Wang ZH, Yuan SL, Wang ZW (2016) Isotopic evidence for distinct sources of mercury in lake waters and sediments. Chem Geol 426:33–44
17. Lee JH, Kwon SY, Yin RS, Motta LC, Kurz AY, Nam S-I (2021) Spatiotemporal characterization of mercury isotope baselines and anthropogenic influences in lake sediment cores. Global Biogeochem Cy 35:e2020GB006904

Chapter 10
Design and Implementation of a New Type of Hydrazine Sewage Treatment System

Feiran Xu, Xiaochun Zhang, Zengguang Li, Xiaobo Zhang, and Yiyong Lin

Abstract Through the comparison of natural purification method, ion exchange method, ozone-ultraviolet-activated carbon combined treatment and other methods at home and abroad, the optimal treatment method of ozone-ultraviolet-activated carbon combined treatment was selected. However, in view of the problem of reducing the capacity of the three-stage combined treatment system of a sewage treatment system, combining the present situation and many factors, the optimization design was carried out from the aspects of process equipment, process flow, software module, etc., which solved the difficult problem of the three-stage combined treatment system, which made the processing capacity double and the discharge water test completely qualified. The design of software control system has greatly improved its related security. Finally, a new method to improve the capacity of hydrazine waste water treatment system by adding small mobile photo catalytic oxidation waste water treatment equipment, corrosion inhibitor and nano-TiO_2 photo catalytic oxidation was also explored.

Keywords Nitrogen tetroxide · Unsymmetrical dimethyl hydrazine · Ozone-ultraviolet-activated carbon combined treatment · Design and implementation

10.1 Introduction

Hydrazine, especially nitrogen tetroxide (N_2O_4) and Unsymmetrical dimethyl hydrazine (UDMH), are toxic substances, but they are widely used because of their good combustion-supporting ability. Untreated discharge can seriously pollute the water source and surrounding environment, thus endangering people's health. At present, hydrazine waste water treatment methods mainly include heating method, physical adsorption method, microbial degradation method, ultraviolet catalytic

---

F. Xu · X. Zhang · Z. Li · X. Zhang · Y. Lin (✉)
Xichang Satellite Launch Center, Xichang, Sichuan, China
e-mail: 799729599@qq.com

K. Y. Foo et al. (eds.), *Water Nexus Approach for Sustainable Development*, Environmental Science and Engineering, https://doi.org/10.1007/978-3-031-89238-7_10

method and chemical oxidation method. These methods can remove hydrazine pollutants in waste water to the maximum extent, but there are some problems such as easy secondary pollution, serious energy consumption, high cost, long treatment period and poor treatment effect. Therefore, it is urgent to seek new technologies and methods for the treatment of hydrazine waste water, so as to realize the non-toxic treatment of hydrazine waste water. At present, there are mainly several kinds of waste water treatment including hydrazine [1–7]:

Activated carbon adsorption method. Activated carbon has strong adsorption, so it is one of the most important methods for the treatment of UDMH waste water. The disadvantage of this method is that the input-output is relatively large, because it is necessary to reprocess the released UDMH when the activated carbon is regenerated.

Natural purification method. UDMH and its harmful decomposers in water can be decomposed under the condition of natural oxygen and light, so that the sewage can be purified. This method is an effective, economical and simple method for waste water treatment. However, the disadvantage is that the treatment time is longer, the requirements for dissolved oxygen and light conditions are higher, and ammonia and hydrazine volatiles can be found in the upper part of the sewage treatment tank.

Ion exchange method. The basic principle of ion-exchange method is to use the adsorption and exchange of ion-exchange resin, through the exchange of metal ions and ion-exchange resin in water, so as to achieve the goal of water treatment. However, the method has more investment at one time, too many cations such as Fe^{3+}, Ca^{2+} and Mg^{2+} in the sewage will affect the resin exchange ability, and the UDMH in the sewage is not destroyed, but transferred to the ion exchange resin, which will produce secondary pollution.

Chlorination. Chlorination method generally includes two methods: hypochlorite method and UV-chlorination method. For hypochlorite method, it is found that the products of the oxidation-reduction reaction of UDMH and calcium hypochlorite are FMH, FDH, DMF and so on. In the reaction of hydrazine with hypochlorite, nitrosamines and chlorinated hydrocarbons are produced. These substances are more toxic, and the secondary pollution will be caused when they are discharged into the water body. In addition, a large amount of heat will be released during the reaction. For UV chlorination, chlorine gas and UDMH can react quickly under UV irradiation. The reaction formula is as follows:

$$(CH_3)2NNH_2 + 2Cl_2 + H_2O \rightarrow 2CH_3OH + 4HCL + N_2 \tag{10.1}$$

The study on the effect of UV light intensity on the reflection rate shows that the light intensity of 1 W/L is 20 times of that of 0.1 W/L. The advantage of UV-chlorination is that the treatment speed is fast and the treated sewage is non-toxic.

Plasma technology. Compared with conventional waste water treatment methods, plasma is a system produced by partial ionization or complete ionization of gas under certain conditions, which can realize thermodynamic chemical reaction at lower energy [8]. At the same time, its secondary pollution to the environment is very small. From the point of view of application, corona discharge, dielectric barrier discharge and sliding arc discharge are easy to be used in large-scale discharge plasma. However, the electrode design is complicated and the energy consumption is high. At present, most of the dielectric barrier discharge reactors are still in the laboratory research stage.

Ozone-ultraviolet-activated carbon combined treatment method. By means of ozone oxidation, activated carbon adsorption and ultraviolet photochemistry, a series of chemical reactions between pollutants and ozone occur in the waste water under ultraviolet irradiation and activated carbon catalysis, resulting in oxidative degradation and becoming toxic or low toxic substances, thus achieving catalytic purposes. The advantages of this method are as follows: the decomposition of harmful substances is thorough; the time-consuming is short, the treatment is simple and easy to operate; the sewage can be discharged directly after treatment. However, the method consumes a large amount of water, and the cost of equipment construction, management and maintenance is high.

Therefore, the combined treatment method of ozone-UV-activated carbon is better, but there is still a lot of room for design and improvement in the use and management of the equipments.

10.2 Analysis of the Present Situation of Sewage Treatment System

10.2.1 Analysis of Water Quality Status

The waste water treated by a sewage treatment system mainly contains toxic substances such as N_2O_4 and UDMH. The main pollutants in sewage are combustion products and washing products such as N_2O_4 and UDMH, which are toxic and complex. In general, the content of UDMH in waste water is 10 mg/L ~ 50 mg/L. Considering the influence of other substances, the content of partial UDMH is 50 mg/L in the design. Therefore, the sewage that this system needs to deal with is different from domestic sewage, marine fishery sewage and so on, mainly is to deal with the toxic sewage which contains the partial UDMH and its combustion product and the washed-out product, as well as the sewage that the laboratory tests N_2O_4 and the partial UDMH and so on. If the content of toxic substances such as UDMH in the effluent is too high, the effluent quality is difficult to reach the standard. The treated wastewater should meet the requirement of < hydrazine fuel and nitro oxidant waste water treatment and discharge >.

10.2.2 Analysis of the Present Situation of Sewage System

Through comparative study at home and abroad, the waste water treatment system adopts a better ozone-ultraviolet-activated carbon three-stage combined treatment method. The system is mainly composed of four parts: sewage collection, ozone production, combined treatment and control system. Using the principles of activated carbon adsorption, ultraviolet light catalysis and ozonation, the three-stage combined treatment equipment is composed of "ozone-UV-activated carbon". The principle is shown in Fig. 10.1.

According to the total design water quantity of the sewage system, the design treatment capacity is 3 m^3/h (In accordance with the treatment of 200 m^3 sewage). However, with the increase of sewage quantity in recent years, the treatment capacity of this sewage treatment system is far from enough. At the same time, there are some problems in sewage treatment system, such as low treatment efficiency, high failure rate and serious pipeline corrosion aging. The maximum processing capacity of the system is reduced to 2.5 m^3/h, and the plasma generator can be automatically shut down due to the instability of pressure and water pressure, and the processing capacity is reduced to 2 ~ 2.5 m^3/h. At the same time, with the decrease of the treatment capacity, the treated sewage also appears the condition that the test is not up to standard, thus pollutes the environment and causes the personal injury. Therefore, the treatment capacity of the three-stage sewage treatment system is under severe test, and it is urgent to improve the design.

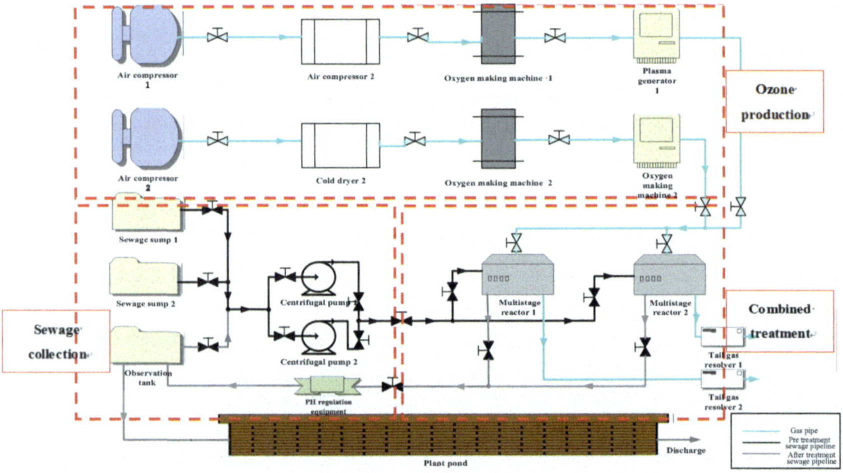

Fig. 10.1 Working schematic diagram of three-stage combined sewage treatment system

10.3 Design and Implementation of New Sewage Treatment System

10.3.1 Technology Design of New Process Equipment

Figure 10.2 shows the current process flow of the sewage treatment system. The specific process of the method is as follows: the sewage in the sewage tank is extracted by the sewage pump, which passes through the mechanical filter tower, the ultraviolet oxidation tower and the activated carbon filter tower, and finally enters the sewage observation tank; at the same time, the air produced by the air compressor is separated by a cyclone and purified by drying, and then enters the ozone generator, and the ozone is produced by discharge under the action of high voltage; Ozone enters the UV oxidation tower to oxidize the waste water at the same time with the UV light. The treated sewage can be discharged directly from the sewage observation tank, and the untreated sewage can be retreated through the process.

At present, with the expansion of the industry, the amount of sewage has doubled (400 m^3 sewage) compared with the previous one, so as to increase the amount of sewage (double), after improving the design, the design treatment capacity needs to reach 6 m^3/h (originally 3 m^3/h). Among them, plasma generator, multi-stage reactor and ozone preparation (Fig. 10.1) are the core parts of the system. If we can improve the design and enhance the capability of the core equipment, the capability of the hydrazine waste water treatment system can be greatly improved.

Plasma generator design improvement.

The plasma generator can convert oxygen into ozone and other active gases in the case of discharge. Ozone is a strong oxidant that oxidizes UDMH into harmless compounds. At present, two plasma generator ionization chamber leakage serious, resulting in automatic shutdown of equipment; sewage treatment room water supply pressure is too low (requirements 0.2 Mpa, actually less than 0.1 Mpa), only barely

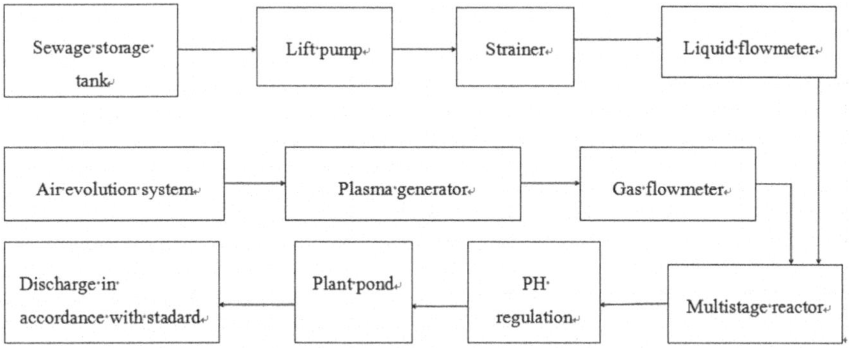

Fig. 10.2 Process flow of sewage treatment system

meet the plasma generator water cooling device water, can not guarantee other work at the same time water. Therefore, it is necessary to replace the plasma generator to improve its equipment performance. At the same time, the water pressure should be raised to more than 0.6 MPa. If the water source is changed from domestic water to working water, it can meet the requirement.

Improvement of multi-stage reactor processing capacity design.

The multi-stage reaction tower is generally divided into three levels. The first-stage reaction tower is designed as a partial UDMH reaction tower, which uses micro-porous aeration tube to distribute gas and is filled with granular activated carbon to enrich oxidation. The second tower is designed as formaldehyde reaction tower, because of the slow reaction of formaldehyde, so it is designed to increase the ultra-violet lamp and catalyst, strengthen the effect of treatment, the design adopts 4 small reaction tanks in series arrangement. The third stage reactor is filled with activated carbon fillers, and the waste water is adsorbed and treated to ensure the water quality is up to standard. At present, the maximum treatment capacity of the multi-stage reactor is less than 3 m^3/h (2 sets), mainly due to the serious aging of the silicon-rubber aeration pipe and gas connection pipe in the multi-stage reactor, which leads to over-load operation, gas leakage and leakage. If the maximum treatment capacity is 2.5 m^3 ~ 3.0 m^3 per hour and the total treatment capacity is higher than 5 m^3/h (2 sets), the current sewage treatment capacity can be satisfied.

Ozone generator (Ozone production part) ozone production capacity design improvement.

The ozone production part consists of air compressor, cold dryer, oxygen generator, plasma generator and tail gas resolver, as shown in Fig. 10.1. The working flow is as follows: air is compressed by air compressor and then cooled and dried by air cooler, then purified into oxygen by oxygen generator, then into plasma generator to produce ozone, ozone is then entered into multi-stage reactor for sewage treatment, and the excess ozone is decomposed by tail gas decomposer and discharged into atmosphere.

Generally speaking, the sewage treatment system with three-stage combined treatment is composed of two sets of identical treatment equipment, each of which has four sub-systems: gas circuit system, waterway system, circuit system and control system. Each sub-system plays an important role: the air system is prepared to provide ozone and deal with excess ozone, the water system extracts sewage, the circuit system supplies power to each equipment, and the control system operates real-time. The ozone generator can convert oxygen into ozone and other active gases in the case of discharge. Ozone is a strong oxidant that oxidizes UDMH into harmless compounds. The experimental results show that the ratio of O_3 to UDMH should reach 6:1 for the waste water treatment quantity which has doubled at present. The concentration of UDMH is calculated at 50 mg/L (taking into account the consumption of other products in the waste water), and the amount of ozone required for the treatment of waste water of 6 m^3/h is 1800 g/h.

In order to increase ozone concentration, oxygen is used as the gas source of the generator. Considering a certain amount of surplus, two sets of ozone generating devices should be selected (1 kg/h per set, 100 mg/L ~ 135 mg/L of ozone concentration), 2 m^3/h of external circulation cooling water flow, and about 10 m^3/h of oxygen are needed to meet the demand.

Before upgrading, the ozone generator used in the system is 80 ~ 100 mg/L with a production capacity of 400 g/h (2 sets); after upgrading, the ozone generator with an ozone output concentration of 100 mg/L ~ 135 mg/L is designed, and the new ozone system is composed of 2 oxygen-enriched 1 kg/h ozone generators, which are equipped with oxygen-making system, water cooler system, feeding system, exhaust gas destruction system, automatic control system and monitoring instrument. The ozone output of this ozone system increased by 2.5 times compared with that of the previous system. As shown in the following Formula (10.2).

$$1\text{kg/h} \times 2 \div (0.4\text{g/h} \times 2) = 2.5 \tag{10.2}$$

According to the operating process of the sewage treatment system, the sewage test is carried out at the operating table for 7 days, 8 h per day, and the treatment of the sewage is 400 m^3. As shown in the following Formula (10.3), the sewage treatment system can reach the treatment capacity of 7.2 m^3/h, exceeding the design capacity of 6 m^3/h, and meet the requirements of < hydrazine fuel and nitro oxidant sewage treatment and discharge requirements > after treatment (Table 10.2).

$$400 \div 7 \div 8 = 7.2\text{m}^3/\text{h} \tag{10.3}$$

10.3.2 Software Control System Design

Control programming.

The whole sewage treatment system consists of manual and automatic control of electric equipment such as waste water self-priming pump, ozone generator, metering pump and so on. Automatic control executes predetermined control logic, manual control, according to the actual process to open the corresponding control equipment.

(1) To determine whether or not it is in the boot state, otherwise it will end directly.
(2) Judging control mode, automatic or manual.
(3) If it is manually controlled, open the corresponding device according to the click instruction.
(4) If it is automatic control, control the action of the equipment according to the logic operation (when it is automatic, click on the start 1 # reaction tower to start the feed water electric valve, self-priming pump, ultraviolet lamp, ozone

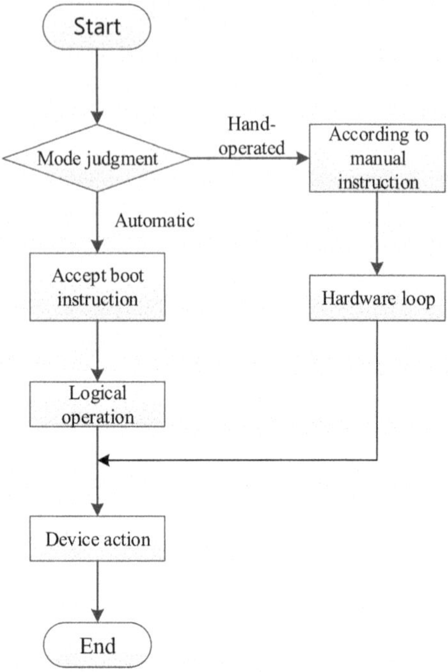

Fig. 10.3 Equipment control flow chart

generator, pH automatic adjustment according to the program setting). As shown in Fig. 10.3.

PH value PID control design.

The module mainly adjusts the speed of metering pump, monitors the pH value of the treated waste water, and realizes the function of controlling metering pump according to the concentration value. According to the output signal of pH secondary instrument, the speed of metering pump can be adjusted by feedback. The automatic program setting function can correspond to different pump flow in different pH range.

(1) To determine whether or not it is in the boot state and under the automatic control condition, otherwise it will end directly.
(2) Calculate the new value of the control quantity.
(3) Comparing the new value of the calculated control quantity with the set value, the cumulative result is calculated.
(4) Feedback the calculation results to the metering pump and adjust its speed. Figure 10.4 shows.

Fault detection and early warning system design.

Aiming at the practical requirement and operation complexity of sewage treatment, this paper aims to design a fault detection and early warning system with clear structure, quick response and high accuracy. The following are detailed design ideas and implementation processes [9].

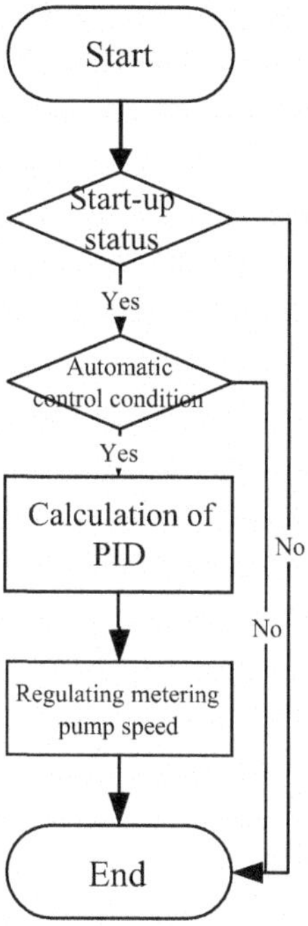

Fig. 10.4 PH metering pump speed flow chart

Ozone generator fault design.

This module mainly carries on the unified monitoring processing to the ozone generator equipment many kinds of fault information, and sends the alarm signal, performs the alarm function. According to the demand of alarm function of ozone generator system, it mainly reads more than ten kinds of fault signals such as temperature switch stop alarm, low flow secondary alarm, power failure stop alarm, scram stop alarm, ozone leakage secondary alarm and so on.

(1) To determine whether or not it is in the boot state, otherwise it will end directly.
(2) To determine whether the failure condition is satisfied.
(3) If the failure condition is satisfied, the secondary failure alarm shall be carried out and the shutdown disposal shall be carried out.
(4) The ozone generator failure and alarm flow chart is shown in Fig. 10.5

Alarm system design.

The alarm design effectively improves the treatment capacity of the sewage treatment system. This module mainly deals with various alarm signals of ozone concentration alarm, observation tank liquid level alarm, sewage tank liquid level alarm, rectifier current value alarm and so on. Figure 10.6 shows.

(1) To determine whether or not it is in the boot state, otherwise it will end directly.
(2) To determine whether the alarm conditions are satisfied.

Fig. 10.5 Trouble and alarm flow chart of ozone generator

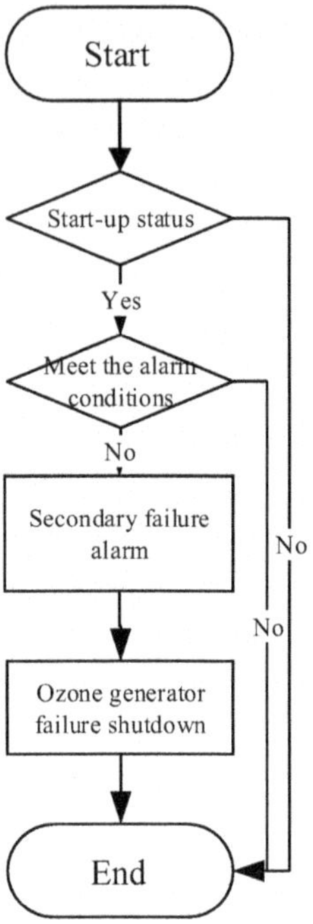

(3) If the failure condition is satisfied, the alarm shall be given and the corresponding disposal shall be carried out. As shown in Fig. 10.6.

10.3.3 New Design, Application and Effect Comparison

After the design optimization of the new technology, new equipment and new software, the treatment capacity of the sewage treatment system has been greatly improved. The following are the results of comparison after the new technology is put into use.

Comparison of sewage treatment capacity enhancement.

After optimized design, the treatment capacity of sewage treatment system was increased from 2.5 m^3/h to 7.2 m^3/h. The ability to meet the design requirements of 6 m^3/h is shown in Table 10.1.

Analysis and comparison of discharge water indicators after optimized design.

The design takes the content of partial UDMH in waste water as 50 mg/L. The treated waste water should meet the requirements of Table 10.2 for treatment and discharge of hydrazine fuel and nitro oxidant. It is pointed out that there are many kinds of substances in the waste water, mainly hydrazine, UDMH, MMH and their

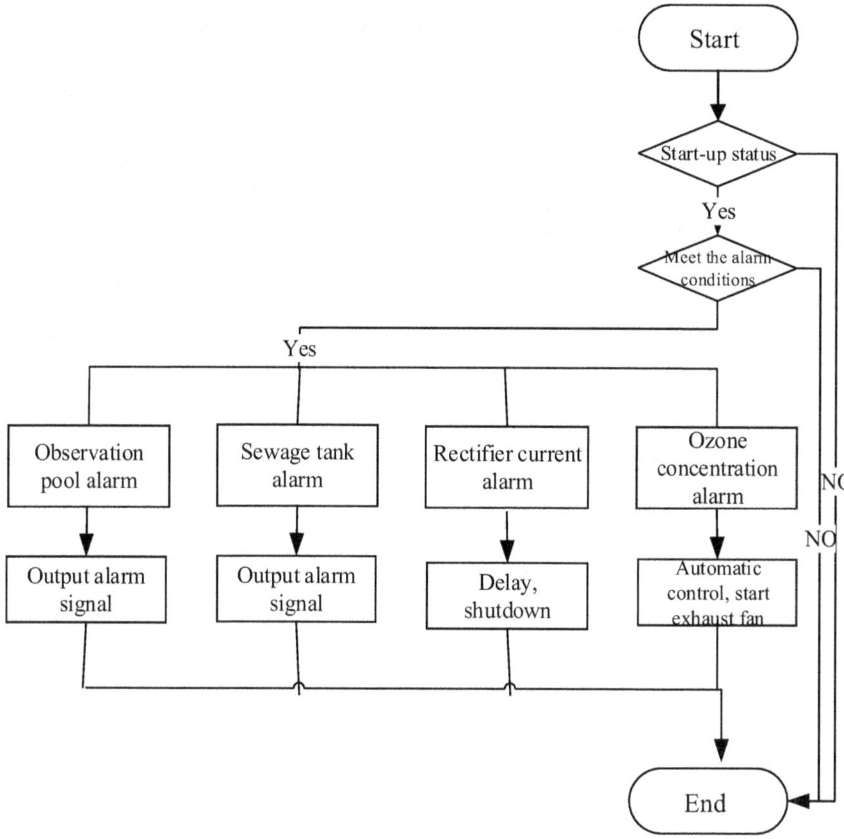

Fig. 10.6 Flowchart of alarm and disposal

Table 10.1 Comparison of system processing capability

Ordinal number	Nomenclature	Processing ability and performance index before upgrading	Design processing ability and performance index	Actual processing capacity and performance index	Improvement rate of effect (%)
1	Processing capacity	2.5 ~ 3 m^3/h	6 m^3/h	7.2 m^3/h	188
2	Ozone production	0.4 kg/h (2 sets)	1 kg/h (2 sets)	1 kg/h (2 sets)	150
3	Ozone concentration	80 ~ 100 mg/L	100-135 mg/L	100–142 mg/L	42

Table 10.2 Sewage discharge standards for hydrazine fuels and nitro oxidants

Ordinal number	Item	Maximum permissible emission standard	Standard number
1	PH	6~9	GB 6920
2	Suspended matter	150 mg/L	GB 11,901
3	Five-day biochemical oxygen demand (BOD_5)	30 mg/L	HJ 505
4	Chemical oxygen demand	120 mg/L	HJ/T 399
5	Cyanide	0.5 mg/L	HJ 484
6	Formaldehyde	2.0 mg/L	GB 13,197
7	Unsymmetrical dimethyl hydrazine (UDMH)	0.5 mg/L	GB/T 14,376
8	Monomethyl hydrazine	0.2 mg/L	GB/T 14,375
9	Hydrazine	0.1 mg/L	GB/T 15,507
10	Ammonia nitrogen	25 mg/L	HJ 535
11	Nitrate nitrogen	30 mg/L	GB 7480
12	Nitrite nitrogen	0.1 mg/L	GB 7493

intermediates, which are reductive substances in chemical properties. The effect of conventional treatment is not ideal, it will form nitrogen tetroxide (melt water into nitric acid), cyanide and formaldehyde, among which formaldehyde is the most difficult to decompose, nitrosamines are more carcinogenic, and pH can not control [10]. According to the discharge requirement, a sewage test item and method as shown in Table 10.2 shall be selected at the time of testing according to the actual situation of composition analysis in the produced sewage. It has been proved that the newly designed ozone-ultraviolet-activated carbon combined treatment of hydrazine waste water has a good effect.

According to the standard requirements, the waste water before and after treatment will be compared after the system capacity optimization design. Among them, the test before sewage treatment is unqualified. And the test indexes after sewage treatment meet the standard requirements. Divide the generated sewage (400 m^3) into three parts, each treated, tested, and the results are shown in Table 10.3.

Security comparison after optimized design.

The newly designed waste water treatment system has a good effect on the treatment of waste water containing hydrazine. Through practical application and water quality detection, it is found that the combined treatment of ozone-ultraviolet-activated carbon has good oxidation effect, and the indexes in the discharged water can meet the discharge requirements without secondary pollution.

Ozone tail gas can do no harm to people's health. Generally speaking, in the process of wastewater oxidation reaction, it will lead to the discharge of active gases such as ozone, which is not fully involved in the reaction. However, after improving the design, by introducing the tail gas into the decomposition of the tail gas, by

Table 10.3 Comparison of test indexes before and after treatment of sewage treatment system

Ordinal number	Item	Maximum permissible emission standard	Test results of sewage before treatment with qualified system	Sewage test results after system treatment(1)	Sewage test results after system treatment(2)	Sewage test results after system treatment(3)
1	PH	6~9	8.2	7.3	7.4	7.7
2	Formaldehyde	2.0 mg/L	0.18 mg/L	0 mg/L	0 mg/L	0 mg/L
3	Suspended matter	150 mg/L	62.4 mg/L	23.2 mg/L	22.1 mg/L	21.8 mg/L
4	Unsymmetrical dimethyl hydrazine (UDMH)	0.5 mg/L	0.04 mg/L	0 mg/L	0 mg/L	0 mg/L
5	Chemical oxygen demand	120 mg/L	47.0 mg/L	8.0 mg/L	8.5 mg/L	9.0 mg/L
6	Nitrite nitrogen	0.1 mg/L	0.56 mg/L (Disqualification)	0.02 mg/L	0.01 mg/L	0.01 mg/L
7	Monomethyl hydrazine	0.2 mg/L	0 mg/L	0 mg/L	0 mg/L	0 mg/L
Conclusion			Disqualification	Qualified	Qualified	Qualified

heating decomposition into oxygen, the concentration can reach 0.1 ppm, completely harmless to the human body.

The ozone detector is installed in the installation room of the multi-stage reactor and the ozone generator. At the same time, the relevant software control alarm system is designed. If the harmful gas in the room exceeds the set limit, the safety measures such as the alarm system, the start of the forced exhaust air and so on are adopted to ensure that the handling system does not cause any hidden danger.

Application conclusion.

Therefore, the treatment capacity of the newly designed sewage treatment system has been increased from 2.5 m^3/h to 7.2 m^3/h after practical application. It can meet the design requirement of 6 m^3/h. At the same time, the test results of discharge water after optimization design can meet the standard requirement and meet the discharge requirement. Especially the highly toxic substances, such as UDMH, formaldehyde and monomethyl hydrazine, are all 0 mg/L after systematic treatment. The addition of control system and alarm system also make the security greatly improved. Therefore, the new design is reasonable, effective and safe and environment-friendly.

10.4 Exploration of New Process and New Methods

10.4.1 Increase of the Latest Mobile Photocatalytic Oxidation Waste Water Treatment Equipment

Mobile photocatalytic oxidation waste water treatment equipment-VU-V coupling advanced oxidation catalytic treatment equipment, the processing capacity of each set can reach 1.5–2 m^2/h. At present, some sewage treatment methods are to use large-scale container to transfer the sewage to the sewage treatment room, the process is cumbersome and often spills in the transfer process. The device is fast and convenient, easy to maintain and efficient to handle. This method can be used as backup and complement each other with the three-level joint processing method.

10.4.2 Study on Antisepsis Design of Sewage Collection System

The corrosion type caused by sewage is mainly nitric acid corrosion, which causes large area corrosion of sewage collection pond, treatment pond and sewage well, and is easy to produce leakage and pollute the environment. Therefore, chemical method can be used to change the PH value or corrosive property of sewage, so as to achieve the purpose of anti-corrosion and environmental pollution.

(1) Addition of corrosion inhibitor. Generally speaking, in nitric acid, the presence of nitrous acid can promote corrosion, so if you add the substances that destroy nitrous acid or choose the compounds that can promote the decomposition of nitrous acid as corrosion inhibitor, it can play a good role in anti-corrosion. The compound containing nitrogen, oxygen and sulfur and the compound which can decompose nitrous acid and the compound inhibitor have strong inhibition effect in nitric acid. Among them, aldehydes, amines and sulfur-nitrogen complexing agents have the best effect.
(2) Adjusting PH value. PH value is an important factor that affects the treatment effect of ozone oxidation. The reaction of ozone with OH− can promote the decomposition of ozone and is beneficial to the production of ·OH, so the oxidation ability of ozone under alkaline conditions is stronger. The pH value has a significant effect on the decomposition reaction of ozone. Under strong acid and alkaline conditions, the decomposition rate of ozone is significantly higher than that under neutral conditions.

10.4.3 Study on the Photocatalytic Oxidation of TiO_2

Nanomaterials as new functional materials have attracted great attention in the academic field. Nanomaterials have a series of new physical and chemical properties; small size effect; surface effect quantum size effect and macroscopic quantum tunneling effect. Nano-scale TiO_2 photocatalytic oxidation is adopted, which is a deep oxidation process. TiO_2 is a kind of non-selective strong oxidant, which can oxidize many kinds of refractory organic compounds. It is reported in the literature that the system composed of H_2O_2, TiO_2 and Cu^{2+} can degrade UDMH very well. Nano-TiO_2 is harmless to human body, and its optical properties accord with Rayleigh light scattering theory, which can pass through visible and ultraviolet light. However, there are still some technical problems in the application of this method, such as the dispersion of nano-particles, etc. [11].

10.5 Conclusions

Through the comparison of natural purification method, ion exchange method, chlorination method, plasma technology, ozone-ultraviolet-activated carbon combined treatment and other methods at home and abroad, the optimal method of ozone-ultraviolet-activated carbon combined treatment was selected. According to the total design water (200 m^3) quantity of the sewage system, the design capacity is 3 m^3/h. However, in view of the increase of the amount of sewage (400 m^3) and the decrease of the capacity of the sewage treatment system, combined with the present situation and many factors, the optimization design was carried out from the aspects of process equipment, process flow, software module, etc., and the difficulty of the three-stage combined treatment equipment was solved, so that the treatment capacity was increased from 2.5 m^3/h to 7.2 m^3/h, and the discharge water test was completely qualified. The addition of software control system has greatly improved its related security. Finally, a new method to improve the capacity of hydrazine waste water treatment system by adding small mobile photocatalytic oxidation waste water treatment equipment, corrosion inhibitor and nano-TiO_2 photocatalytic oxidation was also explored. Thus, the overall operation efficiency of the sewage system is improved, and the development and innovation of hydrazine wastewater treatment technology are promoted.

References

1. Wu Y, Zhao B, Xu ZL (2021) Progress in degradation treatment of dimethyl hydrazine wastewater. Chem Propellant Polym Mater 19(3):32–36
2. Guo Z, Zhang ZJ, Cui HY (2021) Efficient treatment of hydrazine propellant waste water by ozone micro-nano-bubble technology. Chem Propellant Polym Mater 19(4):63–68
3. Hu B, Yang YK, Zhang X (2023) Design of the treatment system for industrial wastewater treatment in Chemical Park. Water Supply Drainage China 9(6):66–70
4. Denise G, Gabrliela MM, Armando Al (2022) Greening urban areas with decentralized waste water treatment and reuse: a case study of Ecoparque in Tijuana, Mexico. Water 14(4):596
5. He YY, Wang YF, Shi JF et al (2022) Incorporating metal–organic frameworks into substrates for environmental applications. Chem Eng J 446:2
6. Zhang M, Sun WQ, Zhou J (2023) Research progress on mechanism of catalytic ozonation and preparation of catalysts. Water Purif Technol 42(8):20–28
7. Domingues E, Silva MJ, Vaz T et al (2022) Sulfate radical based advanced oxidation processes for agro-industrial effluents treatment: a comparative review with Fenton's peroxidation. Sci Total Environ 832:155029
8. Neyts EC, Ostrikov K, Sunkara MK (2015) Plasma catalysis: synergistic effects at the nanoscale. Chem Rev 115:13408–13446
9. Liu YH (2019) Design of automatic control system for chemical waste water treatment based on PLC. China Petrol Chem Stan Qual 43(11):178–180
10. Jia Y, Cui H, Mu XG (2016) Propellant pollution and control. Beijing University of Aeronautics and Astronautics Press, Beijing, pp 19–31
11. Haleh M, Gholamreza MK et al (2022) Catalytic ozonation of acetaminophen with a magnetic, cerium-based metal-organic framework as a novel, easily-separable nano composite. Chem Eng J 434:134614

Chapter 11
A Pattern of Changes in the Water Environment of the Yellow River Under Water-Sediment Regulation

Zhenzhen Yu, Yuxia Yang, Li Yan, Shikun Zhang, Jingyi Deng, and Yanbin Hao

Abstract To evaluate the water quality of the Yellow River under the water-sediment regulation scheme (WSRS) prior to seasonal flooding in 2023, river water was sampled and monitored at the sections of Downstream Xiaolangdi Reservoir, Huayuankou, Jiahetan, and Gaocun, typical sites for monitoring the Yellow River's water quality. The evaluation results reveal overall excellence in water quality from Downstream Xiaolangdi Reservoir to Gaocun, with all monitoring factors meeting the standards for Classes I, II, and III water quality, except for a temporary decrease in dissolved oxygen (DO) when suspended sediment was discharged from high-turbidity water. Notably, there is a correlation between the concentration of water quality factors and that of suspended sediment: higher suspended sediment concentration (SSC) at different times corresponds to elevated levels of chemical oxygen demand (COD), ammonia nitrogen (NH_3–N), permanganate index (CODmn), five-day biochemical oxygen demand (BOD_5), total phosphorus (TP), and manganese at the same section. On the contrary, lower SSC leads to a decrease in the pH and DO levels. No significant difference was made to the Yellow River's water quality along the lower reaches of Xiaolangdi Reservoir under the WSRS, with the impact diminishing from upstream to downstream across the monitoring sections. Overall, the study findings objectively assess the pattern of changes in the Yellow River's water ecosystem under the WSRS before seasonal flooding, providing crucial technical insights into how the Yellow River basin can be preserved and restored.

Keywords Pre-flood water-sediment regulation · Xiaolangdi reservoir · Prototype sampling and monitoring · Water quality factor changes · Suspended sediment concentration

Z. Yu (✉) · Y. Yang · L. Yan · S. Zhang · J. Deng · Y. Hao
Yellow River Water Resources Protection Institute, Zhengzhou, China
e-mail: zzyuhydro@126.com

K. Y. Foo et al. (eds.), *Water Nexus Approach for Sustainable Development*,
Environmental Science and Engineering, https://doi.org/10.1007/978-3-031-89238-7_11

11.1 Introduction

In 2023, to effectively manage the water-sediment dynamics in the Yellow River basin, particularly amid the preparation of midstream reservoirs for seasonal flooding, a pre-flood water diversion and sediment regulation scheme was carried out in a safe, controlled, and coordinated manner. The endeavor lasted 20 days, from 9 a.m. on June 21 to 8 a.m. on July 11, with the maximum discharge of Xiaolangdi Reservoir controlled at approximately 4500 m^3/s at the station of Huayuankou. The scheme hinged upon the coordinated scheduling of Wanjiazhai, Sanmenxia, and Xiaolangdi reservoirs, supplemented by coordination of the dams of Luhun, Guxian, and Hekoucun on the tributaries. This time the sediment discharge from Xiaolangdi Reservoir totaled 2.717 billion tons. As downstream channel scouring continued, main channel contraction was effectively restrained and notably restored. Given that, the minimum bankfull discharge rose back to the pre-flood level of 4600 m^3/s in 2023 from 1800 m^3/s in 2002; the likelihood of floodplains downstream significantly dropped; and the estuary ecosystem effectively improved. Remarkably, this would help prevent flooding, reduce silt, and benefit the environment. Following the relevant requirements of the Yellow River Protection Law, we had river water sampled and monitored on site at the typical midstream and downstream sections of the river to delve into the water quality of these sections prior to seasonal flooding in 2023. Subsequent to that, analysis was conducted to understand the pattern of changes in the water quality of the lower reaches of Xiaolangdi Reservoir before water diversion and sediment regulation, with an aim to provide supportive prototype data for the ecological conservation and restoration of the Yellow River basin.

11.2 Overall Water Environment Reality Across the Yellow River Basin

As China's crucial ecological barrier, demographic center, and economic driver, the Yellow River basin is of strategic importance to the overall development of the country and to socialist modernization [1–4]. The basin's water quality has been improving year on year, as evidenced by a shift in water quality from mildly polluted to good over the five years ending in 2022. That year, the number of sections meeting standards for Classes I, II, and III water quality within the Yellow River basin collectively accounted for 87.5% of the total across the region, up by 21.1% points from 2018. In addition, the sections throughout the main stream of the Yellow River were up to the Class I or II criteria, indicating excellent water quality, and the river water quality across major tributaries was good. The overall water quality changes in the Yellow River basin are shown in Table 11.1.

Table 11.1 Statistical table of overall water quality changes in the Yellow River basin

Year	Waterbody	Proportion/%					
		I	II	III	IV	V	Inferior V
2018	Basin	2.9	45.3	18.2	17.5	3.6	12.4
	Main stream	6.5	80.6	12.9	–	–	–
	Tributary	1.9	34.9	19.8	22.6	4.7	16
2019	Basin	3.6	51.8	17.5	12.4	5.8	8.8
	Main stream	6.5	77.4	16.1	–	–	–
	Tributary	2.8	44.3	17.9	16	7.5	11.3
2020	Basin	6.6	56.2	21.9	12.4	2.9	–
	Main stream	3.2	96.8	–	–	–	–
	Tributary	7.5	44.3	28.3	16	3.8	–
2021	Basin	6.4	51.7	23.8	12.5	1.9	3.8
	Main stream	14	81.4	4.7	–	–	–
	Tributary	5	45.9	27.5	14.9	2.3	4.5
2022	Basin	7.2	57.8	22.4	8.4	1.9	2.3
	Main stream	14	86	0	0	0	0
	Tributary	5.9	52.3	26.8	10	2.3	2.7

Note Statistics based on the *Report on the State of the Ecology and Environment in China* (2018–2022)

11.3 Prototype Observation of Water Quality Under Water-Sediment Regulation

11.3.1 Identification of Monitoring Area and Sites

Based on the changing hydrological and sedimentary dynamics downstream of Xiaolangdi Reservoir during the pre-flood water-sediment regulation period, we concurrently had river water sampled at the sections of Downstream Xiaolangdi Reservoir, Huayuankou, Jiahetan, and Gaocun from June 16 through July 13, which accounted for most of the water-sediment regulation period. Specific monitoring sites are shown in Table 11.2.

11.3.2 Water Function Zoning and Water Quality Targets

The Yellow River's main stream under our research was divided into two primary water function zones—Yellow River Henan Development and Utilization Area and Henan Yulu Development and Utilization Area—and four secondary zones,

Table 11.2 Water quality monitoring sites along the Yellow River's main stream

No.	Monitoring site	Distance downstream of Xiaolangdi Reservoir	Note
1#	Downstream Xiaolangdi Reservoir	2.4 km	Monitored at the Xiaolangdi Yellow River Highway Bridge
2#	Huayuankou	128 km	Monitored at the Zhengzhou Yellow River Highway Bridge
3#	Jiahetan	233.4 km	Monitored at the Kaifeng Yellow River Highway Bridge
4#	Gaocun	312.6 km	Monitored at the Dongming Yellow River Highway Bridge, Shandong province

namely the Jiaozuo Drinking and Agricultural Water Area, Zhengzhou and Xinxiang Drinking and Industrial Water Area, Kaifeng Drinking and Industrial Water Area, and Puyang Drinking and Industrial Water Area, with Class III as the water quality target in these zones. The water function zones and their corresponding water quality targets are detailed in Table 11.3.

Table 11.3 Water function zoning and water quality targets of the Yellow River's main stream under research

Primary zone	Secondary zone	Start of section	End of section	Representative section	Length (km)	Water quality target
Yellow River Henan development and utilization zone	Jiaozuo drinking and agricultural water area	Xiaolangdi Reservoir	Gubozui	Downstream Xiaolangdi Reservoir	78.1	III
	Zhengzhou and Xinxiang drinking and industrial water area	Gubozui	Langchenggang	Huayuankou	110	III
	Kaifeng drinking and industrial water area	Langchenggang	Dongbatou	Kaifeng bridge	58.2	III
Henan Yulu development and utilization zone	Puyang drinking and industrial water area	Dongbatou	Dawangzhuang	Gaocun	134.6	III

11.3.3 Sampling Frequency for Water Quality Monitoring

The Yellow River's water was sampled on site on June 16 (before the WSRS), June 23 and 26(at the stage of clean water discharge), July 10 (at the sediment discharge stage), and July 13 (after the WSRS).

11.3.4 Water Quality Monitoring Factors

The monitoring factors adopted in our research included the pH, DO, COD, NH_3–N, CODmn, BOD_5, TP, and manganese, all of which are typical parameters outlined in the Environmental Quality Standards for Surface Water (GB3838-2002). The sampling was followed by laboratory analysis.

11.4 Water Quality Analysis Under WSRS

Water sampling was conducted at Downstream Xiaolangdi Reservoir, Huayuankou, Jiahetan, and Gaocun between June 16 and July 13. Results from the water quality assessment [5–8] unveiled overall excellence in water quality across the researched stretch, with all monitoring factors meeting the standards for Classes I, II, and III water quality, except for a transient drop in DO at the silt-discharge stage.

During the period when water was discharged and sediment regulated, the maximum SSC at the sections of Downstream Xiaolangdi Reservoir, Huayuankou, Jiahetan, and Gaocun reached 329 kg/m^3 (July 7), 53.3 kg/m^3 (July 9), 48.2 kg/m^3 (July 9), and 33.9 kg/m^3 (July 11), respectively. Levels of the monitoring water quality factors fluctuated over time and space, largely attributed to SSC, which, to be specific, was negatively correlated with the pH and DO. However, there was a positive correlation between SSC and such elements as COD, NH_3–N, CODmn, BOD_5, TP, and manganese.

11.4.1 Basic Factors of Water Quality Monitoring

The pH.

When monitored, the pH across the researched sections averaged 8.13, which met the requirement of the Environmental Quality Standards *for* Surface Water. Having identified the average pH value of all sections during a given period of time, we found that the value peaked at 8.28 on June 16, and hit the lowest point on July 10, at 7.7.

Upon the analysis of the pH level along the researched area, it was discovered that lower SSC led to a decline in the pH of water bodies of the Downstream Xiaolangdi Reservoir, Huayuankou, Jiahetan, and Gaocun sections. When the levels of sediment stayed high on July 10, the pH dropped to 7.4 at the Jiahetan section before bouncing back at Gaocun (Figs. 11.1 and 11.2).

Dissolved oxygen.

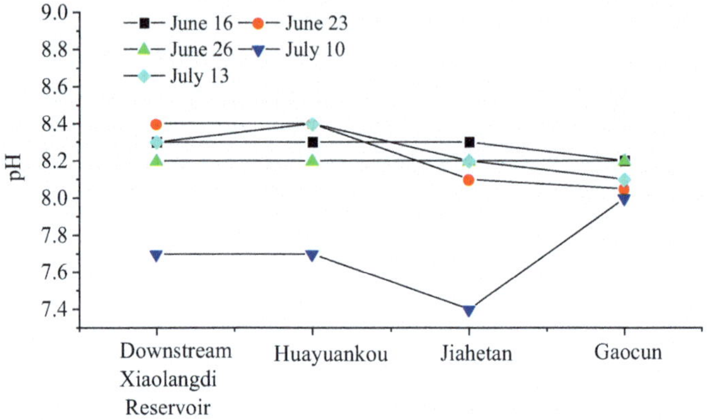

Fig. 11.1 Changes of the pH along the lower reaches of Xiaolangdi Reservoir under WSRS

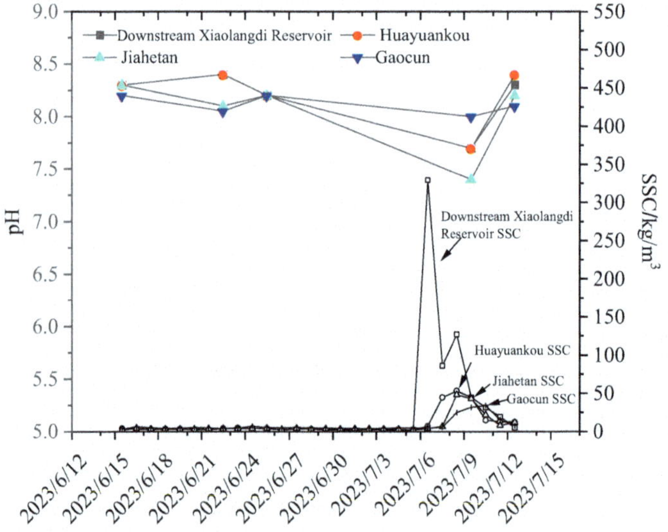

Fig. 11.2 Temporal changes of the pH along the lower reaches of Xiaolangdi Reservoir under WSRS

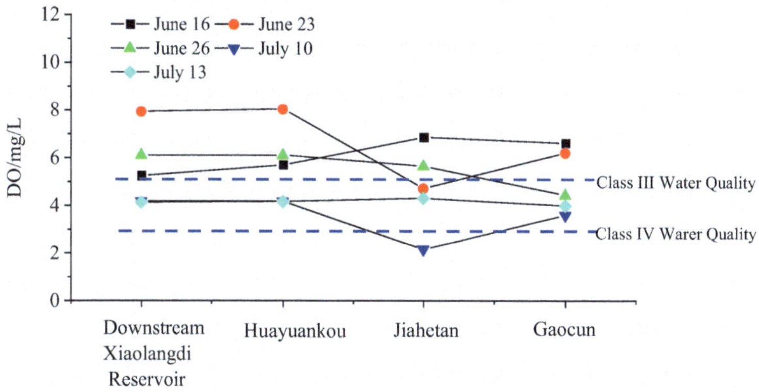

Fig. 11.3 Changes of DO levels along the lower reaches of Xiaolangdi Reservoir under WSRS

The average DO level under monitoring was 5.22 mg/L, which overall satisfied the standards for Class III water quality. Upon the averaging of DO concentrations across all studied sections during a particular period, we found the figure maximized on June 23 and minimalized on July 10, at 6.72 mg/L and 3.54 mg/L, respectively.

According to the analysis of DO distribution along the area under research, the lower the SSC is, the more stable the DO level is in the sections of Downstream Xiaolangdi Reservoir, Huayuankou, Jiahetan, and Gaocun, rating these water bodies Class II or III in terms of water quality. On July 10 which saw higher SSC, the DO level decreased between Downstream Xiaolangdi Reservoir and Gaocun, with the figure dropping to 2.16 mg/L at Jiahetan before a bounce-back at Gaocun. By this parameter, all sections met the criteria for Class IV water quality, except for Jiahetan, which fell under Class V (Figs. 11.3 and 11.4).

Chemical oxygen demand.

Throughout the monitoring period, the COD across various sections averaged 14.53 mg/L, consistently meeting the standards for Class I water quality. For the same monitoring time interval, the median COD level across sections peaked at 17.75 mg/L on July 10, with the minimum value recorded at 12 mg/L on June 16.

The analysis of COD distribution throughout the researched area showed that in the context of low SSC levels, the COD went up from Downstream Xiaolangdi Reservoir and Huayuankou to Jiahetan to Gaocun at the early water-sediment regulation period, while diminishing along the way when clean water was discharged, with water quality ranging between Classes II and III. On July 10 when there were high SSC levels, the COD amount increased across the board. It stood at 18 mg/L at Downstream Xiaolangdi Reservoir and rose to 19 mg/L at Huayuankou and Jiahetan, before dropping to 15 mg/L at Gaocun. By this parameter, all sections were rated as Class III for water quality, except for Gaocun, which was rated as Class I (Figs. 11.5 and 11.6).

Ammonia nitrogen.

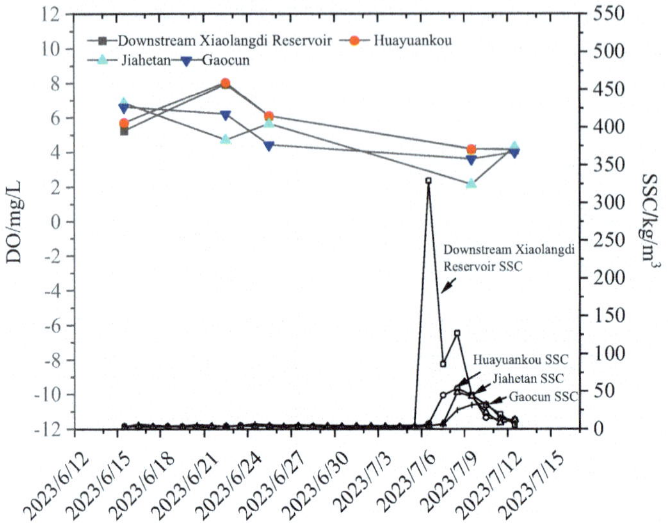

Fig. 11.4 Temporal changes of DO levels along the lower reaches of Xiaolangdi Reservoir under WSRS

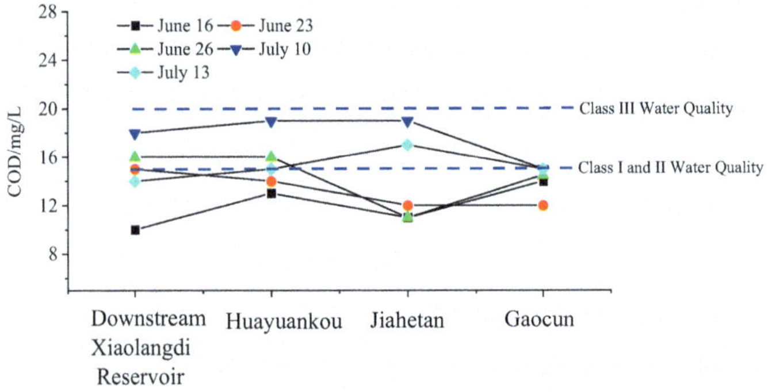

Fig. 11.5 Changes of COD along the lower reaches of Xiaolangdi Reservoir under WSRS

The NH_3–N concentration across different sections maintained an average of 0.29 mg/L throughout the monitoring period, aligning with the Class II water quality standards. During the same monitoring period, its median concentration across sections maximized at 0.6 mg/L on July 10 and minimized on June 16, at 0.07 mg/L.

By analyzing how the NH_3–N concentration was distributed along the researched area, we discovered that when SSC stayed low, the NH_3–N amount tended to stabilize along the sections from Downstream Xiaolangdi Reservoir, Huayuankou, Jiahetan, and Gaocun, ranging between Classes I and II in terms of water quality. During the period of higher SSC, observed on July 10, the NH_3–N levels increased. Specifically,

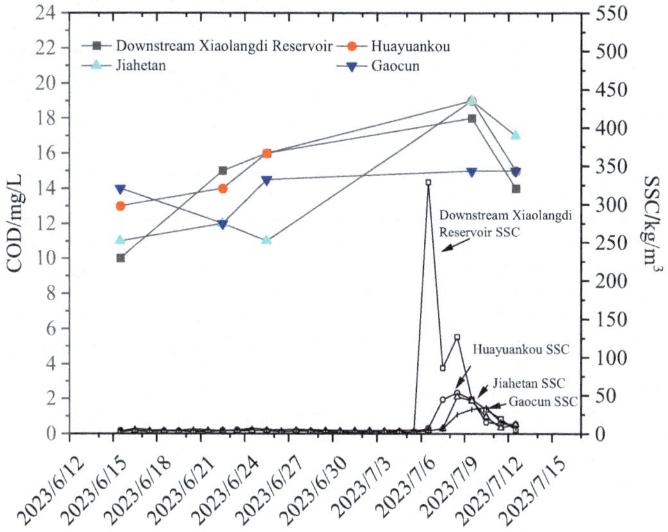

Fig. 11.6 Temporal changes of COD along the lower reaches of Xiaolangdi Reservoir under WSRS

Downstream Xiaolangdi saw an NH_3–N concentration of 0.74 mg/L, which then rose to 0.95 mg/L at Huayuankou. Nonetheless, it fell at Jiahetan and Gaocun to 0.41 mg/L and 0.31 mg/L, respectively. The collective NH_3–N concentration across all sections fell under the Class III water quality standards, with Jiahetan and Gaocun rated as Class II (Figs. 11.7 and 11.8).

Permanganate index.

Across all sections, the CODmn averaged 3.12 mg/L, conforming to the Class II water quality standards. During the same observation period, the mean CODmn

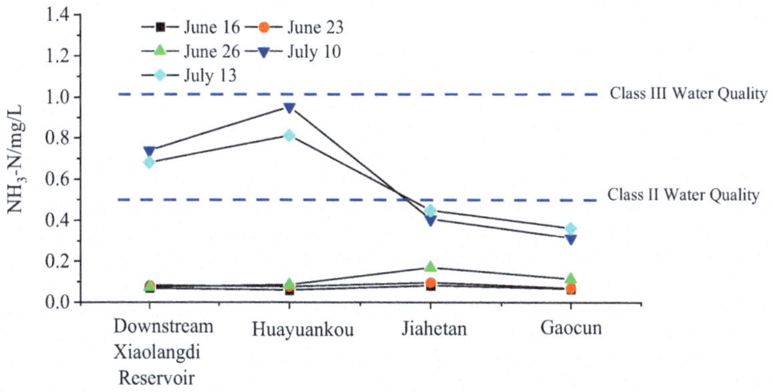

Fig. 11.7 Changes of NH_3–N levels along the lower reaches of Xiaolangdi Reservoir under WSRS

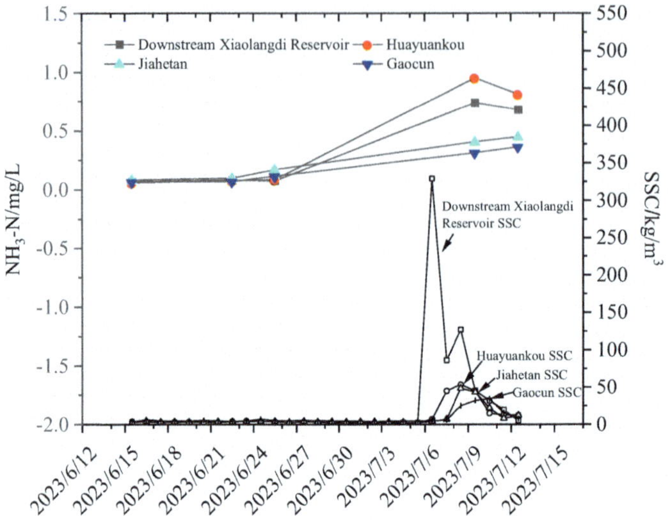

Fig. 11.8 Temporal changes of NH_3–N levels along the lower reaches of Xiaolangdi Reservoir under WSRS

across sections peaked at 3.19 mg/L on July 10, contrasting with a low of 3.05 mg/L recorded on June 16.

An analysis of how the CODmn changed along the studied stretch revealed that during the phases of diminished SSC, NH_3–N levels across the sections of Downstream Xiaolangdi Reservoir, Huayuankou, Jiahetan, and Gaocun tended to stabilize and fell below the Class II water quality threshold. On July 10 which witnessed high SSC levels, the index between Downstream Xiaolangdi Reservoir and Gaocun rose higher than on other dates of monitoring, but remained within the Class II water quality range. To be specific, the CODmn from Downstream Xiaolangdi Reservoir and Huayuankou to Jiahetan to Gaocun exhibited a diminishing trend along the stretch, at 3.5 mg/L, 3.5 mg/L, 3.1 mg/L, and 2.65 mg/L, respectively (Figs. 11.9 and 11.10).

Biochemical oxygen demand.

Under monitoring, its average value across all sections stood at 2.89 mg/L, making it fall under the Class I water quality standards. Having averaged the parameter throughout all sections during the same monitoring period, we established that its mean displayed a zenith of 3.5 mg/L on July 10, juxtaposed with a low of 2.3 mg/L observed on June 16.

By investigating the distribution of the BOD_5 along the researched stretch, it was observed that at the early WSRS stage, its amount went up from Downstream Xiaolangdi Reservoir and Huayuankou down to Jiahetan to Gaocun, but stayed within the Class I water quality range. It, however, stably ranged between the Classes I and III standards amid the discharge of clean water. On July 10 which saw high SSC

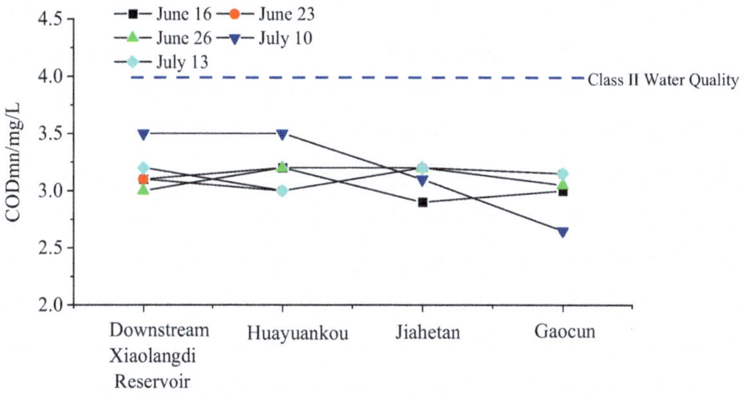

Fig. 11.9 Changes of the CODmn along the lower reaches of Xiaolangdi Reservoir under WSRS

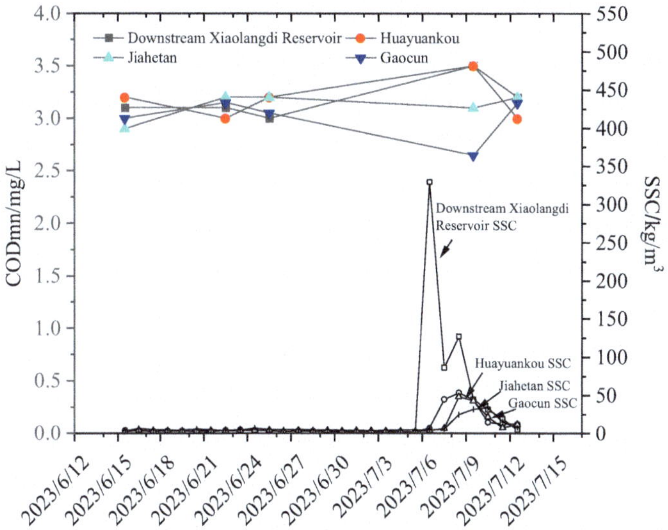

Fig. 11.10 Temporal changes of the CODmn along the lower reaches of Xiaolangdi Reservoir under WSRS

levels, the SSC in the lower reaches of Xiaolangdi Reservoir peaked among all monitoring time intervals, whereas the BOD_5 increased, to a certain extent, from the reservoir downwards. Concretely, the BOD_5 value reached 3.5 mg/L at Downstream Xiaolangdi Reservoir and rose to 3.7 m/L and 3.8 mg/L at Huayuankou and Jiahetan, respectively, before dropping down to 3 mg/L at Gaocun. Overall, the parameter met the criteria for Class III water quality (Figs. 11.11 and 11.12).

Total phosphorus.

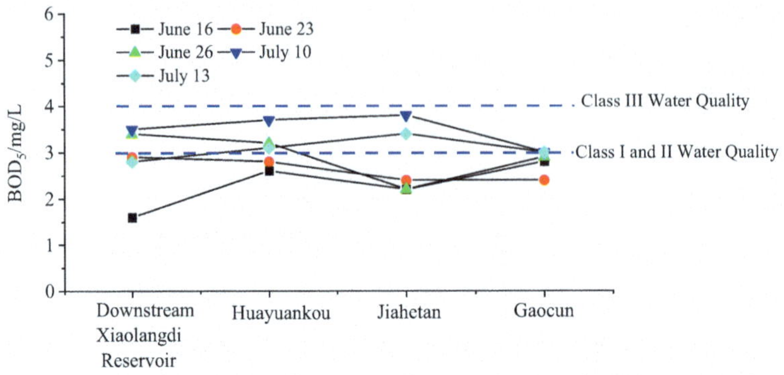

Fig. 11.11 Changes of the BOD_5 along the lower reaches of Xiaolangdi Reservoir under WSRS

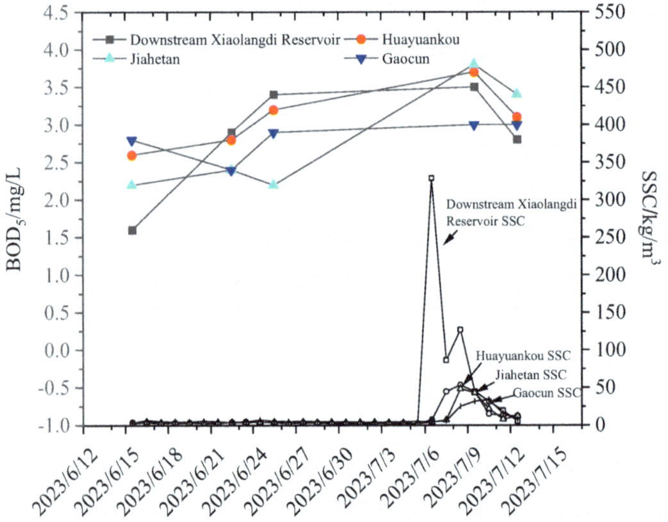

Fig. 11.12 Temporal changes of the BOD_5 along the lower reaches of Xiaolangdi Reservoir under WSRS

Its average level at all sections under monitoring stood at 0.06 mg/L, falling under the Class II water quality standards. Upon the averaging of TP levels across sections during the same period of observation, we discovered its mean peaked at 0.07 mg/L on July 13, with the lowest level recorded on June 16, at 0.04 mg/L.

An analysis of how TP levels were distributed along the studied stretch revealed that when there were lower levels of SSC, the TP amount along the lower reaches of Xiaolangdi Reservoir plateaued out in a way that aligned with the Class II water quality standards. On July 10 which saw elevated sediment levels, the TP amount went up as monitoring sites moved downstream, with the maximum recorded at Jiahetan,

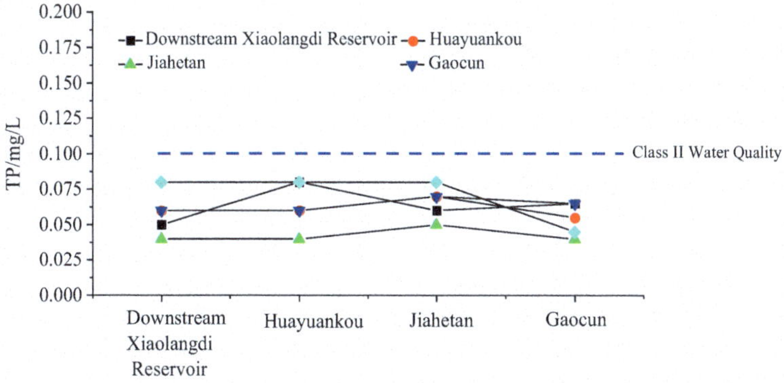

Fig. 11.13 Changes of TP levels along the lower reaches of Xiaolangdi Reservoir under WSRS

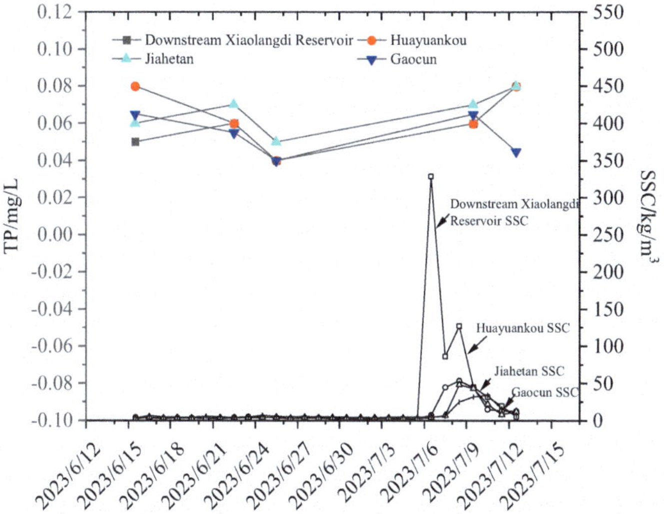

Fig. 11.14 Temporal changes of TP levels along the lower reaches of Xiaolangdi Reservoir under WSRS

at 0.07 mg/L, which stayed within the Class II water quality range (Figs. 11.13 and 11.14).

11.4.2 Representative Heavy Metal Factor

Sediment exerts a significant impact on the water environment and water quality of the Yellow River as it is capable of adsorbing heavy metals, represented by manganese

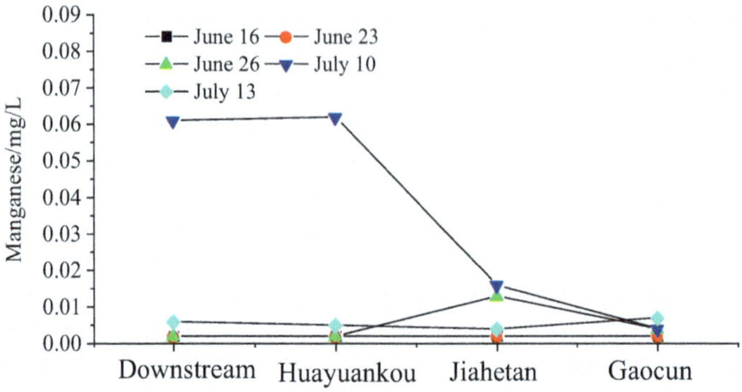

Fig. 11.15 Changes of manganese concentration along the lower reaches of Xiaolangdi Reservoir under WSRS

in our research. Given that, we examined how manganese concentration in the lower reaches of Xiaolangdi Reservoir was affected by varying SSC levels under WSRS.

Throughout the monitoring time, the median manganese concentration across all sections stood at 0.01 mg/L. After the concentration of manganese at various sections during the same period was averaged, we found it peaked at 0.04 mg/L on July 10 and hit the lowest point on June 16 and 23, at 0.002 mg/L. On July 10, the whole researched stretch saw an overall increase in TP, with the zenith observed at Huayuankou, at 0.062 mg/L, below the standard limit of 0.1 mg/L required in the supplementary project on centralized surface drinking water sources (Figs. 11.15 and 11.16).

11.5 Assessment and Conclusions

In this paper, we sampled and monitored river water from the four sections of the Yellow River, namely Downstream Xiaolangdi Reservoir, Huayuankou, Jiahetan, and Gaocun, on June 16, 23 and 26, as well as on July 10 and 13, 2023. These selected dates were representative of the periods before, during, and after the water-sediment regulation of the river before seasonal flooding. The assessment results showed that the water quality of the main-stream section between Downstream Xiaolangdi Reservoir and Gaocun during the time of water diversion and sediment regulation was excellent as a whole. All monitoring factors met the standards for Classes I to III water quality, except for a temporary decrease in DO across all sections when sediment was discharged from high-turbidity water. The concentration of water quality contributors is correlated with that of sediment. Specifically, the levels of COD, NH_3–N, CODmn, BOD_5, TP, and manganese increase as SSC rises within the same section at different times. Meanwhile, a lower concentration of sediment leads to a decline in the pH and DO levels.

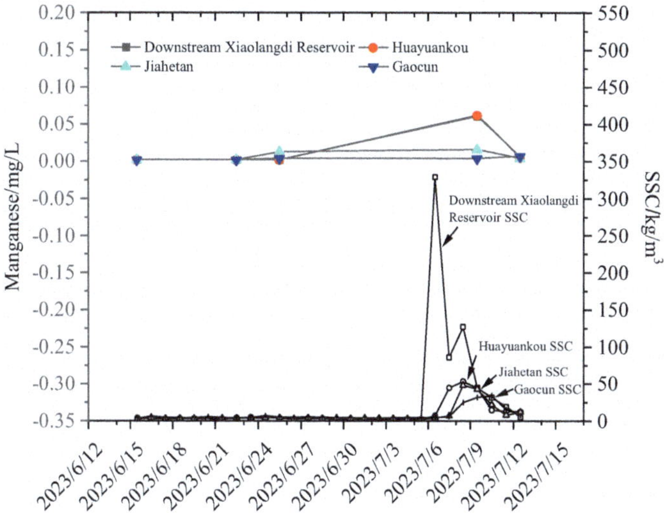

Fig. 11.16 Temporal changes of manganese concentration along the lower reaches of Xiaolangdi Reservoir under WSRS

Overall, sediment as a carrier of pollutants plays an important role in shaping the water ecosystem. On-site inspections revealed that there was no change in the water quality category of water bodies downstream of Xiaolangdi Reservoir upon the regulation of water and sediment in the facility, and no significant difference was made to the water quality of the Yellow River. Also, the impact of such regulation on the river sections of Downstream Xiaolangdi Reservoir, Huayuankou, Jiahetan, and Gaocun diminished as the monitoring site moved downstream.

Acknowledgements The authors acknowledge the support from Outstanding Young Talents Science and Technology Foundation of Yellow River Conservancy Commission (Grant No. HQK-202320), Special Scientific Research Project of Yellow River Water Resources Protection Institute (Grant No. KYY-KYZX-2022-01), National Natural Science Foundation of China (Grant No. 51709126), Major Science and Technology Special Fund of Henan Province (Grant No. 201300311400).

References

1. Liu XY, Wang RL, Zhang YF et al (2020) Threshold of the runoff utilization of the Yellow River. J Hydraul Eng 51(6):631–641. https://doi.org/10.13243/j.cnki.slxb.20190765
2. Yu ZZ, Li Y, Wang RL et al (2020) Study on self-purification water demand flow in the Yiluo River under multi-objective coordination control. Water Resour Prot 36(5):38–42. https://doi.org/10.3880/j.issn.1004-6933.2020.05.006

3. Wang H, Hu P (2020) Key issues of ecological conservation in the Yellow River Basin from a water cycle perspective. J Hydraul Eng 51(9):1009–1014. https://doi.org/10.13243/j.cnki.slxb.20200675
4. Liu CH, Xu LG, Gao HY (2010) Evaluation methods and research progress of river water quality. J Hohai Univ (Nat Sci) 38:290–293. http://www.cnki.net
5. Xu ZX (2005) Comprehensive water quality identification index for environmental quality assessment of surface water. J Tongji Univ (Nat Sci) 33(4):482–488. http://www.cnki.net
6. Yang L, Song J, Song B et al (2015) Primary pollutant water quality identification index method and its application to comprehensive evaluation of river water quality. Environ Sci Technol 38(11):239–245. http://www.cnki.net
7. Peng D, Liu Z, Zhang Y et al (2023) Evaluation methods of aquatic ecological environment quality and their application progress in the Yellow River Basin. Environ Monit China 39(2):41–54. http://www.cnki.net
8. Liu YL, Zheng YA (2022) Water quality assessment and spatial-temporal variation analysis in Yellow River Basin. Environ Sci 43(3):1332–1345. http://www.cnki.net

Chapter 12
Defluoridation of Fluoride Contaminated Water by Electrocoagulation

Fenghao Wu, Zhiqi Long, and Qiyuan Gu

Abstract Currently, a growing population and dwindling water supplies have increased human demand for clean water, but the quality of both surface and groundwater has declined due to the discharge of hazardous chemicals and industrial waste. Diseases caused by contaminated water pose a huge burden and threat to human health and also have a huge economic impact on society. Pollutants from various sources are responsible for changes in the chemical composition of water, including organic pollutants, inorganic pollutants, and bacteria. Among them, the problem of fluoride pollution in water has attracted global attention. Coagulation-flocculation method, precipitation method, biological method, membrane technology, adsorption method, ion exchange method, and electrocoagulation technology are all commonly used methods for fluoride removal. Electrocoagulation technology has become a current application because of its high efficiency of in-situ flocculation and fluoride removal and stable fluoride removal effect. Prospective technology. This article reviews methods for removing fluoride from water and focuses on the mechanism, advantages, and fluoride removal effects of electrocoagulation fluoride removal technology.

Keywords Electrocoagulation · Defluoridation · Wastewater treatment

F. Wu · Z. Long (✉) · Q. Gu
National Engineering Research Center for Environment-Friendly Metallurgy in Producing Premium Non-Ferrous Metals, China GRINM Group Corporation Limited, Beijing, China
e-mail: longzhiqi@vip.163.com

GRINM Resources and Environment Technology. Co., Ltd., Beijing, China

General Research Institute for Nonferrous Metals, Beijing, China

Beijing Engineering Research Center of Strategic Nonferrous Metals Green Manufacturing Technology, China GRINM Group Corporation Limited, Beijing, China

K. Y. Foo et al. (eds.), *Water Nexus Approach for Sustainable Development*,
Environmental Science and Engineering, https://doi.org/10.1007/978-3-031-89238-7_12

12.1 Introduction

Fluoride contamination is prevalent in surface water and groundwater in many parts of the world, especially in Asia and Africa, such as India, Pakistan, and other countries. The average concentration of fluoride in groundwater exceeds 5 mg/L [1, 2]. The World Health Organization specifies that the concentration of fluoride ions in drinking water must be below 1.5 mg/L. If the human body consumes excessive fluoride, it will cause lesions in multiple organs such as the thyroid gland, cranial nerves, bones, and liver, causing serious health problems [2, 3], if excessive fluorine accumulates in the plant, it will seriously affect photosynthesis, eventually leading to leaf yellowing and necrosis [4]. The main sources of fluorine pollution are natural fluorine-containing rocks and volcanic eruptions, as well as pollutants emitted by human metal smelting, coal burning, etc. Fluorosis is a global problem that must be monitored and mitigated, and appropriate measures taken to reduce fluoride concentrations in drinking water in affected areas to within permissible limits. This article will summarize the currently commonly used fluoride removal technologies and focus on the application of electrocoagulation in treating fluoride in wastewater, providing ideas for the subsequent development and promotion of fluoride removal technologies.

12.2 Overview of Fluoride Removal Methods

There are currently many fluoride removal methods, but most of them cannot be safe and cost-effective at the same time and are widely used. Currently, only a few reliable fluoride removal methods are available for practice, including coagulation-flocculation, precipitation, biological methods, membrane technology, adsorption, ion exchange and electrocoagulation.

12.2.1 Coagulation-Flocculation

The coagulation-flocculation method and the precipitation method both remove the fluoride in the wastewater in the form of insoluble aggregates or insoluble solid precipitation by adding chemicals. These two methods have a small dosage of chemicals, a large volume of water to be treated, and a high processing speed. It has the advantages of speed and low cost, but it is difficult to make the fluorine content of treated wastewater meet discharge standards, and a large amount of sludge will be produced after treatment and requires additional treatment.

12.2.2 Biological

Biological methods allow pollutants to be oxidized or degraded into safe intermediate products under aerobic or anaerobic conditions and converted into energy [5]. This method is usually used in combination with physical or chemical methods to remove fluorine to a certain extent. For example, Liu et al. [6] studied a new type of facultative anaerobic denitrification copper-tungsten bacteria that can cause calcium precipitation and remove fluoride from wastewater. The biological method has high treatment efficiency under low-temperature conditions and is easy to maintain a consistent pH, but the treatment process is time-consuming and often needs to be used in conjunction with other methods to meet emission standards.

12.2.3 Membrane Technology

Membrane technology includes reverse osmosis, nanofiltration and electrodialysis technology. This technology is mainly driven by concentration differences or electricity. It uses the selective permeability of the membrane to trap chloride ions on one side of the membrane and separate them from other ions in the wastewater to achieve removal. For fluoride, this technology does not require the addition of additional chemicals, has a high efficiency of fluoride removal and low subsequent contamination. It is an effective fluoride removal technology, but it still has problems such as high investment cost, membrane degradation, membrane fouling and blockage, and electrode passivation.

12.2.4 Adsorption

The adsorption method is one of the commonly used fluoride removal technologies. Fluoride ions are removed from wastewater through adsorption, ion exchange, complexation, etc. Commonly used adsorbents include aluminum-based adsorbents, calcium-based adsorbents, and oxides/hydroxides. Adsorbents, carbon-based adsorbents, natural material adsorbents and agricultural, biomass-based adsorbents, building material adsorbents and nanomaterial-based adsorbents. This method has high fluoride removal efficiency, good cost-effectiveness and simple operation. However, due to the limited number of membrane regenerations, it cannot be regenerated for a long time and can easily cause secondary pollution.

12.2.5 Ion Exchange

The ion exchange method is to exchange certain anions in the ion exchange resin with fluoride ions, usually the fluoride ions in the water replace the hydroxyl ions or any other anionic groups of the metal lattice. This method has high fluoride removal efficiency and a stable fluoride removal effect. However, due to the high price of ion exchange resin and high regeneration cost, this method is cost-effective, and a large amount of wastewater will be produced during the resin regeneration process, causing secondary pollution.

12.2.6 Electrocoagulation

Electrocoagulation technology uses an electrochemically dissolved sacrificial electrode (soluble anode) immersed in the solution to generate metal coagulant in situ under the action of an external electric field [7], that is, the metal cations generated by the soluble electrode are hydrolyzed in water, polymerized into a series of multinuclear hydroxyl complexes and hydroxides with aggregation and adsorption functions, aat the same time, the cathode undergoes a reduction reaction, and hydrogen bubbles with tiny structures are produced. This bubble has good adhesion properties and floats on the surface. The process can play the role of air flotation, bringing suspended matter to the water surface and removing it there. Figure 12.1 shows a simple electrocoagulation device and a schematic diagram of typical reactions in electrocoagulation. The entire electrocoagulation process can be described as [8, 9]:

(1) The oxidation reaction at the anode produces metal cations (Formula 12.1).
(2) The cathode Reduction reaction occurs to generate hydrogen gas and hydroxyl ions (Formula 12.2).
(3) Metal cations interact with hydroxyl ions to generate metal hydroxides (Formula 12.3).
(4) Harmful pollutants are oxidized to form safe intermediates.
(5) The interaction between negatively charged pollutants and metal hydroxides is neutralized.
(6) The pollutants are neutralized and adsorbed onto the surface of metal hydroxides, forming flocs.
(7) Hydrogen bubbles are adsorbed on the surface of pollutants and flocs, carrying them to the surface of the liquid in the electrolyzer. The flocculation and air flotation work together to remove pollutants from the water.

$$M\,(n) \rightarrow M^{n+} + ne^- \tag{12.1}$$

$$nH_2O + 2ne^- \rightarrow n/2H_2 + nOH^- \tag{12.2}$$

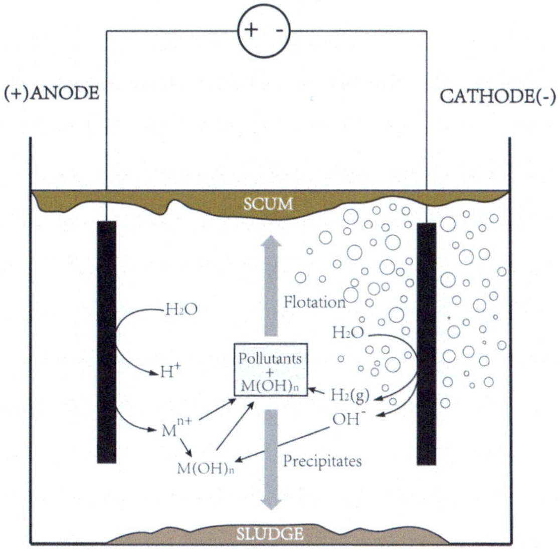

Fig. 12.1 Simple electrocoagulation device and schematic diagram of typical reactions in electrocoagulation

$$Mn + +nOH- \rightarrow M(OH)n \qquad (12.3)$$

The mechanism of action in the electrocoagulation process mainly includes electrochemical dissolution, electrochemical redox, flocculation, and electrolytic air flotation. The process of metal ions forming flocs through hydrolysis and polymerization is similar to the chemical flocculation process, including compressing the electric double layer. It has the functions of catching with a sweep net, adsorbing bridging, and adsorbing electricity. In addition, electroflocculation also includes electric field-induced particle depolarization and flocculation and electroabsorption on the electrode surface.

The biggest advantage of electrocoagulation is that it will produce flocculant in situ under the action of an electric field, without adding additional chemicals, and will not increase the Cl^-, SO_4^{2-} plasma content in the water [7, 10], which is conducive to subsequent processing, and at the same time, it will be produced at the cathode. The H_2 and OH^- provide pH regulators for the solution in situ without additional addition [11, 12], compared with chemical flocculation, the flocs produced by electrocoagulation are larger, more stable, acid-resistant, and contain less water. It is beneficial to subsequent separation [9]. At the same time, the uniform metal dissolution of electrodes in electroflocculation technology reduces the need for strong mixing machinery, reduces energy consumption, and makes the flocculation process more effective, the sludge generated during the electroflocculation process is mainly metal oxidation Materials are easy to settle and dehydrate, and the entire electrocoagulation process is low-sludge-generating. In addition, electrocoagulation technology has a wide range of adaptability to wastewater quality. Parameter settings can be adjusted in time according to changes in wastewater quality or volume to improve effluent quality. Stable and good processing effect. However, due to the serious electrode passivation problem during the electrocoagulation process, the passivation film generated on the electrode surface will reduce the treatment efficiency and increase energy consumption, seriously affecting the economic benefits of this method and making it impossible to achieve large-scale engineering applications. Electrode passivation has become a bottleneck problem in the widespread application of electrocoagulation technology. At the same time, the formation of flocculant during the electrocoagulation process is easily affected by hydraulic conditions, especially pH, making its operation more difficult [13].

Electrocoagulation technology is considered to be one of the promising technologies in the field of fluoride removal because of its versatility, high efficiency and environmental compatibility. As shown in Table 12.1, many scholars have studied it in recent years and have obtained good results in fluoride removal.

Table 12.1 Optimal operating parameters for electrocoagulation defluoridation

Water type	Electrode type (electrode spacing/cm)	Current density (mA/cm^2)	Fluoride removal rate (%), initial concentration	References
Groundwater	Fe//Al (2)	12	95, F^- = 4.93 mg/L	Alimohammadi et al. [14]
Wastewater	Fe//Al// Stainless (1)	60	90.1, F^- = 42 mg/L	Aoudj et al. [15]
Synthetic water	Al//Al (3)	16.7	94.5, F^- = 25 mg/L	Behbahani et al. [16]
Groundwater	Al//Al (0.46)	7	80.29, F^- = 4.08 mg/L	Castañeda et al. [17]
Synthetic water	Al//Al (1)	1.85	81.7, F^- = 30 mg/L	Drouiche et al. [18]
Synthetic water	Al//Al (0.5)	2	98, F^- = 20 mg/L	Hashim et al. [19]
Synthetic water	Al//Al (2) Fe// Fe (2)	7.14	60, F^- = 10 mg/L	Govindan et al. [20]
Groundwater	Al//Al (0.5)	10	85.5 >, F^- = 8.98 mg/L	Mena et al. [21]
Synthetic water	Al//Al	2	97.6, F^- = 15 mg/L	Un et al. [22]

12.3 Experiment

12.3.1 Material

Use KXN-1530D DC regulated power supply, self-made PVC electric flocculation tank, stainless steel plate and aluminum plate (12 cm * 5 cm * 0.2 cm), weigh 0.0459 g of sodium fluoride solid (AR) into a 1000 ml volumetric flask, add Deionized water to the mark, configure simulated wastewater ($c(F^-) = 0.0219$ g/l), refer to the international standard for preparation, cleaning, and evaluation of corrosion specimens ASTMG1-03 (2017) to process the electrode plate [23], and handle the electrode for the first time: use Dip cotton into 75% ethanol to clean the processing residue on the surface of the electrode plate. Use 600-grit sandpaper to wipe the surface of the electrode evenly and polish off the oxide layer on the surface of the electrode. After polishing, the surface should have a metallic luster and should not have too many pits, the electrode should be used again. Treatment: Soak in 3% mass fraction of hydrochloric acid for 20 min to detach surface oxides, then rinse the surface acid with deionized water. After rinsing, place it in an oven, adjust the temperature to 60 °C and dry for 5 min.

12.3.2 Experimental Procedure

The dissolution rate of the anode increases with the increase of current density, and the amount of metallic hydroxide flocs also increases, thereby improving the efficiency of decontamination. When the current density exceeds the optimal current density, it will not affect the removal of pollutants, because there are enough metal hydroxide flocs for pollutant settling. Higher current densities will cause the solution temperature to rise, causing the coagulation process to be unstable, and limiting the removal effect. At the same time, higher current density will increase the economic cost. The experiment determined that the distance between the plates was 1 cm, the pH was 8, and NaCl was 0.75 g/L. The design was designed to examine the fluorine removal effect when the current density was 5 mA/cm^2 and 10 mA/cm^2. The reaction was conducted for a total of 50 min, and samples were taken regularly (at intervals of 10 min).

12.3.3 Analysis

The calculation formula of fluoride ion removal efficiency in water is shown in Formula (12.4).

$$\eta\,(\%) = (C_i - C_o)/C_i * 100\% \tag{12.4}$$

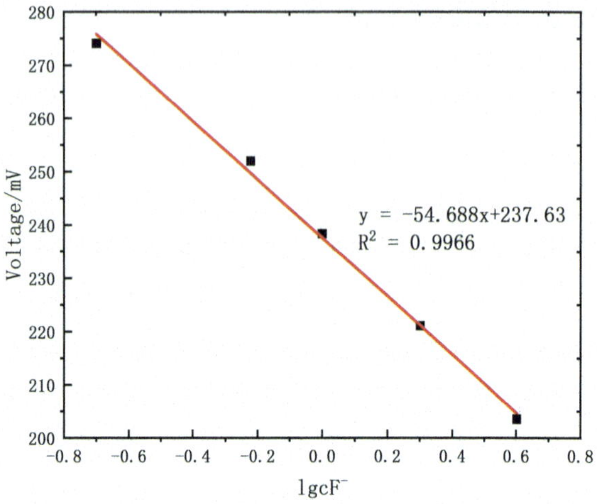

Fig. 12.2 Fluoride ion standard curve

In the formula, η is the removal efficiency, C_i represents the initial concentration of fluoride ions in the water, and C_o represents the concentration of fluoride ions in the water after treatment.

The PXSJ-216F ion meter is used to detect fluoride ions. After activating and cleaning the electrode, the fluoride ion concentrations are tested at 0.2 mg/L, 0.6 mg/L, 1.0 mg/L, 2.0 mg/L, and 4.0 mg/L. potential, draw the lgc(F^-) potential standard curve (Fig. 12.2), and obtain the curve expression $y = -54.688x + 237.63$.

12.3.4 Result

Under the experimental conditions (the distance between the plates is 1 cm, pH = 8, NaCl = 0.75 g/L, initial c(F^-) = 9 mg/L), electrocoagulation has a relatively obvious effect on fluoride removal and the current density increases, the fluoride removal effect increases, but the voltage will increase and the energy consumption will increase (Fig. 12.3). Subsequent optimization should be further combined with other operating conditions, and according to the experimental phenomenon, brown flocs appeared in the experiment, considering the presence of iron in the stainless-steel plate, the experiment Pungent smelling gas appeared during the process, but it was not obvious. It is considered that chlorine gas appeared during the treatment process.

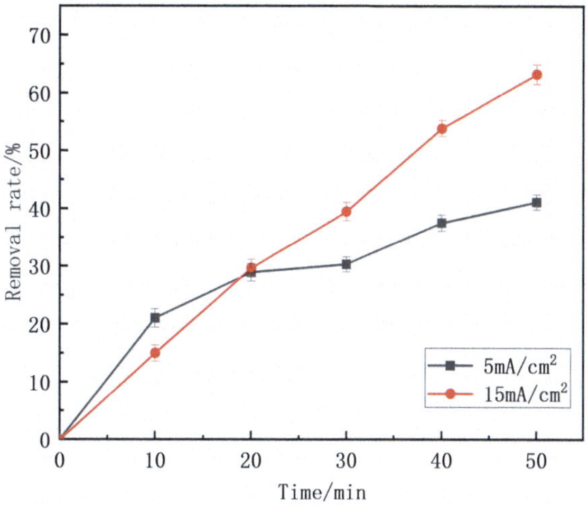

Fig. 12.3 Fluoride removal efficiency at different current densities

12.4 Conclusions and Recommendation

The problem of fluoride pollution in water has attracted global attention. The fluoride content in water has a significant impact on the environment and human health, depending on its concentration. Therefore, it is very important to find methods and technologies to deal with fluoride pollution in water reasonably and effectively. Currently, commonly used fluoride removal methods include coagulation-flocculation, precipitation, biological methods, membrane technology, adsorption, ion exchange and electrocoagulation. Although each method can achieve fluoride removal to a certain extent, there are still many challenges.

At present, electric flocculation technology has gradually become an industrial application technology for water treatment and is also widely used in the field of fluoride removal. From the single-factor experimental results, it can be seen that the electric flocculation technology has a better fluoride removal effect, but due to its severe electrode bluntness Chemicalization problems prevent large-scale promotion and application. In-depth research is still needed to further improve the electroflocculation defluoridation technology.

References

1. Alabdulaaly AI, Khan MA (2013) Occurrence of fluoride in ground waters of Saudi Arabia. Appl Water Sci 3:589–595. https://doi.org/10.1007/s13201-013-0105
2. Ahmad S, Singh R, Arfin T, Neeti K (2022) Fluoride contamination, consequences and removal techniques in water: a review. Environ Sci Adv 5:620–661. https://doi.org/10.1039/D1VA00039J

3. Aggarwal A, Mehta S, Gupta D et al (2012) Clinical & immunological erythematosus patients characteristics in systemic lupus Maryam. J Dent Educ 76(11):9–1532. https://doi.org/10.4103/ijmr.IJMR_1356_15
4. Elloumi N, Abdallah FB, Mezghani I et al (2005) Effect of fluoride on almond seedlings in culture solution. Fluoride 38(3):193. http://homepages.ihug.co.nz/~spittle/fluide-journal.htm
5. Rezai B, Allahkarami E (2021) Wastewater treatment processes—techniques, technologies, challenges faced, and alternative solutions. Soft computing techniques in solid waste and wastewater management. Elsevier 2:35–53. https://doi.org/10.1016/B978-0-12-824463-0.00004-5
6. Liu J, Su J, Ali A et al (2022) Potential of a novel facultative anaerobic denitrifying *Cupriavidus* sp. W12 to remove fluoride and calcium through calcium bioprecipitation. J Hazard Mater 423:126976. https://doi.org/10.1016/j.jhazmat.2021.126976
7. Ingelsson M, Yasri N, Roberts EP (2020) Electrode passivation, faradaic efficiency, and performance enhancement strategies in electrocoagulation—a review. Water Res 187:116433. https://doi.org/10.1016/j.watres.2020.116433
8. Das PP, Sontakke AD, Purkait MK (2023) Electrocoagulation process for wastewater treatment: applications, challenges, and prospects. Development in wastewater treatment research and processes, pp 23–48. https://doi.org/10.1016/B978-0-323-95684-0.00015-4
9. Mondal S, Purkait MK, De S (2018) Advances in dye removal technologies. Springer, Singapore, p 323
10. Ölmez T (2009) The optimization of Cr(VI) reduction and removal by electrocoagulation using response surface methodology. J Hazard Mater 162(2–3):1371–1378. https://doi.org/10.1016/j.jhazmat.2008.06.017
11. Wang CT, Chou WL, Kuo YM (2009) Removal of COD from laundry wastewater by electrocoagulation/electroflotation. J Hazard Mater 164(1):81–86. https://doi.org/10.1016/j.jhazmat.2008.07.122
12. Tartakovsk B, Kleiner Y, Manuel MF (2018) Bioelectrochemical anaerobic sewage treatment technology for Arctic communities. Environ Sci Pollut Res 25(33):32844–32850. https://doi.org/10.1007/s11356-017-8390-1
13. Nasrullah M, Singh L, Krishnan S et al (2020) Electrocoagulation treatment of raw palm oil mill effluent: effect of operating parameters on floc growth and structure. J Water Process Eng 33:101114. https://doi.org/10.1016/j.jwpe.2019.101114
14. Alimohammadi M, Mesdaghinia A, Shayesteh M et al (2019) The efficiency of the electrocoagulation process in reducing fluoride: application of inductive alternating current and polarity inverter. Int J Environ Sci Technol 16:8239–8254. https://doi.org/10.1007/s13762-019-02297-4
15. Aoudj S, Khelifa A, Drouiche N et al (2015) Simultaneous removal of chromium(VI) and fluoride by electrocoagulation–electroflotation: application of a hybrid Fe–Al anode. Chem Eng J 267:153–162. https://doi.org/10.1016/j.cej.2014.12.081
16. Behbahani M, Moghaddam MA, Arami M (2011) Techno-economical evaluation of fluoride removal by electrocoagulation process: optimization through response surface methodology. Desalination 271(1–3):209–218. https://doi.org/10.1016/j.desal.2010.12.033
17. Castañeda LF, Coreño O, Nava JL, Carreño G (2020) Removal of fluoride and hydrated silica from underground water by electrocoagulation in a flow channel reactor. Chemosphere 244:125417. https://doi.org/10.1016/j.chemosphere.2019.125417
18. Drouiche N, Palahouane B, Aoudj S et al (2016) Defluoridation of post treated photovoltaic wastewater using aluminum electrodes: optimization of operating parameters and sludge characteristics. J Solution Chem 45(11):1571–1579. https://doi.org/10.1007/s10953-016-0507-5
19. Hashim KS, Shaw A, Khaddar R et al (2017) Defluoridation of drinking water using a new flow column-electrocoagulation reactor (FCER)-Experimental, statistical, and economic approach. J Environ Manage 197:80–88. https://doi.org/10.1016/j.jenvman.2017.03.048
20. Govindan K, Raja M, Maheshwari SU et al (2015) Comparison and understanding of fluoride removal mechanism in Ca^{2+}, Mg^{2+} and Al^{3+} ion assisted electrocoagulation process using Fe and Al electrodes. J Environ Chem Eng 3(3):1784–1793. https://doi.org/10.1016/j.jece.2015.06.014

21. Mena V, Betancor-abreu A, González S et al (2019) Fluoride removal from natural volcanic underground water by an electrocoagulation process: parametric and cost evaluations. J Environ Manage 246:472–483. https://doi.org/10.1016/j.jenvman.2019.05.147
22. Un UT, Koparal AS, Ogutveren UB (2013) Fluoride removal from water and wastewater with a bach cylindrical electrode using electrocoagulation. Chem Eng J 223:110–115. https://doi.org/10.1016/j.cej.2013.02.126
23. Gl A (2003) Standard practice for preparing, cleaning, and evaluation corrosion test specimens. ASTM international. https://www.astm.org/g0001-03.html

Chapter 13
Design and Application of Intelligent Tide Observation System

Ying Ye, Zhi Kuo Sun, and Yong Mao Huang

Abstract In coastal areas, disasters caused by extreme weather risk occur frequently, posing a serious threat to these regions. In order to improve the frequency of tide data observation during disasters such as storm surges, an intelligent tide observation system was designed based on unmanned tide gauge stations. The existing data collection system and observation mode have been upgraded and redesigned. The observation system can automatically increase the observation frequency during weather disasters based on the set threshold, without the need for remote commands. The system can also automatically conduct data quality control based on set thresholds, providing more useful real-time observation data for ocean warning and disaster reduction efforts.

Keywords Observation system · Intelligent data collection · Self-adjusting sampling frequency

13.1 Introduction

In the construction of national economy and the enhancement of maritime disaster prevention and mitigation capabilities, the combination of storm surge and astronomical tide has caused significant water level rise, inundation, and overflow in coastal areas, resulting in major losses. Especially in coastal cities such as Zhuhai in southern China, this region has unique geographical characteristics for disastrous weather and climate events, making it a frequent area for extreme weather and maritime disasters. The frequency of sudden disasters is also the highest in the country, with strong typhoons exacerbating storm surge disasters, posing a serious threat to the lives of coastal residents, maritime development, and economic growth. In 2017, the cities of Zhuhai, Jiangmen, Zhongshan, and Guangzhou suffered significant losses due to the storm surge caused by Typhoon Hato. The affected area of aquaculture in

Y. Ye (✉) · Z. K. Sun · Y. M. Huang
Southern Marine Science and Engineering Guangdong Laboratory (Zhuhai), Zhuhai, Guangdong, China
e-mail: yeying@sml-zhuhai.cn

K. Y. Foo et al. (eds.), *Water Nexus Approach for Sustainable Development*, Environmental Science and Engineering, https://doi.org/10.1007/978-3-031-89238-7_13

the province reached 18,238.80 ha, with a total loss of 22,800 tons of aquaculture production. In 2018, Typhoon Mangkhut made landfall in the Pearl River Estuary area, causing severe erosion to the 5 km-long tourist beach in Xichong, Shenzhen. Coastal tourist facilities and hotels were severely damaged, with the storm surge becoming one of the important factors constraining the economic development of the Pearl River Delta [1, 2].

Tides are periodic fluctuations in sea level caused by the gravitational forces of the ocean [3]. Tide data is of great importance for navigation safety, port management, and coastal protection [4]. Tide data can be measured using tide gauges, tide stations, or satellite remote sensing. The collection and analysis of tide data elements can provide a detailed description of the marine environment, serving as a crucial basis for marine predictions and decisions. Therefore, in the work of storm surge warnings, comprehensive and rapid tide level data undoubtedly serves as a powerful technical support to improve the accuracy of warning reports [5, 6].

There are various methods for tidal observation, ranging from the earliest water level measurements, float-type tide gauges, acoustic water level gauges, high-frequency shore-based radars, to the combination of unmanned aerial vehicles and other unmanned intelligent technologies [7]. Currently, there are many unattended tidal stations with flexible and low-cost installation, which can be used for emergency observation. These stations adopt mature automated devices such as pressure tide gauges, radar tide gauges, and acoustic tide gauges. However, in the face of severe convective weather, the observation frequency of once per minute is still insufficient. It is necessary to increase the observation frequency according to the task requirements in order to provide more powerful data support for early warning reports. Traditional operational tide gauges operate according to a fixed process of collection, calculation, storage, and transmission, and cannot change the collection frequency [8]. Moreover, in traditional marine hydrological and meteorological data collection, storage, and analysis systems, there are problems such as low data collection frequency and insufficient data quality analysis capabilities. Therefore, this paper proposes a tide intelligent real-time observation system, with an ARM processor as the core hardware platform, integrating various sensors and software to achieve the collection, storage, and analysis of hydrological and meteorological data. According to the program setting of the observation system, the system can intelligently switch the appropriate observation frequency, which can meet the demand for increasing the frequency of data collection in disaster weather such as storm surge.

13.2 Observational System Composition

The intelligent tide real-time monitoring system collects environmental observation elements in real-time. The designed software algorithm can automatically determine whether to increase the observation frequency based on the set threshold. This system resolves the need for obtaining fine-grained observation data in adverse weather

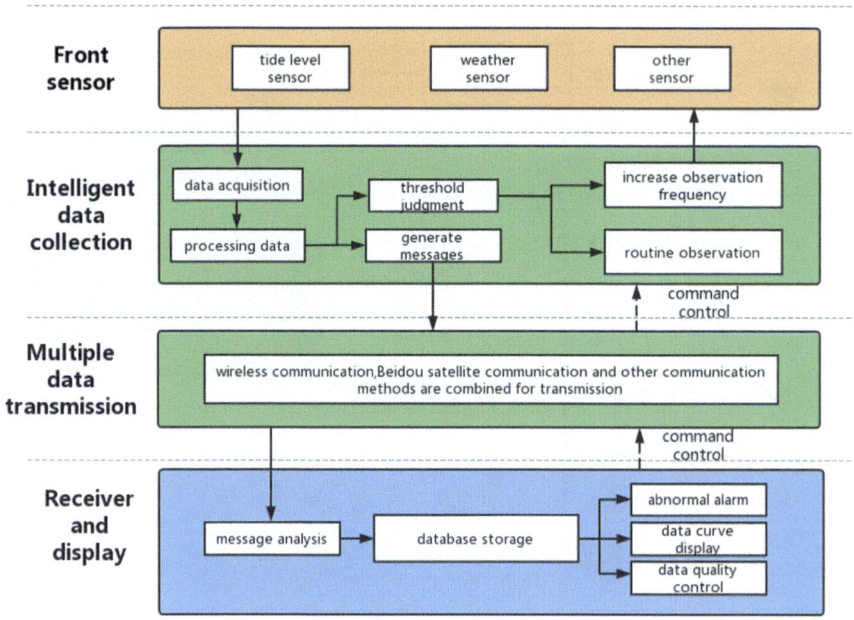

Fig. 13.1 The overall framework of the intelligent tide real-time monitoring system

conditions such as storm surges, and provides important data and assists decision-making for the forecasting and mitigating work.

Figure 13.1 illustrates the overall framework of the tide intelligent observation system. Various hydrological and meteorological observation devices such as tide gauges, air pressure sensors, wind direction and wind speed sensor are integrated into the data collector through intelligent collection methods. The data collector reads, interprets, and quality controls the raw sensor data to generate data packets. These data packets are then transmitted in real-time through a multi-channel transmission system to the data server on the data receiving terminal for standard storage and management.

13.3 Composition of Tide Level Observation Equipment

As shown in Fig. 13.2, the observation equipment of the intelligent tide observation system includes hydrological and meteorological observation sensors and video devices. Hydrological data, meteorological data and videos information are transmitted in real-time to the data receiving center.

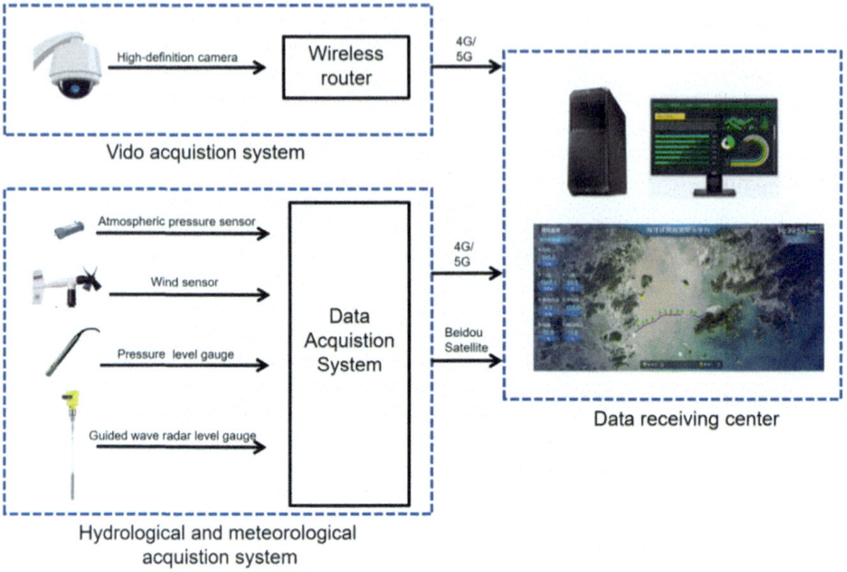

Fig. 13.2 Composition of the intelligent tide observation system

13.3.1 Hydrological and Meteorological Observation Sensors

This article selects devices that are suitable for nearshore and low power consumption, powered by solar energy and batteries to ensure the stable operation of the system.

Radar observation tide gauge.

Radar tide gauges use radar echo ranging principles to measure water levels [8]. Microwaves can penetrate vacuum and are not affected by environmental factors such as atmosphere, volatile substances, or space. Therefore, radar beams can measure the distance between the water surface and the tide gauge, thus achieving accurate monitoring of the tide level. Radar water level measurement devices are generally divided into contact radar tide gauges and non-contact radar tide gauges. This article chooses the contact radar tide gauge, equipped with a small tide gauge well, to focus microwave energy without being disturbed by ocean waves, resulting in high equipment accuracy and low infrastructure costs.

Air pressure sensors, wind direction and wind speed sensor.

The air pressure sensors uses an electronic barometer. The measurement of wind speed and wind direction data is carried out by a propeller-type anemometer. The sensor's stainless steel and precision conductive plastic potentiometers are suitable for harsh environments with high temperatures, humidity, and salinity.

13.3.2 Video Devices

The video device consists of a high-definition camera, solar panel, battery, and wireless router, primarily used to assist in the tide observation. It can monitor, capture images of the site, and transmit the images back to the data receiving center. Users can also remotely monitor the site in real time through their mobile phones and desktop computers. In order to ensure the stability of tide level data observation, the video device has a separate transmission system and power supply system, which are not shared with other observation devices.

13.3.3 Multi-path Data Transfer

The tide monitoring system is required not only to observe tidal levels and weather data, but also to store and manage a variety of marine observation data from multiple equipment sources, multiple time sequences, and multiple accuracy requirements. Therefore, from the perspective of data transmission, the intelligent tidal monitoring system uses multi-channel data transmission to achieve high-speed transmission of system observation data and observation images.

The system utilizes Data Transmission Unit (DTU) for real-time transmission of small data such as real-time messages. At the same time, using Beidou satellite communication, small data is backed up to ensure the integrity of observation data. For large data such as videos, within areas covered by 5G, the system utilizes wireless routers with built-in 5G transmission modules for real-time transmission of image and video files. The 5G transmission speed can reach 20 Gb/s, which is over 10 times faster than 4G network transmission speed, allowing users to watch live videos of tide gauge stations in real-time through computers or mobile phones.

13.3.4 Data Receiving Center

The data receiving center receives real-time data sent by data transmission devices, serving as an essential component of the tide intelligent observation system. It is responsible for receiving, storing, and managing observation data. The data receiving center can send remote commands to the front-end collection system to increase the observation frequency in harsh weather conditions. Moreover, it is responsible for managing various station information, such as latitude and longitude of tide observation sites, observation elements, communication methods, and real-time management of device operation status. It also provides users with device alarm information.

13.4 Intelligent Data Collection Mode

Most domestic tide data collectors use SCM or PC/104 platforms to to collect multiple analog signals, frequency signals, and digital I/O signals. Although SCM has low power consumption and is cost-effective, their processing speed, storage capacity, and software and hardware expansion are limited, which cannot meet the requirements of intelligent automatic control. The hardware of the intelligent data observation system adopts the ARM embedded core board, which has low power consumption and can be built-in with Ubuntu16.04 operating system. It is powerful and can be embedded with various algorithms.

In the intelligent tide observation system, ARM architecture was chosen as the core processor of the system due to its high efficiency, low power consumption, and easy portability, and is widely used in embedded systems. The ARM embedded core board used in this system features a six-core processor with dual Cortex-A72 big cores and four Cortex-A53 small cores, with a high frequency of up to 1.8 GHz. The core board is installed with the Ubuntu system, and the display screen software is compiled using Python language and PYQT5 graphic framework. In addition, it is equipped with a 4G module for real-time transmission. The real-time data files generated by the built-in software system can be directly transmitted in real-time through the main control board without the need for connecting data transmission devices or modules such as DTU.

According to the original workflow, the tidal intelligent observation system has made improvements in "data collection", "data transmission", and "data reception." The intelligent data collection mode is divided into "regular collection mode" and "self-adjusting frequency mode mode." The "regular collection mode" is for daily observations, with an observation frequency of 10 min, that is, a message is generated every 10 min and sent in real time to the data receiving end; the "automatic identification mode" is a mode that increases the observation frequency, with an observation frequency of 1 min or 30 s. The front end of the data collector sets thresholds based on observed factors, and if the observed values exceed or fall below the set thresholds, the observation frequency is either increased or decreased. In addition, the data receiving end can also remotely send instructions to the front end data collection system to increase or decrease the observation frequency. The system program first collects data, then compares the data with the set thresholds, and ultimately determines whether to change the current collection mode (Fig. 13.3).

13.5 Application of the System

In order to verify the applicability of the intelligent observation system, a selected location at the Geili Coast in Zhuhai was chosen to compare the existing data collection system with the intelligent observation system. Various hydrological and meteorological observation devices were installed, such as front tide gauge, wind speed

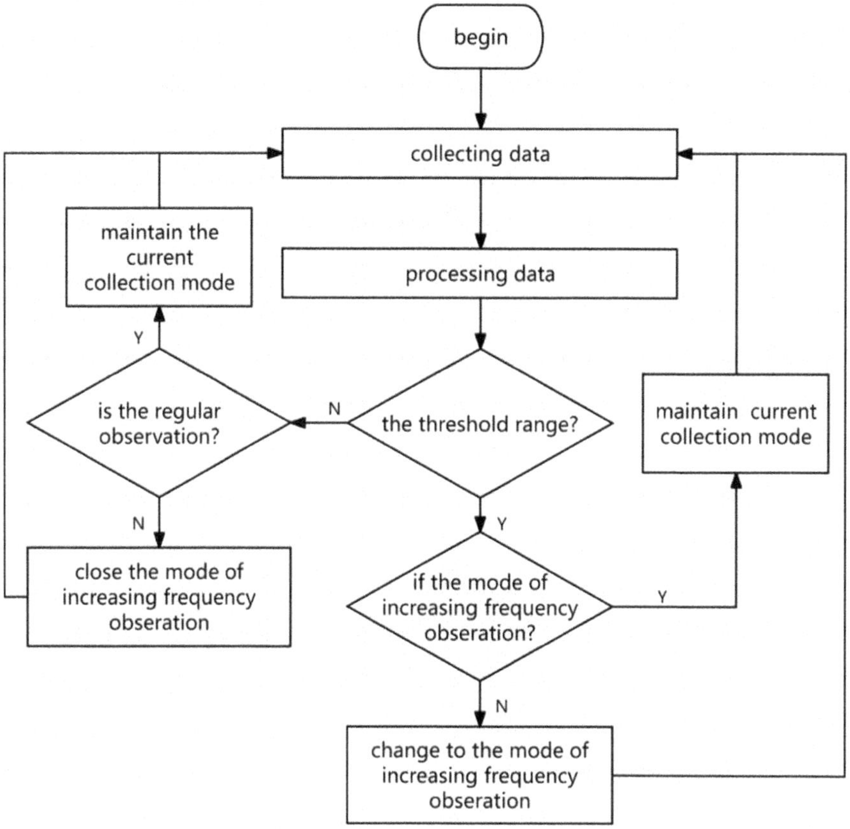

Fig. 13.3 Flow chart of "self-adjusting frequency mode" for data intelligence collection

and direction sensors, air pressure sensors. A 100 W solar panel and 100 AH battery were used for continuous power supply. The observation data was processed using the moving average filtering method. By comparing the instantaneous wind speed data observed by the existing data collection system and the intelligent data collection system (Fig. 13.4), it can be seen that the intelligent collection system can observe instantaneous wind speeds of up to 14 m/s when the sampling frequency is increased to once per minute, providing more useful information. By increasing the observation frequency, one can obtain faster and more detailed observational data.

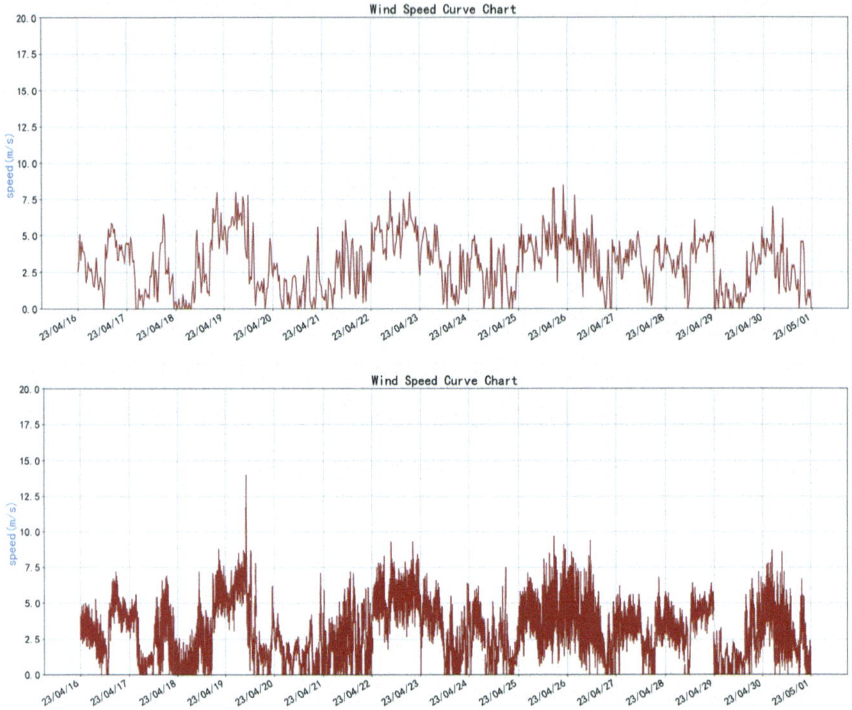

Fig. 13.4 Data from existing collection system and intelligent collection system

13.6 Conclusions

This article presents a tidal intelligent real-time observation system with a new feature design of "self-adjusting frequency". Compared to existing tidal observation systems, this system does not require manual operation on-site, is cost-effective, and highly efficient. Not only does it adjust the observation frequency quickly, but it also reduces the operational errors and inconvenience caused by manually adjusting the observation frequency. This system can achieve more accurate sea level height and meteorological observation data under disaster weather conditions such as storm surges.

This system can be applied to marine meteorology, marine transportation and other maritime departments in coastal areas, by constructing tide gauge stations to obtain fine-grained sea level height and meteorological observation data, enriching basic survey data in maritime areas, and thereby improving the simulation accuracy of high-precision numerical models. It provides scientific support for China's marine disaster prevention and reduction work, basic support for navigation safety and emergency rescue, reduces economic losses caused by marine disasters, and fully leverages the important role of marine observation in promoting economic development.

References

1. Zhi JL, Huang XX (2020) Comparion and cause analysis of high wind processes in No. 1713 Typhoon Hato and No. 1822 Typhoon Mangkhut. Guizhou Meteorol
2. Tu J, Luo Q, Liu H (2021) Investigation and countermeasures analysis of damages of coastal seawalls by Typhoons Hato (1713) and Mangkhut (1822). Guangdong Water Res Hydropower
3. Wei H (2016) Design and implementation of low cost marine environmental monitoring equipment. Ship Sci Technol 52(36):41–55
4. Kleinosky LR (2007) Vulnerability of Hampton Roads, Virginia to storm-surge flooding and sea-level rise. Nat Hazards (1):323–329
5. Benlu Z (2010) Business application of Fujian storm surge warning auxiliary decision system. Mar Forecasts 27(5):16–19
6. Guo Y, Zhang J, Zhang L (2009) Computational investigation of typhoon-induced storm surge in Hangzhou Bay, China. Estuar Coast Shelf Sci 85(4):530–536
7. Zhao J, Shen W, Wu Y et al (2007) Modern marine surveying and charting. Wuhan University Press, Wuhan, pp 2–3
8. Lixia S, Yuanyuan H (2018) Analysis and storage of meteorological observation data based on Python. Ningxia Eng Technol 15(55):91–95
9. Xu Z, Zhang J (2007) Application of measurement technology of radar liquid level in the marine engineering. Ship Eng 29(5):83–85

Chapter 14
Facile Synthesis of Ferrocene Modified Graphene and Application for Acid Fuchsin Removal in Water

Jinsheng Cheng, Sheng Zhong, Haoceng Wang, Jing Li, Wenying Sun, Weihong Wan, Xiaowei Zhou, and Lanzhao Zhong

Abstract In this work, a interesting ferrocene modified graphene (FMG) were synthesized via facile chemical methods under simple and mild conditions. The structure and composition of the prepared FMG were characterized by transmission electron microscopy (TEM), high-resolution transmission electron microscopy (HRTEM), energy dispersive X-Ray spectroscopy (EDX), raman spectra, fourier-transform infrared spectroscopy (FTIR) and cyclic voltammogram (CV) etc. Further investigations showed that the prepared FMG nanocomposites possess superior purification advantages foracid fuchsin, a typical azo dye, which fosters the prepared FMG nanocomposites a special candidate for acid fuchsin removal in water.

Keywords Ferrocene modified graphene · Acid fuchsin · Removal

14.1 Introduction

With the development of industries, the increase in population, the intensification of urbanization, and the increase in the use of fertilizers and pesticides, the water as the source of life has been severely polluted and has become a serious issue that affects human survival and development in the 21st century [1–3]. Among them, extensive nitrogen dyes are important sources of pollution of water body pollution [4, 5]. Large nitrogen dyes are discharged into the water environment and will cause

J. Cheng (✉) · J. Li · W. Sun · W. Wan · X. Zhou
Shaoguan University, Shaoguan, China
e-mail: chengjins@gmail.com

S. Zhong
School of Environment, China University of Geosciences, Wuhan, Wuhan, China

X. Zhou · L. Zhong
Technical Center, Zhuhai Jingwu Tech. Co. Ltd., Zhuhai, China

H. Wang
Shaoguan, China

K. Y. Foo et al. (eds.), *Water Nexus Approach for Sustainable Development*, Environmental Science and Engineering, https://doi.org/10.1007/978-3-031-89238-7_14

severe water pollution. Traditional puppet nitrogen dye treatment methods include chemical adsorption method, membrane treatment method etc. [5, 6]. These treatment methods have the advantages of simple process and low threshold, but the processing efficiency still has certain limitations, and it is also easy to cause secondary pollution. Some sewage treatment agents are more expensive and cannot be used in circular, which could limit the industrial application of such purification methods.

Graphene nanomaterial has a large surface and good physical and chemical properties, and has been widely used in the study of or dye degradation or removal of puppet nitrogen dyes [7–10]. In the present study, a novel FMG nanocomposites were synthesized readily under mild conditions. The structure and composition of the prepared FMG were characterized by transmission electron microscopy (TEM), high-resolution transmission electron microscopy (HRTEM), energy dispersive X-Ray spectroscopy (EDX), raman spectra, fourier-transform infrared spectroscopy (FTIR) and cyclic voltammogram (CV) etc. Further investigations showed that the prepared novel FMG nanocomposites could combine both photocatalytic advantages of ferrocene and graphene nanosheets foracid fuchsin, a typical azo dye, which fosters the prepared FMG nanocomposites a special candidate foracid fuchsin removal in water. In this study, the acid fuchsin in the water sample is the template pollutant. The removal efficiency of the acid fuchsin was studied under different conditions. It also examines the effects and feasibility of temperature, addition amount of FMG, and the starting concentration of dyes of the acid fuchsin in this work.

14.2 Results and Discussion

14.2.1 Preparation and Characterization of FMG Nanocomposites

To date, few reports have delved into FMG preparation and application, which might combine both superior photocatalytic properities of graphene and Ge nanocubes. The work illustrated here presented a novel FMG nanocomposites preparation procedures, which involving reaction and loading effect between aminoferrocene containing Fe (II) cation and graphene oxide with abundant COOH groups, together with successively reduction treatment by sodium borohydride powder and hydrazine hydrate step by step. After vacuum dried, the featured FMG nanocompositess were finally afforded (Fig. 14.1).

The obtained FMG nanocomposites were characterized by TEM, HRTEM, EDX etc. The TEM images illustrated in Fig. 2a displayed a view of the prepared FGM nanocomposites, highly uniform crystalline iron cation structures of ferrocene on graphene nanosheets were observed. HRTEM images showed in Fig. 2b indicated clear evidence of a homogeneous distribution of crystalline iron cation (10–30 nm in size) on few-layered graphene nanosheets. The presented image in Fig. 14.1c also showed energy-dispersive X-ray spectroscopy (EDX) analysis and CV (Fig. 14.3)

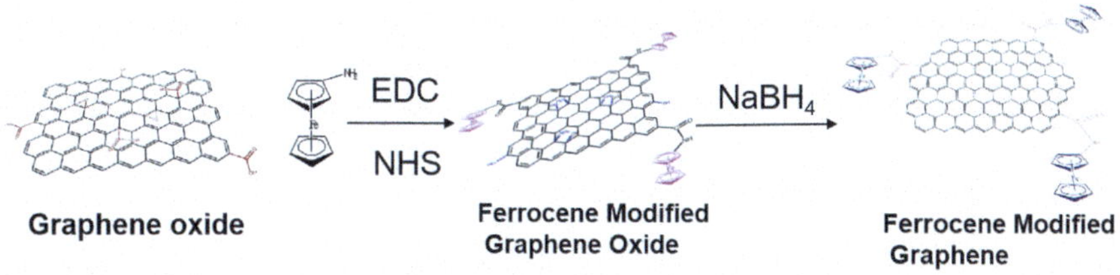

Fig. 14.1 Synthesis of the prepared FMG nanocomposites

etc. of the prepared FMG nanocomposites, further confiming few-layed graphene nanosheets supported FGM nanomaterials.

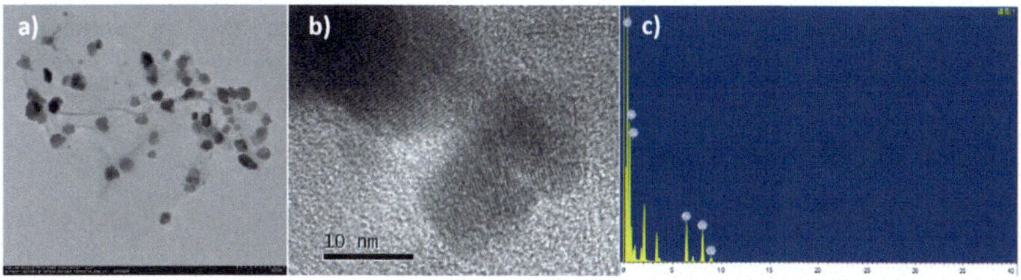

Fig. 14.2 TEM **a** and HRTEM **b** images of FMG nanocomposites and **c** EDX analysis of the prepared FMG nanocomposites

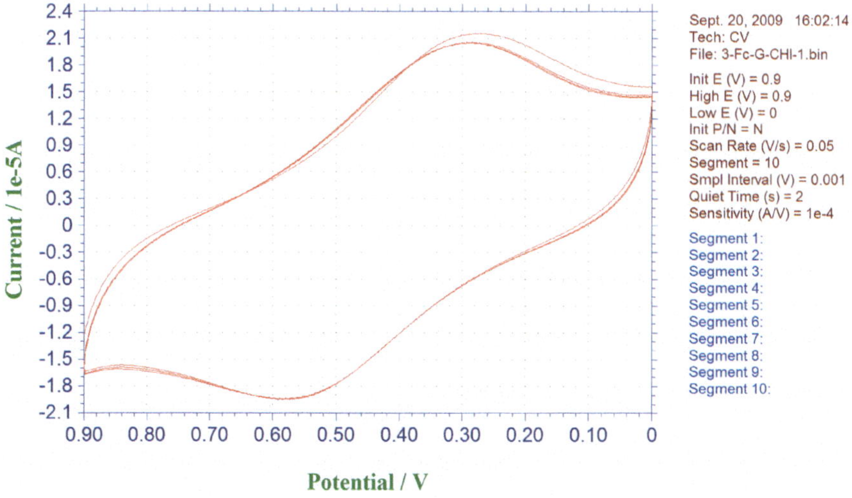

Fig. 14.3 Cyclic voltammogram of the prepared FMG nanocomposites

14.2.2 Removal Performance of Acid Fuchsin by the Prepared FMG Nanocomposites

Effect of removal time on acid fuchsin removal rate.

The removal rate curve of acid fuchsin as a function of removal time is shown in Fig. 14.4. It can be seen from the figure that at each time period, the removal rate of acid fuchsin by FMG was significantly higher than that of graphene nanosheets without amino modification, showing that the removal rate of acid fuchsin by FMG was better than that of graphene without ferrocene modification. Under the conditions of 20 min, 50 min and 120 min, the removal rates of acid fuchsin by FMG were 34.2%, 64.8% and 97.1% respectively, which was obvious higher than the removal rate under the same time conditions when graphene nanosheets are used as the removal agent. The rates were 29.2%, 55.3% and 76.1% respectively. The study found that the removal process of acid fuchsin by FMG can be divided into two stages: the first 60 min was the rapid removal stage, and the next 60 min was the slow removal stage.

The effect of FMG dosage on acid fuchsin removal rate.

Under the condition of 120 min conditions, the FMG use of the undertakings of the acid fuchsin was shown in Fig. 14.5. It can be seen from the figure that within the range of 0–1.5 mg, the removal rate of the acid fuchsin was gradually increased as the amount of FMG increases. The removal effect is the best, the removal time was up to 97.1% under the condition of 120 min conditions; when the amount of FMG was higher than 1.5 mg, the removal rate of the acid fuchsin was declining. This might due to the excessively high nano-dehydration agent added, the reunion effect of nanomaterials will be produced, which reduces the dehydent's slightly reduced surface than the surface, and some FMG surface is covered with it, leading to poor removal efficiency.

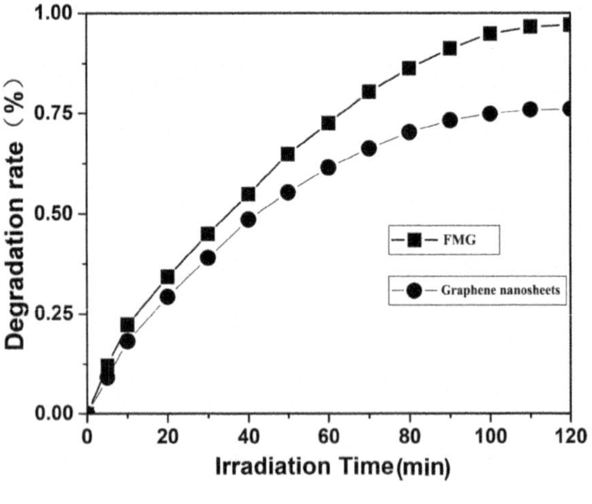

Fig. 14.4 Influence of irradiation time on the degradation efficiency of acid fuchsin

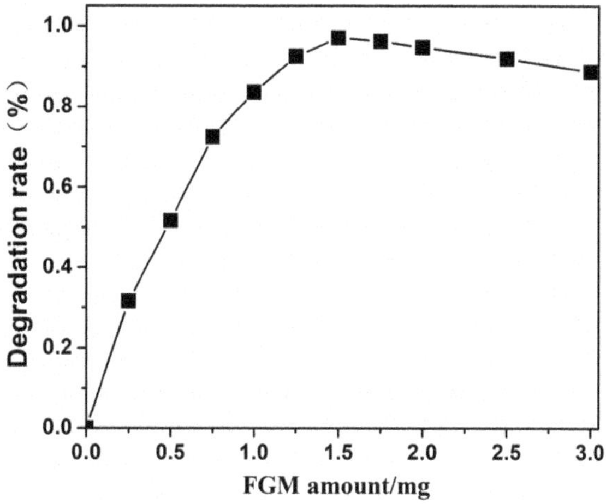

Fig. 14.5 Influence of dosage of acid fuchsin on the degradation efficiency of acid fuchsin photocatalyzed by the prepared FMG nanocomposites

The impact of the initial concentration of acid fuchsin on removal rate.

In this study, different initial concentration of acid fuchsin removal curve is shown in Fig. 14.6. It can be seen from the figure that the initial concentration of the acid fuchsin gradually increased from 0.20 mg/ml to 1.0 mg/ml conditions, and the removal rate of FMG on the acid fuchsin increased. This might due to that under higher conditions, according to the molecular collision model, the collision probability of the acid fuchsin molecule and the FMG molecule is higher, which is conducive to the full contact between the acid fuchsin and FMG, thereby achieving a higher removal efficiency. When the initial concentration reaches a certain degree, ranging from 2.0 mg/ml to 4.0 mg/ml conditions, and the FMG removal rate of acid fuchsin was not significantly improved. When GR dye molecules are surrounded, further improving the impact of the acid fuchsin concentration on the final removal efficiency does not have a great impact; the initial concentration of the acid fuchsin is 1.0 mg/ml, the amount of FMG is 1.5 mg, and the removal time is 120 min conditions. FMG has the best removal effect on the acid fuchsin, and the removal rate reaches the maximum of 97.1%.

Due to the introduction of graphene with ferrocene, the surface-rich ferrocene groups enables the prepared FMG to better purification agents through photocatalytic degradation, hydrogen bonds or static electricity. GR dye molecules. On the other hand, the FMG and acid fuchsin structure contain a polyenzene ring structure, which is easy to form a strong π-π bond, making the adsorption and removal effect of FMG on the adsorption and removal of acid fuchsin.

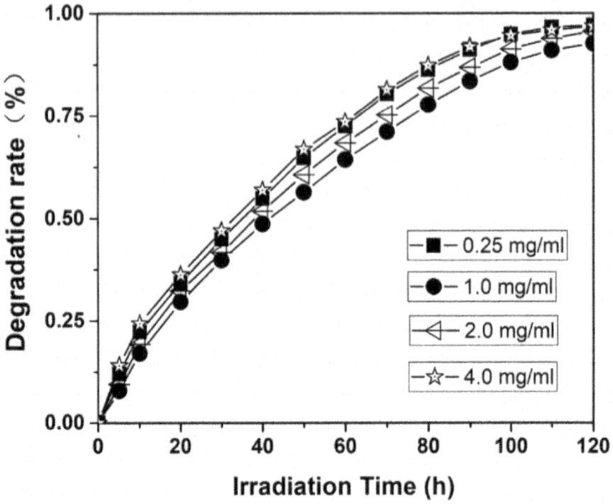

Fig. 14.6 Influence of starting concentration of acid fuchsin on the degradation

14.3 Conclusion

In this work, highly uniform FGM nanocomposites were prepared via covalent bond formation and reduction procedures step by step. The structure and composition of the prepared FMG nanocomposites were systematicly characterized by TEM, HRTEM, EDX and CV analysis. Further investigations revealed that the prepared novel FGM nanocomposites could combine both photocatalytic advantages of ferrocene and graphene nanosheets for acid fuchsin photodegradation in wide concentration range (0.25–4.0 mg/mL) under visible light. At initial acid fuchsin concentrations of 1 mg/L at room temperature, the removal efficiency can achieved to 97.1% and the solution can be decolorized to nearly colorless, obvious high than the result of comercial candidates. Morever, the prepared FGM nanocomposites photocatalyst can be used for even five runs without much loss of photodegradation efficiencies, which fosters FGM nanocomposites a special candidate for azo dye removal in water.

Acknowledgements This work was financially supported by the National Science Foundation of China (No. U22A20152), Guangdong University Innovation Team Project (2019KCXTD017)

References

1. Scholz M (2023) Wetlands for water pollution control. Elsevier Science, Amsterdam. https://www.google.com.hk/books/edition/_/MdPKEAAAQBAJ?hl=zh-CN&kptab=overview
2. Kordbacheh F, Heidari G (2023) Water pollutants and approaches for their removal. Mater Chem Horiz 2(2):139–153. https://doi.org/10.22128/MCH.2023.684.1039
3. Al-Nuaim MA, Alwasiti AA, Shnain ZY (2023) The photocatalytic process in the treatment of polluted water. Chem Pap 77(2):677–701. https://doi.org/10.1007/s11696-022-02468-7

4. Zafar S, Bukhari DA, Rehman A (2022) Azo dyes degradation by microorganisms–an efficient and sustainable approach. Saudi J Biol Sci 29(12):103437. https://doi.org/10.1016/j.sjbs.2022.103437
5. Ganesh R, Boardman GD, Michelsen D (1994) Fate of azo dyes in sludges. Water Res 28(6):1367–1376. https://doi.org/10.1016/0043-1354(94)90303-4
6. Samiee S, Goharshadi EK (2014) Graphene nanosheets as efficient adsorbent for an azo dye removal: kinetic and thermodynamic studies. J Nanopart Res 16:1–16. https://doi.org/10.1007/s11051-014-2542-8
7. Pérez-Molina Á, Morales-Torres S, Maldonado-Hódar FJ, Pastrana-Martínez LM (2020) Functionalized graphene derivatives and TiO_2 for high visible light photodegradation of azo dyes. Nanomater 10(6):1106. https://doi.org/10.3390/nano10061106
8. Guo YF, Deng J, Zhu JY, Zhou CY, Zhou XJ, Bai RB (2016) Removal of anionic azo dye from water with activated graphene oxide: kinetic, equilibrium and thermodynamic modeling. RSC Adv 6(46):39762–39773. https://doi.org/10.1039/C6RA03423C
9. Cheng JS, Du J, Zhu W (2012) Facile synthesis of three-dimensional chitosan–graphene mesostructures for reactive black 5 removal. Carbohyd Polym 88(1):61–67. https://doi.org/10.1016/j.carbpol.2011.11.065
10. Zhang H, Lv X, Li Y, Wang Y, Li J (2010) P25-graphene composite as a high performance photocatalyst. ACS Nano 4:380–386. https://doi.org/10.1021/nn901221k

Part II
Security on Drinking Water

Chapter 15
Research on the Clean Water Diversion Scheme of Banfu Based on Quality Target Accessibility Analysis

Bin Wang, Xinyao Jiang, Wenbin Lv, and Weilin Luo

Abstract In order to improve the problems of Insufficient hydrodynamic force, lack of water environment capacity, and serious black and odorous water in Banfu, this paper proposes a water diversion plan to improve the water quality of the tidal river network in the basin, based on the tidal power of the Modaomen and the tail water of the Wastewater Treatment Plant. The results show that using Modaomen tidal power to divert water can effectively improve the water quality of rivers in Banfu. During the dry season, the water quality of most rivers in Banfu can reach Class IV or Class V standards. After water diversion is implemented for 48 h after a rain, the water quality of rivers in the basin can be basically restored.

Keywords Tidal river network · Water cycle · Black and odorous water · Guangdong-Hong Kong-Macao Greater Bay Area

15.1 Introduction

With the rapid development of urbanization, the ecological environment of urban water systems is increasingly affected by human activities. which has caused a series of urban water environment problems and has become a serious challenge for water environment management work [1–3]. The phenomenon of Black and odorous water in urban rivers is an important water environment problem, seriously restricting the development of urban economic development [4]. The generation of black and odorous water in urban rivers is affected by many factors, the insufficient hydrodynamic power of urban rivers is one of the important influencing factors [5, 6]. The river networks in plain areas is affected by tidal dynamics, the water in the inner

B. Wang (✉) · W. Lv · W. Luo
Pearl River Water Resources Research Institute, Guangzhou, China
e-mail: 278273213@qq.com

X. Jiang
Beijing Urban Construction Design & Development Group Co., Limited, Beijing, China

K. Y. Foo et al. (eds.), *Water Nexus Approach for Sustainable Development*,
Environmental Science and Engineering, https://doi.org/10.1007/978-3-031-89238-7_15

river flows back and forth, causing the continuous accumulation of pollutants, which further aggravates the black and odorous condition.

There are many studies on clean water diversion scheme technology aimed at controlling the water quality of urban rivers and treating black and odorous water bodies at present. The gray water footprint theory and the MIKE21 water quality model was implemented by Han et al. [7], a comprehensive evaluation of the effect of urban lake water environment management was carried out; In view of the water environment problems in the Zhongshun Dawei area of the Pearl River Delta, a joint dispatching plan for diversion from the west to the east is proposed by Du et al. [8]. Shi et al. [9] takes Tiepai River in Baoan District, Shenzhen City as an example, the ecological water supply management plan for rivers in highly built-up areas was studied.

This paper takes the internal river network in Banfu, Zhongshan City as an example, combines the tidal dynamic characteristics of estuary, implements water diversion from Modaomen to improve the water quality of the river network, and builds a one-dimensional hydrodynamic water quality model through MIKE 11 to analyze the accessibility of water quality targets. The clean water diversion scheme under dry season and rainy season conditions was analyzed in order to provide scientific reference for the treatment of river water pollution in cities.

15.2 Overview

Banfu is located in the south of Zhongshan City, Wugui Mountain is located in the east, and adjacent to Hong Kong and Macao, as shown in Figs. 15.1 and 15.2. The town covers 79.7 km^2. The river network in Banfu is complex. There are 34 rivers in the town area with a total length of 56.57 km. The Qijiang River runs through the town, dividing the town into Hexi area and Hedong area. Hedong area is the main town, surrounded by mountains (Jinzhong Mountain) and rivers (Qijiang River), and has prosperous industry, Hexi area has developed flower seedlings and aquaculture industries, The overall topography of the Banfu watershed is higher in the east and lower in the west. Except for the Modaomen Waterway, Qijiang River, mayongchong and Silianchong, which flow in the north–south direction, the other rivers all flow in the east–west direction and eventually flow into the Qijiang River and the Modaomen Waterway.

In recent years, Banfu Town, Zhongshan City has gradually increased its efforts to control water environment problems in the basin. The water quality has been significantly improved, but there are still some stubborn problems: (1) Domestic wastewater, domestic garbage, factory wastewater and other pollution sources in the river can easily pollute the river water quality and increase the risk Risk of black and odorous water. (2) Although there are drainage pipes in old communities, natural villages, they are scattered and unsystematic. And most of them are combined rainwater and sewage system, all kinds of sewage are discharged into nearby rivers

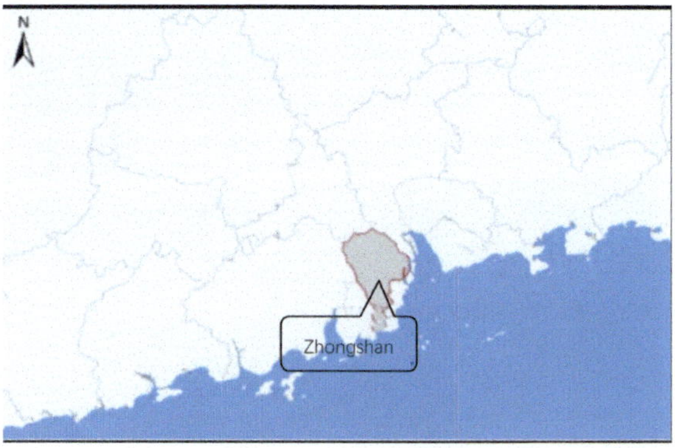

Fig. 15.1 Location of Zhongshan

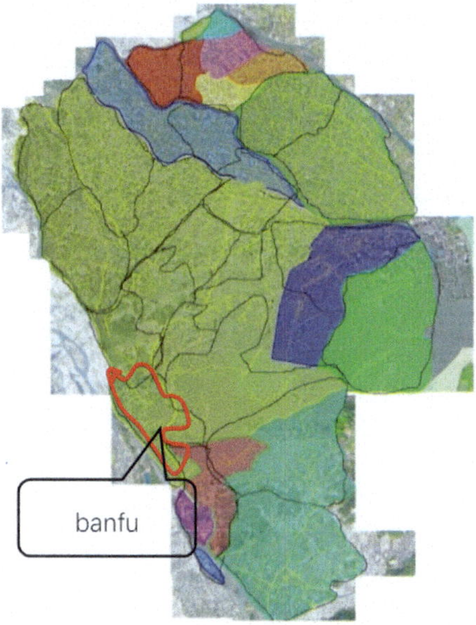

Fig. 15.2 Location of Banfu basin

or ponds. (3) Some domestic waste is dumped directly into rivers and pipes, causing blockage of rivers and pipes.

In order to solve the problem of black and odorous water in Banfu, the Zhongshan Water Bureau has carried out a comprehensive improvement project for black and odorous rivers in Zhongshan City. Through sewage interception projects, river non-point source pollution control projects, dredging projects and silt disposal projects, water system circulation and other project contents. By 2023, all black and smelly rivers in this basin will be eliminated during the construction period, and 80% of rivers will be eliminated as inferior Class V water bodies during the operation period.

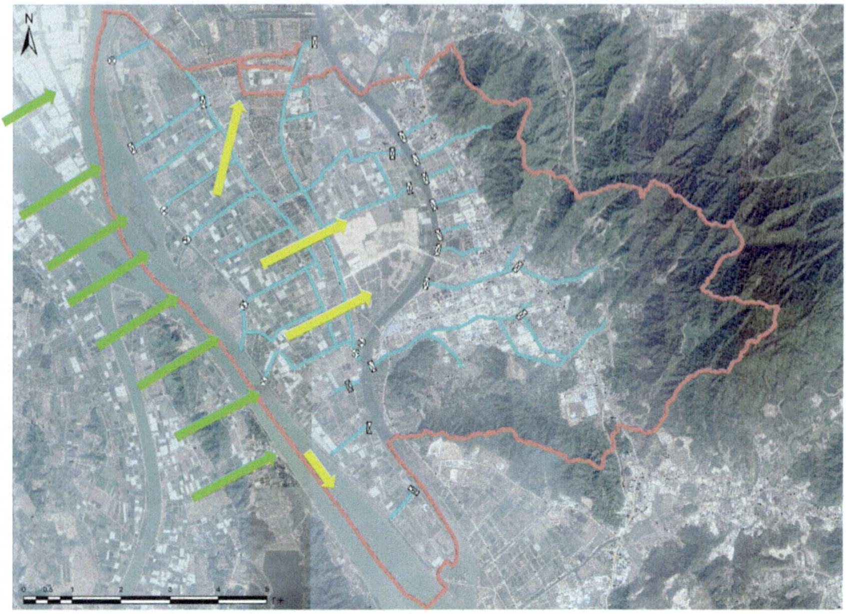

Fig. 15.3 Schematic diagram of flow direction in Hexi district

15.3 Clean Water Diversion Scheme

15.3.1 Hexi Area

Hexi area belongs to Zhongshun Dawei, the highest operating water level of Zhongshun Dawei is 2.0 m. (85 national elevation basis, the same below) Taking into account the actual topography and possible sudden flood drainage needs, the clean water diversion scheme in the Hexi area is controlled to a maximum water level of 1.4 m during daily operation and a maximum water level of 1.6 m during drainage operation. Figure 15.3 presents the flow direction in Hexi district.

The specific method is: the sluice on one side of Modaomen are opened to divert water during high tide. And these sluice are closed when the water level reaches the maximum tide level or the highest control level (1.4 m). When the Qijiang River water level ebbs, the sluice near the Qijiang River drains water.

15.3.2 Hedong Area

Hedong area will treat the tail water from the Banfu Sewage Treatment Plant through upgrading standards and pump it to Gongqing Chung, Lixi Yuejiao Chung, Jinzhong

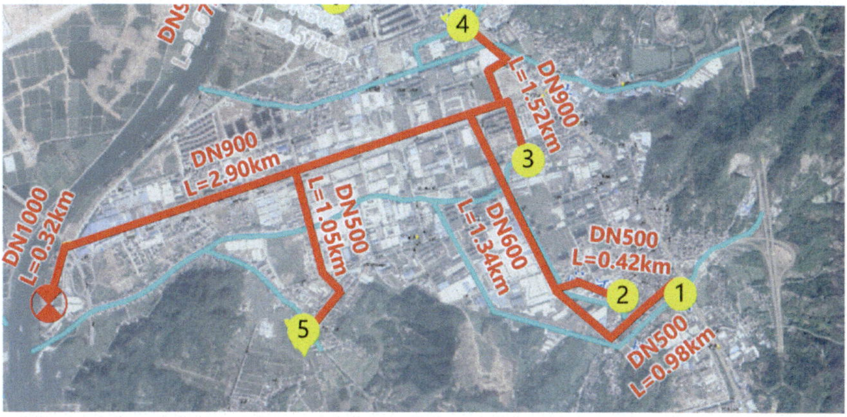

Fig. 15.4 Schematic diagram of water supply pipeline in Hedong district

Table 15.1 Comprehensive land use surface erosion simulation parameters of different pollutants

Pollutants	Erosion coefficient	Washoff coefficient
COD	0.49	1
TN	0.51	1
TP	0.47	1

Chung, Shenwan Central River, and the upstream of the Shenwan Flood Drainage Channel. And using the Qijiang River as a backup water source. Construction of 1 water supply pumping station and 7.75 km of pipeline. The Ermenkan Reservoir and the existing idle water supply pipelines will be used to replenish water in Shapuyong and Sankexunyong. 700 m of DN400PE pipes and 600 m of DN300PE pipes will be added, as shown in Fig. 15.4.

The Jinzhong sluice gate in Hedong area opens to divert water during high tides. The gate is closed after the water level in the inland river reaches the maximum tide level or the maximum control water level (1.7 m), Opening gates for drainage during spring and ebb tides; When the tidal power is weak during small and medium tide, the sluice gate will be closed and the Jinzhong pump station will be opened to introduce more water into Jinzhongchong, and the sluice gate will be opened at low tide for drainage (Table 15.1).

15.4 Analysis of Pollutants Entering the River

There are many point source and non-point source pollution sources entering the river in Banfu. In order to truly evaluate the accessibility of the engineering measures of this project. Domestic, industrial and endogenous pollution use local statistical data. Surface runoff non-point source pollution adopts pipe network-river coupling model,

Simulates the non-point source pollution process of surface pollutants entering rainwater pipes through erosion under rainfall conditions and finally flowing into rivers, as shown in Table 15.2.

Initial rainwater pollution in Banfu is calculated using the SWMM pipe network model. The hydrological analysis module of ArcGIS was adopted to extract subcatchments in the 90 m DEM of the study area. Then the modification tools was used to correct the boundaries based on the street scene and municipal pipes.

The rainfall adopts the Chicago rain pattern, and the rainfall is calculated using the Zhongshan rainstorm intensity formula recommended by the Zhongshan Water Bureau [10]. The water quality indicators select three main pollutants: COD, NH_3–N, and TP. The pollutant wash off module uses an exponential function model, and the input parameters include Erosion coefficient and Wash off coefficient was shown in Table 15.1. It is assumed that the initial amount of pollutants on the surface is linearly related to the amount of local dust fall pollutants. Due to the lack of on-site initial rain measurement data, the simulation parameters used the monitoring results

Table 15.2 Statistical table of parameter sources for the total amount of pollutants entering the river

Category	Subcategories	Data sources	Value
Point source	Industrial	"Investigation Report on Drainage Conditions of Industrial Enterprises in Banfu Town, Zhongshan City" (2018)	"Guangdong Province Local Standard Water Pollutant Discharge Limits"
	Domestic pollution	Population data from "Zhongshan Statistical Yearbook 2022"	"Code for design of outdoor water supply engineering" (GB50013-2021)
	Water treatment plant	Treatment scale (130,000 m^3/d)	Water purification plant effluent standards (Grade A)
Non-point source	Aquaculture	"2022 Zhongshan Statistical Yearbook", fish pond area	"The First National Census of Pollution Sources, Aquaculture Industry Pollution Sources and Emission Coefficients Manual"
	Farmland source	"2022 Zhongshan Statistical Yearbook", farmland area	"National Water Environment Capacity Assessment Technical Guidelines", correction coefficients
	Initial rainwater pollutant	SWMM model calculation	Loss patterns of non-point source pollutants in typical small watersheds in Guangdong Province
Endogenous	Endogenous	River surface area	Yu Guangwei et al. studied the sediment pollution release situation in heavily polluted tidal rivers in cities in the Pearl River Delta

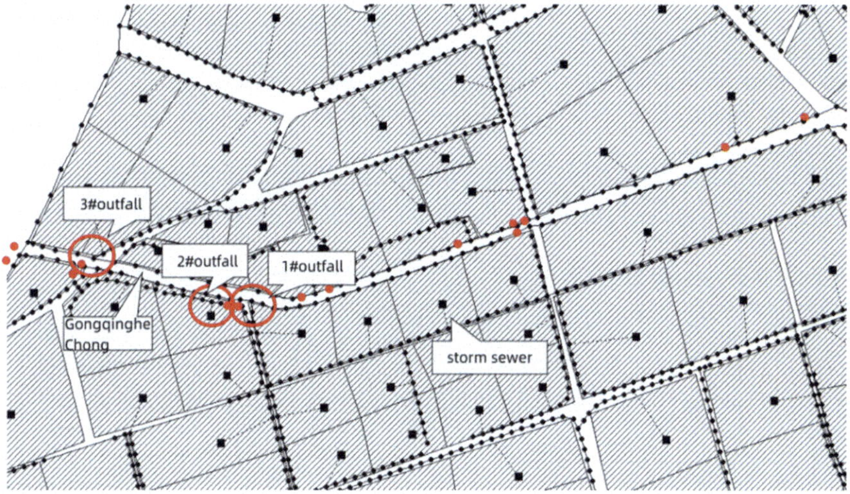

Fig. 15.5 Schematic diagram of the Gongqing river regulation project

of Chen Yanping [11]. The initial surface pollution values of COD, NH_3–N, and TP are 9.65, 0.83, and 0.049 kg/ha respectively.

Figures 15.5 and 15.6 shows the rainwater outlet of the Gongqing River in the typical river section of this project. During a typical rainfall process with a return period of 0.2 years, the flow and pollutant processes of 1#, 2# and the rainwater outlet of the river are as follows As shown in the figure, the simulated initial rain pollutant concentration process has obvious initial effects.

15.5 Model Building Scope

The rivers in Zhongshun Dawei are extremely complex. This paper builds a one-dimensional hydrodynamic water quality model of the river network in the basin based on MIKE 11 software. The model constructed 61 river channels in the main river network of Zhongshun Dawei, as shown in Fig. 15.7. The main rivers involved include Qijiang River, Fuzhou River, Central Drainage Canal, Shijiao River, Western Drainage Canal, Northern Drainage Canal, Gongbei River, Chizhou River, Qianshui Lake, Jinhong River, Wakou Central River, etc. Add 34 water bodies in Banfu Town that do not meet the standard as the generalized river network in Zhongshun Dawei. The model involves 40 sluices, 25 outer sluices and 15 inner sluices.

In this simulation, the diversion water source is river near Modaomen: Shatou Chong, Shisan Qing Chpng, Jiamao Chong, Jiamaojiehe Chong, Luwei Chong, Shouwei Chong, Guiwei Chong, Guangfu Chong, and Huayiwei Chong, Fuling-weichong, Fuweijiaochong and Shiqi river, The downstream boundary is assigned

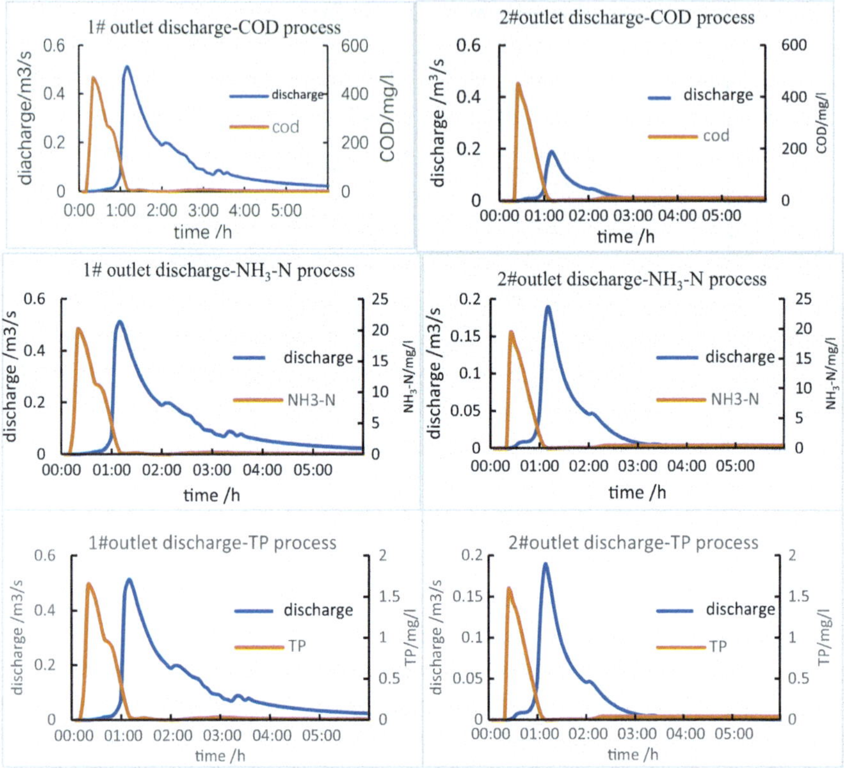

Fig. 15.6 Discharge and each pollutant process diagram for 1#, 2# outlet

a value based on surface Class IV water standards (COD is 25 mg/L, NH_3–N is 1.25 mg/L, TP is 0.25 mg/L).

15.6 Goal Accessibility Analysis

15.6.1 Current Water Quality Analysis

Figure 15.8 shows the concentration diagrams of COD, NH_3–N and TP in various rivers in Banfu. Under the current pollution load conditions entering the river, water quality lower than Class V rivers are: Ludaokeng Chong, Xialuojing Chong, Shagou Chong, Siqing Chong, Baitan Chong, Sankeliang Chong, Gongqinghe Chong, Shapu Chong, Shenwan Flood Drainage Channel, Shenwan Central River, Lixi Yuejiao chong, Banwei chong, Heweiheng chong, Machong and Jinzhong chong. Some rivers in the Hexi area are Class IV water bodies.

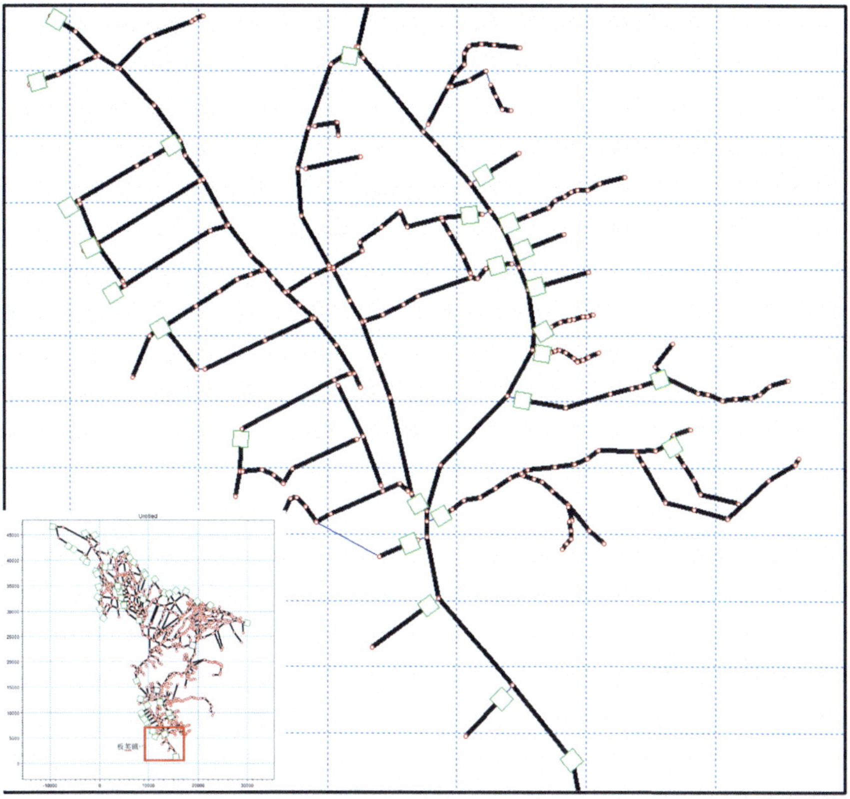

Fig. 15.7 Hydraulic model diagram of Banfu

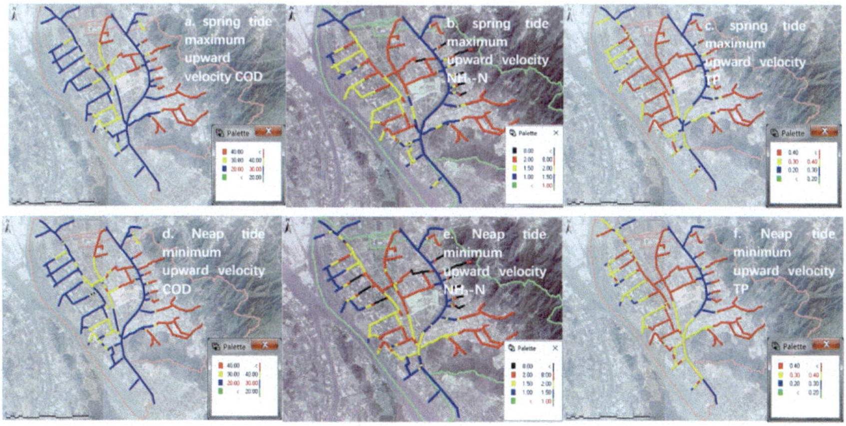

Fig. 15.8 Concentration distribution of each pollutant at the characteristic moment

The simulation results show that in the Hexi area is dominated by fish ponds and farmland, with low development intensity and less pollutants entering the river in current situation. In addition, the Hexi area is close to Modaomen, and part of the water body is able to exchanged under the action of tidal fluctuations, which made the water quality in Hexi area is relatively good. In the Hexi area, the rivers close to the Shiqi River, such as Banwei Chong, Miaojiao Chong and Hewei Chong, are affected by the constant closing of Muwei Sluice, Miaojiao Sluice and Hewei Sluice. The water accumulated in rivers cannot be exchanged with the clean water bodies of Shiqi River, making the water quality of these river water bodies poor. The rivers in the Hedong area are all deadend. Although the sluices are always open, they can only exchange water in the river. The water bodies in the upper reaches of the river cannot be exchanged, making the rivers in Hedong area water quality lower than Class V river.

The changing trends of COD, NH_3–N and TP pollutant concentrations during the spring tides and neap tides are basically similar.

15.6.2 Analysis of Water Quality in Dry Season After Water Diversion

Figure 15.9 shows the concentration diagrams of COD, NH_3–N and TP after the implementation of water diversion in the Banfu. Except for the Minxi Chong, Ludaokeng Chong, Xialuojing Chong, Shagou Chong, Siqing Chong, and Baitan Chong in the Hedong area, the upper reaches of the rivers are still in lower than Class V water bodies, and the rest river in Banfu has reached Class V. The changing trends of COD, NH_3–N and TP pollutant concentrations are basically similar.

Among them, some rivers in the Hexi area, such as Minxi Chong, Guangfu Chong, Fuwei Chong, Luweijie River, Luwei Chung, Kamojie River, Jiamao Chong, Shisanqing Chong and Shatou Chong can meet Class IV. Through the clean water diversion scheme, river hydrodynamic conditions can be effectively improved, making most river water bodies reach Class V.

15.6.3 Analysis of Water Quality in Rainy Season After Water Diversion

Figure 15.10 shows the concentrations of COD, NH_3–N and TP in the Banfu at the beginning of rainfall and 48 h after rainfall. The water quality in the Hexi area recovered faster than that in the Hedong area. The main reason is that the Hexi area is mainly used for fish ponds and farmland, and there is less pollution entering the river. Under rainfall conditions with a return period of 0.2 years, surface source pollutants enter the river with rainfall runoff. After rainfall, river COD pollutants in the Hexi

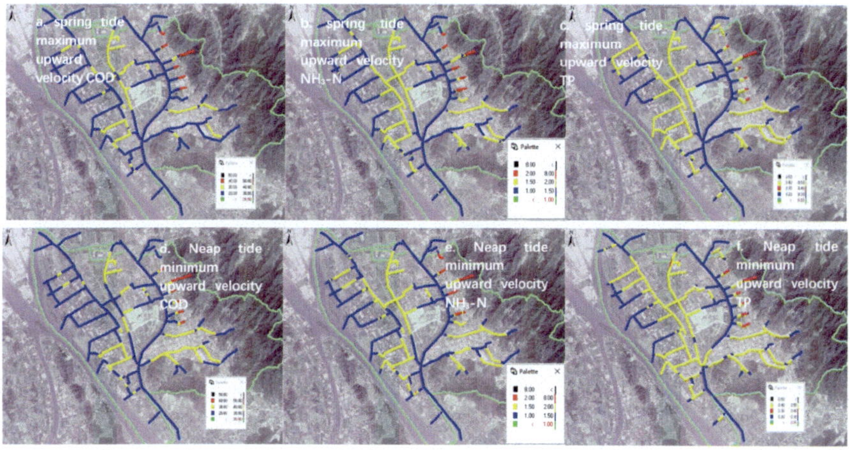

Fig. 15.9 Concentration distribution of each pollutant at the characteristic dry seasons moment after water diversion

area can reach the standard within 6 h. The pollutants with higher concentrations are TP and NH_3–N.

48 h after rainfall, black and odor water can be eliminated in the upstream water bodies of Minxi Chong, Ludaokeng Chong, Xialuojing Chong, Shagou Chong, Siqing Chong and Baitan Chong. Tau Chung, Banwei Chung, Shisan Qing Chung, Si Lian Chung, Fu Lin Wai Chung, Fu Wei Kiao Chung, San Wu Qing Chung and Guayin Chung can meet the standards of Class V water bodies. Sankeliang Chong, Hewei Chong, HeweiHeng Chong, Shatou Chong, Banwei Chong, Shisanqing Chong, Silian Chong, Furenwei Chong, Fuweijiao Chong, Shaluojing Chong and Guayin Chong

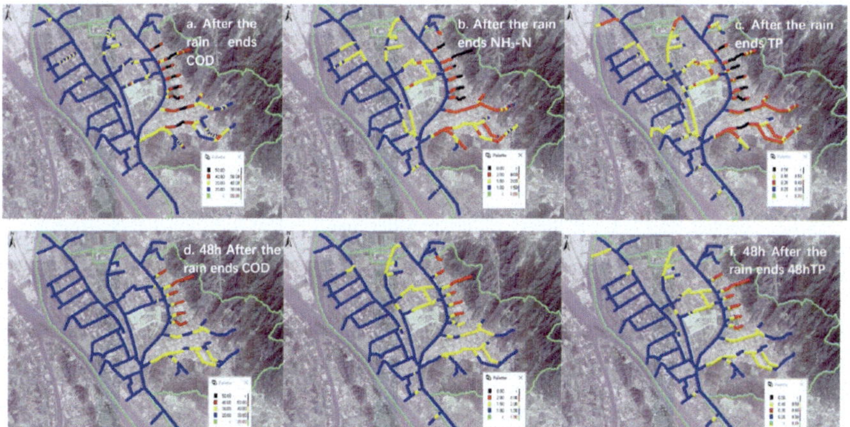

Fig. 15.10 Concentration distribution of each pollutant at the characteristic rainy seasons moment after water diversion

can reach class V. All other rivers can meet Class IV. COD reaches the standard at the earliest in each time period after rain, and the changing trends of NH_3–N and TP pollutant concentrations are basically similar.

15.7 Conclusions

This paper takes the internal river network in Banfu, Zhongshan City as an example, take the tidal dynamic water and the tail water of water treatment plant as the source, to Introduce clean water bodies to Banfu. The conclusions are as follows:

a. After the implementation of the clean water diversion scheme, the average flow rate of the rivers has increased significantly, effectively reducing the water exchange cycle and significantly improving the water quality of the rivers.
b. After the implementation of sewage interception projects, river non-point source pollution control projects, dredging projects and silt disposal projects in Banfu, the total amount of pollutants entering the river can be effectively controlled. However, due to the influence of the reciprocating flow in the river network, pollutants in the river are easy to accumulate, causing the river to swell. After the clean water diversion scheme, most rivers in Banfu can reach Class IV or V standards during the dry season.
c. Under the rainfall conditions of light rain in the rainy season, after adopting the clean water diversion scheme and ramaining for about 48 h, the water quality of the rivers in Banfu can basically be restored.

Acknowledgements Fund project: Guangzhou Science and Technology Plan Project (201904010335).

References

1. Xia J, Shi W (2016) Perspective on water security issue of changing environment in China. J Hydraul Eng 47(03):292–301
2. Xu Z, Cheng T (2019) Basic theory for urban water management and sponge city—review on urban hydrology. J Hydraul Eng 50(01):53–61
3. Wang H, Wang J, Liu J et al (2021) Analysis of urban water cycle evolution and countermeasures. J Hydraul Eng 52(01):3–11
4. Wang X, Wang Y-G, Sun C-H et al (2016) Formation mechanism and assessment method for urban black and odorous water body: a review. Chin J Appl Ecol 27(04):1331–1340
5. Feng Q, Yi J, Liu S et al (2020) The pollution situation, treatment techniques and countermeasures of urban black and odors water in China. Environ Eng 38(08):82–88
6. Sun L, Ma W, Wu J et al (2021) Research on the progress of urban black-odorous water governance and water conservancy measures. China Rural Water Hydropower (08):23–28
7. Han Y, Guo Y (2020) Study on the effect evaluation of comprehensive treatment of water environment in urban lake. Water Conservancy Sci Technol Econ 26(10):10–14

8. Du J, Chen X, Chen Z et al (2012) Water diversion and control for water environment in tidal network of Pearl River Delta. J China Hydrol 32(04):16–21+96
9. Shi G, Tong X et al (2020) Study on ecological compensation scheme of rivers in urban built-up area. Yangtze River 51(4):75–80, 134
10. Ma K, Tu X, Chen X (2021) Simulation of water quantity and quality response of sluice operation in sluice controlled river network area. J China Hydrol 41(06):66–72
11. Chen YP, Huo PS, Tang DD et al (2022) Proposal of ecological water supplement for urban rivers based on quality target reachability analysis. Water Purif Technol 41(5):102–111,149

Chapter 16
Quantitative Evaluation of the Risks of Soil Pb and Cd to Human Health in the Youjiang River Basin, China

Zhengli Mao, Huamei Zhao, and Zhuoping Qin

Abstract This study aimed to quantitatively evaluate the risk of soil toxic metals in the Youjiang River Basin to human health. Shallow soil samples (n = 71) were collected and their concentrations of Cd and Pb were measured. The United States Environmental Protection Agency (EPA) risk to health evaluation model was used to quantitatively evaluate the risk of toxic metals to human health in this basin. The results showed that: (1) All non-cancer-causing risk to health indices were below 1, indicating a small non-cancer-causing risk to health in the study area, mainly due to Pb. Among three exposure pathways investigated, ingestion through contamination of the hands was shown present the highest risk. Exposure to non-cancer-causing risk to health in children was shown to exceed that of adults by a factor of 10. (2) The maximum, minimum, and average total cancer-causing risk to health index values for children were $> 10^{-4}$, 10^{-6}, and 10^{-4}, respectively. This result indicated the presence of a certain cancer-causing risk to health by soil toxic metals to children in most regions, with significant localized risks. The total cancer-causing risk to health index for adults was between 10^{-6} and 10^{-4}. Most cancer-causing risk to health occurred though ingestion from contaminated hands. (3) There were higher non-cancer-causing and cancer-causing risk to health indices in the eastern part of the region, whereas those in other regions remained relatively stable. The higher risks to health in the east could be attributed to the production, development, and utilization of bauxite resources.

Keywords Toxic metals · Soil · Evaluate the risk to health · Spatial distribution · Youjiang River Basin

---

Z. Mao (✉) · Z. Qin
Key Laboratory of Regional Ecological Environment Analysis and Pollution Control in Western Guangxi, School of Civil Engineering and Architecture, Baise University, Guangxi Baise, China
e-mail: zhlmao@163.com

H. Zhao
Library, Baise University, Guangxi Baise, China

K. Y. Foo et al. (eds.), *Water Nexus Approach for Sustainable Development*,
Environmental Science and Engineering, https://doi.org/10.1007/978-3-031-89238-7_16

16.1 Introduction

The population explosion and unprecedented socioeconomic development of China have emphasized the increasing need for conservation of land resources. Land resource conservation is a key part of strategies for rejuvenating rural areas and is needed to ensure the food security and the protection of natural ecosystems. Global changes pose risks to the conservation of agricultural land, and the objectives of conserving agricultural land are to maintain land quantity, quality, and ecological functions. Pollution of soil by toxic metals is a key factors affecting soil ecology [1]. There are both human and natural sources of soil toxic metals. Natural sources mainly include the weathering of parent rock, and these sources determine the background toxic metal concentrations of soil. Anthropogenic sources include industrial and agricultural production, such as mining waste water and combustion engine exhaust fumes. Metals from these sources enter the soil by settlement [2, 3], after which they can contribute to disrupting the soil chemical balance. The pathways under which soil toxic metals can enter humans include skin contact, respiratory inhalation, ingestion through exposure to contaminated hands (hand-mouth intake), and other exposure channels. These toxic metals pose risks to human health by potentially disrupting the internal circulation system and damaging to the central nervous system [4–6].

Most previous studies on soil toxic metal contamination in Guangxi were conducted at a regional scale. The comprehensive study by Deng et al. [7] concluded that soil Cd content in Guangxi was higher than the standard in various types of soil. Li et al. [8] arrived at similar conclusions in their study of farmland of the Lame River Basin in Hechi. Works by and Mao et al. [9] on the spatial distribution, pollution status, and drivers of Pb and Cd in the soil of the Youjiang River Basin concluded that soil Cd contamination far exceeded that of Pb. They also identified mineral mining, urbanization, and thermal power generation to be the main drivers of soil Cd and Pb contamination. These studies focused on the evaluation of the extent of contamination and distribution of toxic metals in this region. However, to date, an evaluation of the risks of toxic metals to the health of humans has not been conducted in this region. The present study utilized the risk to health evaluation model of the Unites States Environmental Protection Agency (USEPA) to evaluate the cancer-causing and non-cancer-causing risks posed by soil Cd and Pb to human health. The spatial distributions of these risks were identified and distinguished between adults and children. The outcomes of the current study can provide a scientific basis for maintaining human health, environmental management, agricultural management, and sustainable development.

16.2 Study Area

The present study chose west Guangxi as a study area. This area along the Youjiang River is a narrow strip extending from east to the west includes the main urban areas of Baise, Tianyang District, Tiandong County, and Pingguo City, with a length and width of ~ 120 km and a 2–20 km, respectively. The study area is bounded by mountains to the north, south, and west, with a clear boundary. An exit with a width of ~ 2 km connects with Pingguo City in the east. The eastern portion of the study area has a relatively flat terrain with some hills with a height no more than 50 m. The western part has a more complex terrain in which the height of hills can react 100 m. The main crop types in the basin are mango, rice, and corn. There is a large bauxite production base at Pingguo to the east of the study area. This area contains thermal power plants, oil refineries, coal mines, and industrial parks (Fig. 16.1).

16.3 Materials and Methods

16.3.1 Collection and Preparation of Samples

Eleven sampling lines were established along the elongated shape of the study area (east by south) at ~ 10 km intervals. The sampling positions along every line in the valley center were distributed evenly at an interval of 3 km. The beginning and end of the sampling lines were at the southern and northern valley edges and on the mountainside and mountain top. These sites were moved when conditions for sampling were not suitable. For each sample, 500 g of shallow soil (depth of 10–20 cm) was collected from the four corners of a rectangle (10 m × 10 m) and center (total of five points). The five samples were mixed after collection. Of each sample, 500 g was kept for analysis in accordance with the quartile method. The coordinates and elevation of each sampling site according to a handheld GPS were recorded. Figure 16.1 shows the locations of the 71 collected soil samples.

Testing of the samples for toxic metals was conducted by the Qingdao Hengli Testing Co., LTD. Sample preparation and testing for Pb and Cd content followed the national standard (GB/T17140-1997) in which the metal concentrations were determined using flame atomic absorption spectrophotometry. The detection limit, limit of quantitative analysis, range for linearity, correlation coefficient R, and regression equation of Cd were $0.05\ \text{mg}\cdot\text{kg}^{-1}$, $0.15\ \text{mg}\cdot\text{kg}^{-1}$, $0.15\text{–}2.0\ \text{mg}\cdot\text{L}^{-1}$, 0.9997, and $Y = 0.3035x - 0.0027$, respectively, whereas those of Pb were $0.20\ \text{mg}\cdot\text{kg}^{-1}$, $0.60\ \text{mg}\cdot\text{kg}^{-1}$, $0.60 - 200.00\ \text{mg}\cdot\text{L}^{-1}$, 0.9990, and $Y = 0.0214x - 0.0013$, respectively. Determination of Cd and Pb concentrations was through an atomic absorption spectrophotometer (model: TAS-986AFG).

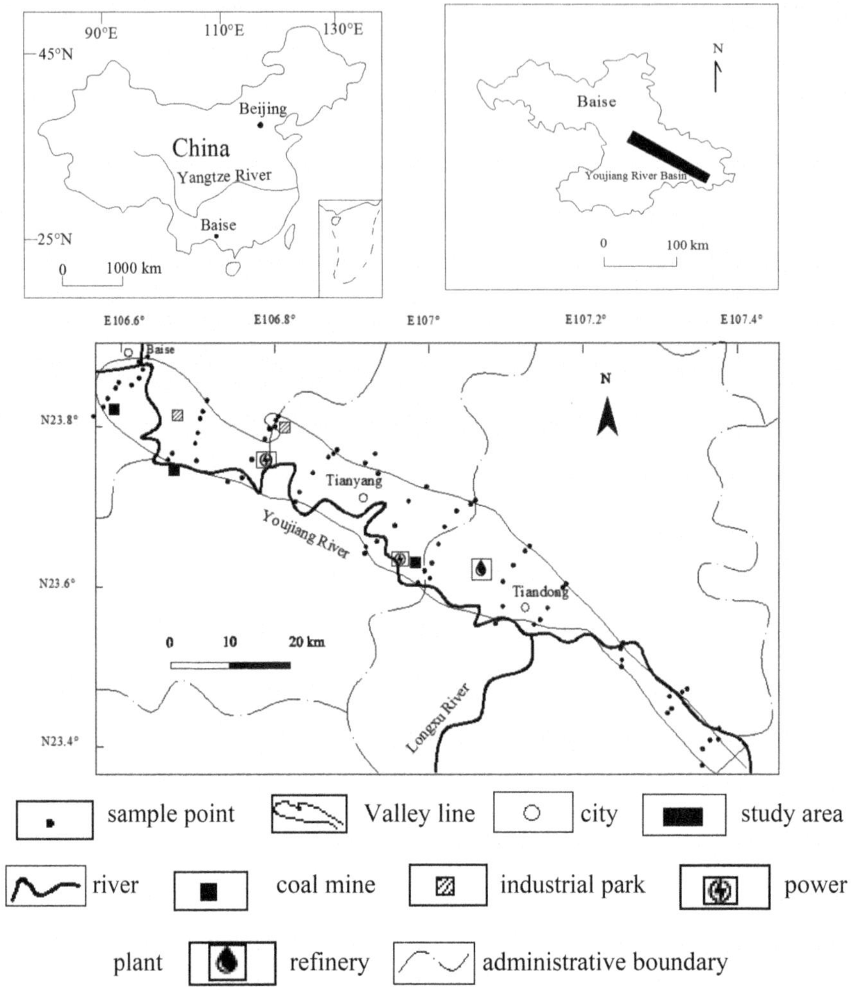

Fig. 16.1 Map of the study area and locations of sampling sites

16.3.2 Human Health Risk Evaluation Model

The present study used the USEPA risk to health evaluation model. This model assumes the existence of three exposure pathways under which soil toxic metals enter humans: (1) hand-mouth intake (ADDing); (2) skin contact (ADDderm) and; (3) respiratory intake (ADDinh). The equations used to calculate the human intake of soil toxic metals in the study area according to:

Daily absorbed dose through hand-mouth intake:

$$\mathrm{ADD_{ing}} = \frac{c \times IngR \times CF \times EF \times ED}{BW \times AT} \tag{16.1}$$

Daily absorbed dose through skin contact intake:

$$\mathrm{ADD_{derm}} = \frac{c \times SA \times CF \times AF \times ABS \times EF \times ED}{BW \times AT} \tag{16.2}$$

Daily absorbed dose through respiratory intake:

$$ADD_{inh} = \frac{c \times InhR \times EF \times ED}{PEF \times BW \times AT} \tag{16.3}$$

Table 16.1 provides the definitions and values of parameters used in Eq. (16.1)–Eq. (16.3).

The equations shown below estimate the risk to human health through the three exposure pathways.

The non-cancer-causing risk (HQi) of a single toxic metal in soil to human health:

$$\mathrm{HQ}_i = \sum \frac{ADD_{ij}}{RfD_{ij}} \tag{16.4}$$

Table 16.1 Parameter values of the model used in the current study to evaluate the risk of toxic metals to human health

Parameter	Meaning	Unit	Value
c	Content measured value of toxic metal	mg/kg	actual measured
IngR	Frequency of hand-mouth intake	mg/d	100 (adults) 200 (children)
InhR	Frequency of respiratory intake	m^3/d	15 (adults) 7.5 (children)
CF	Transfer coefficient	mg/kg and m^3/cm^3	1×10^{-6}
EF	Exposure frequency	d/year	350
ED	Exposure time	year	24 (adults) 6 (children)
BW	Average weight	kg	62 (adults) 15.9 (children)
AT	Average exposure time	d	ED × 365 (non-cancer-causing) 70 × 365 (cancer-causing)
PEF	Dust emission factor	m^3/kg	1.36 × 109
SA	Exposed skin area	cm^2/d	5700 (adults) 2800 (children)
AF	Skin adhesion	mg/($cm^2 \cdot$ d)	0.20 (adults) 0.07 (children)
ABS	Skin absorption factor	–	0.001

The total non-cancer-causing risk (HI) of a single toxic metals in soil to human health:

$$\mathrm{HI} = \sum HQ_i \tag{16.5}$$

The carcinogenic risk (CRi) of a single toxic metal in soil to human health:

$$\mathrm{CR}_i = \sum \mathrm{ADD}_{\mathrm{ij}} \times SF_{ij} \tag{16.6}$$

The total cancer-causing risk (TCR) of toxic metals in soil to human health:

$$\mathrm{TCR} = \sum \mathrm{CR}_i \tag{16.7}$$

In Eq. (16.4)–Eq. (16.7), RfD is the reference dose of toxic metals under the different exposure pathways, representing their threshold intake (per unit body weight and per unit time) below which toxic reactions will not take place in the human; SF represents the slope coefficient (probability of cancer-causing health effects) of cancer-causing risk of toxic metals to human health under the different pathways of exposure. Table 16.2 shows the values of Rfd and SF.

A non-cancer-causing risk index (HQi) of single toxic metal ≤ 1 or the total non-cancer-causing risk index (HI) ≤ 1 indicates a small non-cancer-causing risk to humans. In contrast, a non-cancer-causing risk is considered to exist when their values exceed 1, with the risk considered chronic when their values exceed 10. The acceptable cancer-causing risk index value of a single toxic metal element (CRi) or total cancer-causing risk index value (TCR) ranges from 10^{-6} to 10^{-4}. Values of the cancer-causing risk index of $\leq 10^{-6}$, 10^{-6} to 10^{-4}, and $\geq 10^{-4}$ indicate is no obvious cancer-causing risk, cancer-causing risk, and significant cancer-causing risk, respectively.

Table 16.2 Exposure to soil toxic metals under different pathways and cancer-causing rate

Element	RfDing [mg/(kg · d)]	RfDderm [mg/(kg · d)]	RfDinh [mg/(kg · d)]	SF[mg/(kg · d)]
Cd	1.00E − 03	1.00E − 03	1.00E − 06	6.3E + 00
Pb	3.50E − 03	3.50E − 03	5.20E − 03	

16.4 Results and Discussion

16.4.1 Analysis of the Statistical Characteristics of the Risk to Health Index

Analysis of the characteristics of the non-cancer-causing risk to health index.

Table 16.3 shows the calculated values of the non-cancer-causing risk to health indices for soil Cd and Pb in the study area. All non-cancer-causing risk to health indices were below 1, indicating a small non-cancer-causing risk to human health in this area. The total non-cancer-causing risk of soil Cd and Pb to the health of children generally exceeded that to adults by a factor of 10. The maximum total non-cancer-causing risk to health index for children was 0.845. Soil Pb was the dominant contributor to soil toxic metal non-cancer-causing risk in the study area, accounting for ~ 70% and > 90% of the non-cancer-causing risk to health of children and adults, respectively. The hand- mouth intake route contributed most to the non-cancer-causing risk to health among the three exposure pathways. Most of the non-cancer-causing risk to health by Pb intake to children was through the hand- mouth route. The hand- mouth intake rout also accounted for most of the non-cancer-causing risks to health resulting from Cd ingestion in both children and adults.

Analysis of the characteristics of the cancer-causing risk to health index.

Table 16.4 shows the values of the parameters of the cancer-causing risk to health index for soil Cd in the study region. The cancer-causing risk to health index was only calculated for Cd since data for the cancer-causing risk slope coefficient (SF) of Pb were not available. The total cancer-causing risk to health of children exceeded that of adults by a factor of 10. The maximum, minimum, and average total cancer-causing risk to health index values for children were > 10^{-4}, 10^{-6}, and 10^{-4}, respectively. These results indicated a significant cancer-causing risk to the health of children in the study area. The total cancer-causing risk to health index value for adults ranged from 10^{-6} to 10^{-4}, indicating no significant cancer-causing risk posed by Cd to the health of adults in this area. The hand-mouth exposure route accounted for 95% of total cancer-causing risks to the health of children and adults.

16.4.2 Analysis of the Spatial Distribution of the Risk to Health Index

Spatial locations of non-cancer-causing risks to health.

Figure 16.2 shows the spatial distributions of the non-cancer-causing risk to health index for Pb and Cd and the total non-cancer-causing risk to health index for children and adults. The locations of non-cancer-causing risks to health of Pb were the same

Table 16.3 Table of parameter values for the non-cancer-causing health risk index

Element	Class	Children				Adults			
		HQing	HQderm	HQinh	HQi	HQing	HQderm	HQinh	HQi
Pb	Maximum value	6.00E − 01	3.96E − 04	1.65E − 05	6.00E − 01	7.69E − 02	5.90E-04	8.48E − 06	7.75E − 02
	Minimum value	4.86E − 02	3.21E − 05	1.34E − 06	4.86E − 02	6.23E − 03	4.78E − 05	6.87E − 07	6.28E − 03
	Average value	1.70E − 01	1.12E − 04	4.69E − 06	1.70E − 01	2.18E − 02	1.67E − 04	2.41E − 06	2.20E − 02
Cd	Maximum value	2.23E − 01	2.19E − 02	6.15E − 06	2.45E − 01	8.17E − 04	6.27E − 06	9.02E − 08	8.24E − 04
	Minimum value	2.29E − 02	2.25E − 03	6.32E − 07	2.52E − 02	8.40E − 05	6.44E − 07	9.26E − 09	8.46E − 05
	Average value	6.62E − 02	6.48E − 03	1.82E − 06	7.27E − 02	2.42E − 04	1.86E − 06	2.67E − 08	2.44E − 04
HI	Maximum value				8.45E − 01				7.83E − 02
	Minimum value				7.38E − 02				6.36E − 03
	Average value				2.43E − 01				2.22E − 02

Table 16.4 Table of parameters values for the cancer-causing health risk index

Element	Class	Children				Adults			
		CRing	CRderm	CRinh	CRi	CRing	CRderm	CRinh	CRi
Cd	Maximum value	1.41E – 04	1.38E – 07	3.88E – 09	1.41E – 04	1.80E – 05	2.05E – 07	1.99E – 09	1.82E – 05
	Minimum value	1.44E – 05	1.41E – 08	3.98E – 10	1.45E – 05	1.85E – 06	2.11E – 08	2.04E – 10	1.87E – 06
	Average value	4.17E – 05	4.09E – 08	1.15E – 09	4.17E – 05	5.35E – 06	6.09E – 08	5.90E – 10	5.41E – 06

for children and adults. Higher index values with a narrow range were distributed in the eastern part of the study area. A second area with high index values occurred in the northern-central part. These results indicated that bauxite production in the east of study area was the main source of non-cancer-causing risks to health due to Pb. In contrast, the operation of coal mines and thermal power plants and the activities of industry had no significant impact on non-cancer-causing risk. The spatial distribution of non-cancer-causing risks by Cd to the health of children and adults was generally the same as that of Pb. The spatial distribution of non-cancer-causing risks by Cd showed regions with high values in the eastern, central, and western parts. These regions contained thermal power plants. This result indicated that thermal power generation is a driver of Cd non-cancer-causing risk. There were similar distributions of the total non-cancer-causing risk to health index for Pb between children and adults.

In summary, the primary drivers of soil toxic metal non-cancer-causing risks to human health in this region were the exploitation of bauxite resources in the east, with non-cancer-causing risk to health index values in the east far exceeding those in other regions.

Spatial distribution of cancer-causing risks to health.

Figure 16.3 shows the spatial distribution of cancer-causing risk to health in children. Areas of high cancer-causing risk to health were mainly in the eastern part of the study area. High risk to health index values exceeding 10^{-4} and with a small range were found in the southeast corner of the study area. A second area of high risk to health index values was identified near the central and western thermal power plants. The spatial distributions of the cancer-causing risk to health index for children and adults where basically identical (Fig. 16.3). However, the cancer-causing risk to health index did not exceed 10^{-4}, indicating the source of the risk originated predominantly from the exploitation of bauxite resources in the east and the operation of thermal power plants.

16.5 Conclusions

(1) Soil toxic metals in the study area posed overall small non-cancer-causing risks to the health of children and adults in the study area. However, the cancer-causing risks to the health of children were significant in some areas. The hand-mouth intake route contributed to > 95% of the risks to health.
(2) Children and Adults showed similar spatial distributions of non-cancer-causing and cancer-causing risks to health, with higher values in the east and relatively stable values in other regions. The midwestern part of the study area was characterized by relatively high cancer-causing risk to health indices for children and adults.

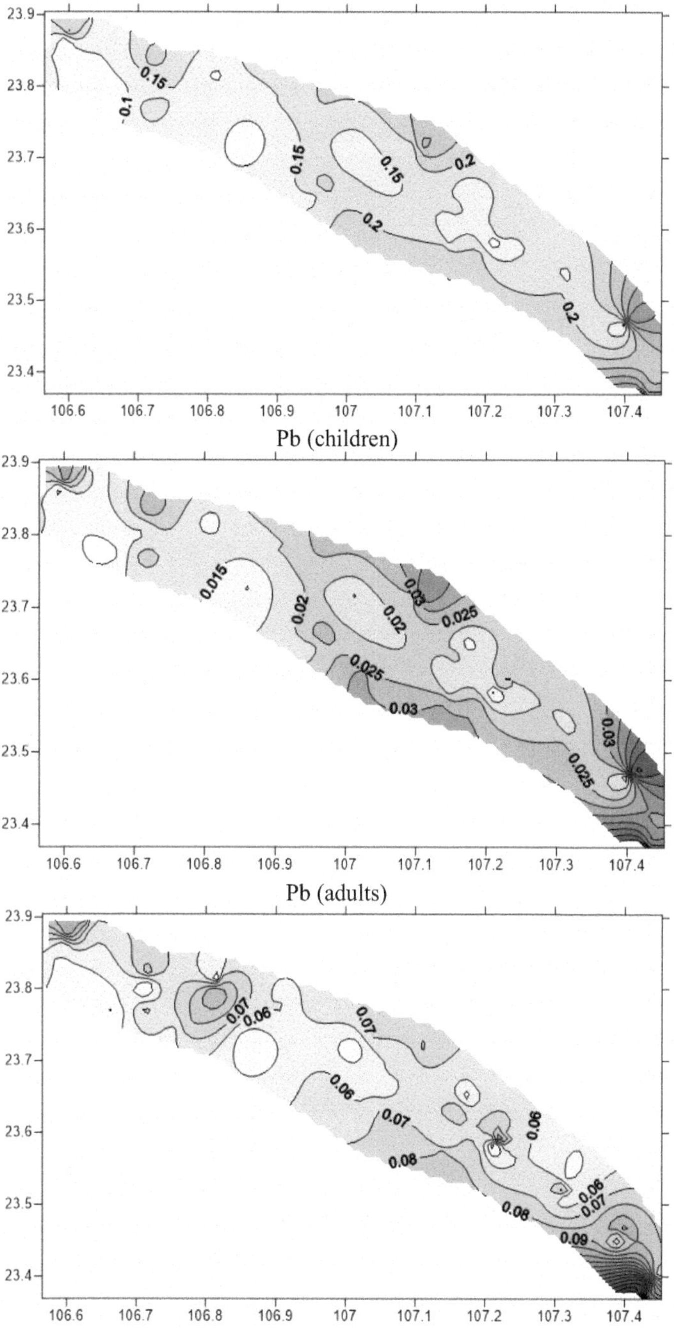

Fig. 16.2 Locations of non-cancer-causing health risks

CCd (children)

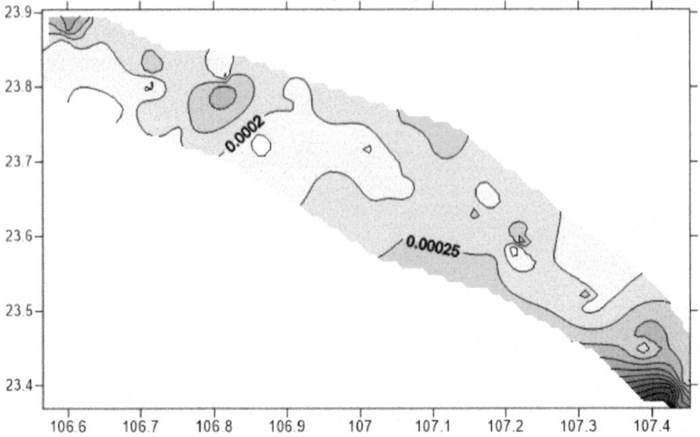

Cd (adults)

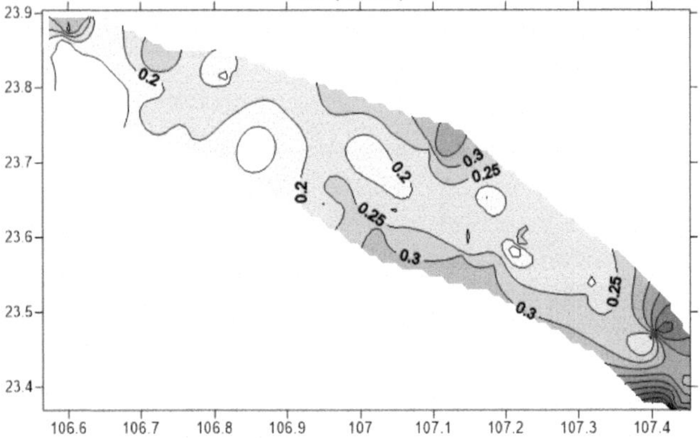

total non-cancer-causing (children)

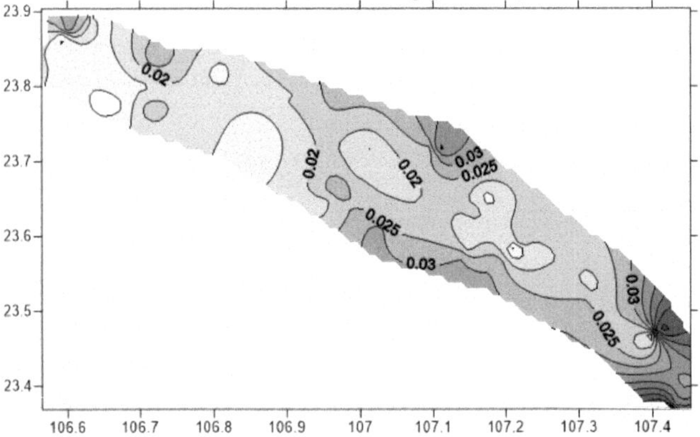

total non-cancer-causing (adults)

Fig. 16.2 (continued)

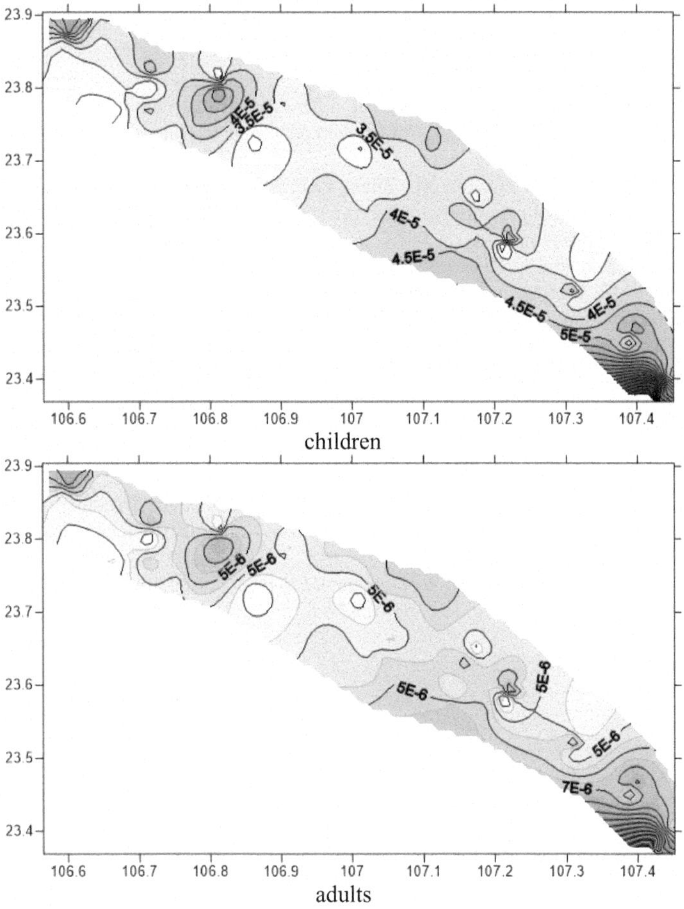

Fig. 16.3 Spatial distribution of cancer-causing health risks

(3) The main drivers of soil toxic metal risks to health in this region were the exploitation of bauxite resources in the east, followed by the operation of thermal power plants.

References

1. Proshad R, Kormoker T, Islam S (2021) Distribution, source identification, ecological and health risks of heavy metals in surface sediments of the Rupsa River, Bangladesh. Toxin Rev 40(1):77–101

2. Chen G, Yang Y, Liu X, Wang MJ (2021) Spatial distribution characteristics of heavy metals in surface soil of Xilinguole coal mining area based on Semivariogram. ISPRS Int J Geo-Inform 10(5):290
3. Sun X, Sun L, Zhang X, Li G, Yu M (2021) The spatial distribution of soil heavy metals and ecological restoration of Datong Abandoned coal mine in Huainan City. J Phys: Conf Ser 1732:012101
4. Lin SX, Zhang ZL, Xiao ZQ, Liu XL, Zhang QH (2022) Atmospheric deposition fluxes and health risk assessment of potentially toxic elements in Caohai Lake (Guizhou Province, China). J Mt Sci 19(4):1107–1118
5. Pachoulis M, Maggos T, Panagopoulos P, Dasopoulou M, Balla D, Stamatelopoulou A, Manousakas ML, Eleftheriadis K, Saraga D (2022) Population health risks assessment from air pollution exposure in an industrialized residential area in Greece. Atmosphere 13(4):615
6. Huang CS, Zhang L, Meng J, Yu Y, Qi J, Shen P, Li X, Ding P, Chen M, Hua G (2022) Characteristics, source apportionment and health risk assessment of heavy metals in urban road dust of the Pearl River Delta, South China. Ecotoxicol Environ Saf 236:113490
7. Deng QY, Zhao YJ, Lin Q, Miao YQ, Xie QY, Jiang WY, Yang N, Wang Y, Chen G (2019) Regional distribution characterstics of cadmium and evaluation of soil pollution situation in Guangxi. Environ Eng 37(1):164–172 (in chinese)
8. Li CB, Huang KX, Wei RB, Huang H, Zhang CL (2018) Study on metals contamination characteristic of heavy in agricultural soil near Lamo River in Hechi city of Guangxi. J Guangxi Univ (Nat Sci Edn) 43(5):2088–2094 (in chinese)
9. Mao ZL, Zhao HM, Qin ZP (2020) Assessment of Cd-Pb pollution in soils of the Youjiang river basin, South China. Eurasian Soil Sci 53(6):829–837

Chapter 17
Exploration of the Application Effect of Vertical Microwave Ultraviolet Sterilization Equipment in Drinking Water Treatment

Hongyi Cao, Zhe Yu, Wenhai Gao, Yanju Zhang, Yiran Shen, and Jinsong Li

Abstract Preliminary experiments were conducted to explore the application effects of vertical microwave ultraviolet (UV) sterilization equipment in the treatment of drinking water. The results indicate that the disinfection efficiency of this vertical sterilization equipment is influenced by the diameter of the tank and the irradiation time. Under the same power, equipment with a smaller tank diameter exhibited better sterilization effects. As the duration of the irradiation of the device increased, the sterilization rate began to rise, but entered a slow growth or stable phase after 60 min. Moreover, this disinfection technology had minimal impact on the water quality indicators of the effluent; the hardness, UV254 absorbance, nitrate nitrogen, and nitrite nitrogen levels did not undergo significant changes after disinfection. Under the experimental conditions, the use of equipment with a power of 1 kW and a diameter of 680 mm individually did not achieve desirable disinfection effects. However, when two devices with a diameter of 450 mm and a power of 1 kW were used in series, the effluent water quality met the "Standards for Drinking Water Quality" (GB5749—2022). The optimal disinfection effect was achieved when a sterilization device with a diameter of 800 mm and a power of 3 kW was used in series with two devices of 450 mm diameter and 1 kW power; the sterilization rate reached approximately 99%, and the removal rate of coliform bacteria was 100%.

Keywords Vertical microwave ultraviolet sterilization equipment · Disinfection effect · Diameter of the tank · Duration of the irradiation · Coliform

---

H. Cao · Z. Yu · W. Gao · Y. Zhang · Y. Shen · J. Li (✉)
Jinan Water Group Co, Ltd, Jinan, Shandong, China
e-mail: c17860639621@163.com

K. Y. Foo et al. (eds.), *Water Nexus Approach for Sustainable Development*,
Environmental Science and Engineering, https://doi.org/10.1007/978-3-031-89238-7_17

17.1 Introduction

The presence of pathogenic bacteria, viruses, or parasites in water can lead to the infection and transmission of human diseases, thus making it one of the primary health risks associated with water-related applications [1, 2]. Therefore, disinfection is a crucial step in the treatment of drinking water. UV disinfection technology has increasingly gained popularity in water treatment due to its high disinfection efficiency, absence of disinfection by-products, and the lack of risk associated with overdosage [3]. The microwave electrodeless UV light source, developed in recent years, is a new type of UV light source generated through microwave excitation [4]. It boasts advantages such as high irradiation intensity and long lifespan [5]. Microwave electrodeless UV radiation has been proven to effectively degrade aquatic pollutants [6, 7].This experiment aimed to evaluate the applicability and disinfection efficacy of a vertical microwave electrodeless UV sterilization device in drinking water treatment by assessing the removal rates of coliform groups and total colony count, as well as identifying factors affecting the disinfection performance, and ultimately selecting efficient disinfection equipment and combination methods.

17.2 Materials and Methods

17.2.1 Test Equipment

Test equipment was installed at the Quehua Water Plant in Jinan city, using the water post-filtration as the experimental source water. Given that the turbidity of the source water was less than 1 NTU, the impact of suspended particles on UV disinfection efficiency could be considered negligible [8], The total colony count of the filtered water ranged between 102–103 CFU/mL, and total coliforms ranged between 1–600 MPN/100 mL. The first phase of the experiment employed devices G680 and G600 in series, both with a power of 1kw and diameters of 680 mm and 600 mm, respectively, as depicted in Fig. 17.1. The second phase involved the use of device G800, with a diameter of 800 mm and power of 3 kw, in series with two devices of the same model, G450-1 and G450-2, each with a diameter of 450 mm and power of 1 kw, as shown in Fig. 17.2. The disinfection equipment, vertical microwave electrodeless UV disinfection devices, operates by generating high-intensity UV light waves through a microwave source-magnetron-activating electrodeless lamp tubes, effectively destroying the DNA/RNA of pathogens for disinfection purposes. The equipment's main body is made of 316L stainless steel to prevent corrosion, featuring a compact footprint and easy, flexible installation. Additionally, with inlets and outlets at opposite ends, water flows from the inlet upwards in a laminar flow, preventing short-circuiting [9], ensuring safe and reliable operation of the equipment.

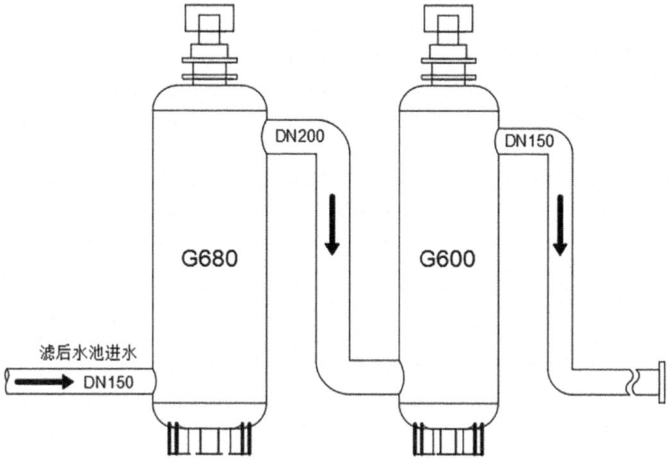

Fig. 17.1 Phase I test system diagram

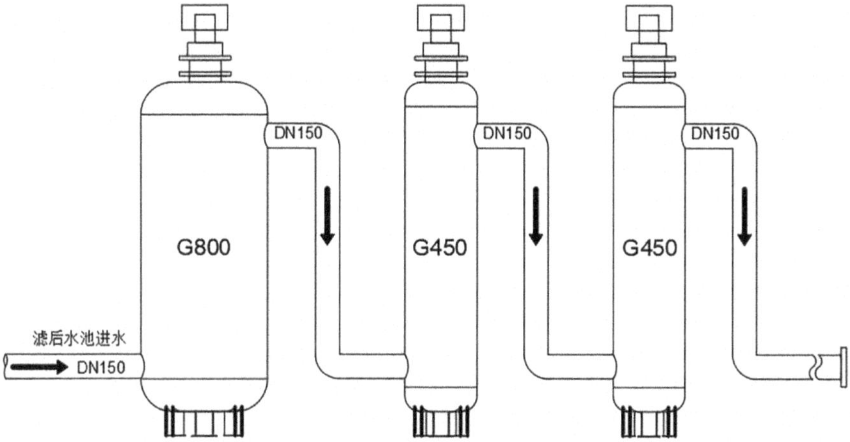

Fig. 17.2 Phase II test system diagram

17.2.2 Experimental Design

During the experiment, samples of the inflow and outflow water were collected before and after disinfection using the UV equipment. The first phase of the experiment investigated the effects of tank diameter and irradiation time on disinfection efficacy, as well as the individual and combined disinfection performance of G680 and G600, and their impact on the quality of the treated water. The second phase of the experiment explored the instantaneous disinfection effects of operating G450-1, G450-2, and G800 individually and in various serial combinations. Due to uneven microbial distribution, multiple sets of experiments were conducted for different equipment combinations to ensure experimental accuracy and eliminate sampling errors.

17.2.3 Test Items

Turbidity: Scattering method—Formazine standard, Aerobic plate count: plate count method, Total coliforms: Enzyme substrate method, Hardness: Titrimetric method, UV254: Ultraviolet–visible spectroscopy, Nitrate nitrogen: Ion exchange chromatography, Nitrite nitrogen: Diazamine coincidence spectrophotometry.

17.2.4 Experimental Methods

To evaluate the disinfection effect of the vertical microwave electrodeless UV equipment, water quality before and after treatment is sampled and analyzed. The sterilization rate is expressed using the following formula (17.1).

$$v = \frac{N_0 - N}{N_0} \tag{17.1}$$

v is Sterilization rate, N_0 is the total number of colonies in raw water, and N is the total number of colonies in water after disinfection. The The units of N and N_0 are CFU/mL.

Total colony count and total coliforms are crucial indicators for assessing the sanitary quality of drinking water [10]. According to China's "Standards for Drinking Water Quality" (GB5749—2022), total coliforms must not be detectable, and the total colony count must be less than 100 CFU/mL. These criteria, along with sterilization rates, are used to evaluate whether the disinfection effect meets the standard, whether the output water complies with drinking water sanitary standards, and whether it can be applied in actual production.

17.3 Results and Discussion

17.3.1 First Phase of the Experiment

The disinfection effects of G600 and G680 when activated separately and together

Bactericidal effect of G680 and G600 when used separately shown in Figs. 17.3, 17.4, and 17.6, the instantaneous sterilization effects of G680 and G600 are relatively poor, with the highest sterilization rate of G680 reaching 70.57%, and G600 achieving a maximum of 76.84%. Additionally, both devices frequently exhibited negative sterilization effects, with the removal rate of coliforms not reaching 100%. This is possibly because the lamp tubes in these vertical devices are concentrated in the center of the tank, resulting in a thicker layer of water being irradiated. The UV light is absorbed by the water layer, leading to a weaker inactivation effect [11], When the dose of irradiation is insufficient, some microbial groups cannot be fully and evenly exposed and thus are not inactivated. As depicted in Figs. 17.5 and 17.6, when G680 and G600 are operated simultaneously, the sterilization rate fluctuates between 50%–60%, and the total colony count in the effluent is sometimes higher than 100 CFU/L. Although the removal rate of coliform bacteria reaches 100%, it still does not consistently meet the "Standards for Drinking Water Quality" (GB5749—2022).

The impact of tank diameter and irradiation duration on disinfection efficacy

In each set of experiments, the valves before and after the disinfection equipment were closed, and testing was conducted at intervals of 0 min, 15 min, 60 min, 120 min, and 150 min. As shown in Figs. 17.7, 17.8, and 17.9, with the increase in irradiation duration, the disinfection effects of the three test groups gradually strengthened. When the irradiation duration exceeded 60 min, the sterilization rate generally exceeded 80%, entering a slow growth or even stabilization phase, where the sterilization rate no longer changed significantly with time, and 100% removal of coliforms could be achieved. Moreover, as seen in Figs. 17.7 and 17.8, with the increase in disinfection duration, the disinfection effect of G600, which has a smaller tank diameter, tended to be superior to that of G680. At 15 min of disinfection, the removal rate of coliforms could reach 100%, and the total colony count in the effluent was reduced, meeting the "Standards for Drinking Water Quality" (GB5749—2022). As the disinfection duration increased, the total colony count in the effluent from G600 stabilized at around 9 CFU/mL, never reaching 0 CFU/mL, possibly due to the presence of a small fraction of bacteria in the water that have strong resistance to microwave UV radiation or the shielding protection effect of impurities in the water on some bacteria [12].

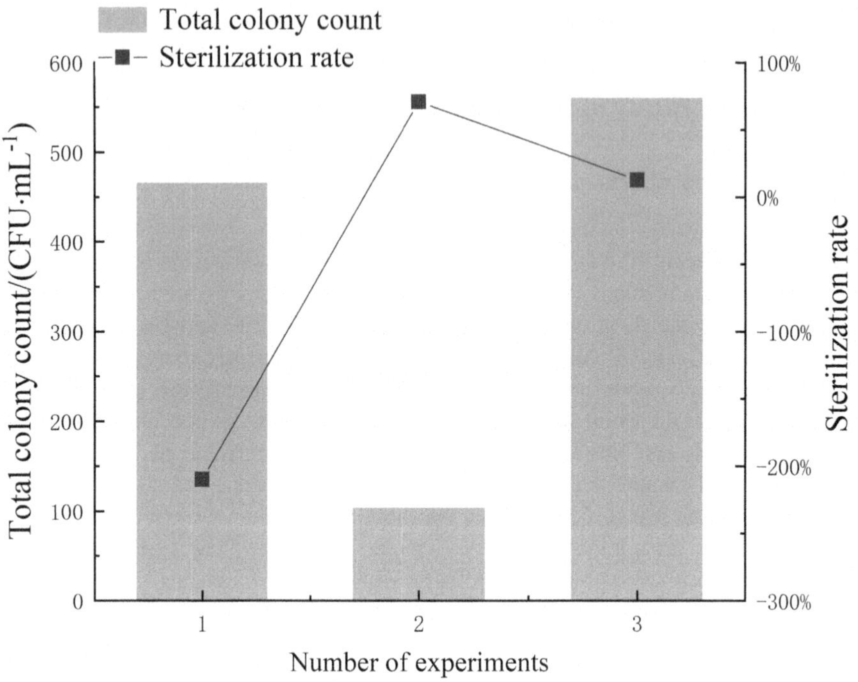

Fig. 17.3 Bactericidal effect of G680 when used alone

The impact of microwave electrodeless ultraviolet disinfection on the water quality

This experiment measured the water quality of effluent under different UV irradiation duration, reflecting the impact of microwave electrodeless UV disinfection on effluent parameters such as nitrate nitrogen, nitrite nitrogen, hardness, and UV254. Photolytic reactions occur in water after UV irradiation, generating hydroxyl radicals. These active hydroxyl radicals degrade organic pollutants almost non-selectively [13], leading to a reduction in humic substances and other large molecular organic materials in the water. As shown in Fig. 17.11, UV254 slightly decreased from 0.018 to 0.016. Additionally, after irradiation, nitrogen-containing substances in water undergo a series of chemical reactions, gradually converting into nitrites [14], resulting in a slight increase in nitrite nitrogen content in the effluent, as illustrated in Fig. 17.13, where nitrite nitrogen increased from 0.0002 mg/L to 0.022 mg/L. With the increase in irradiation time, as shown in Figs. 17.10 and 17.12, hardness fluctuated around 250 mg/L, and nitrate nitrogen around 2.3 mg/L. However, overall, the changes in effluent parameters such as hardness, UV254, nitrate nitrogen, and nitrite nitrogen were minimal, indicating that the physicochemical properties of the water were essentially unchanged.

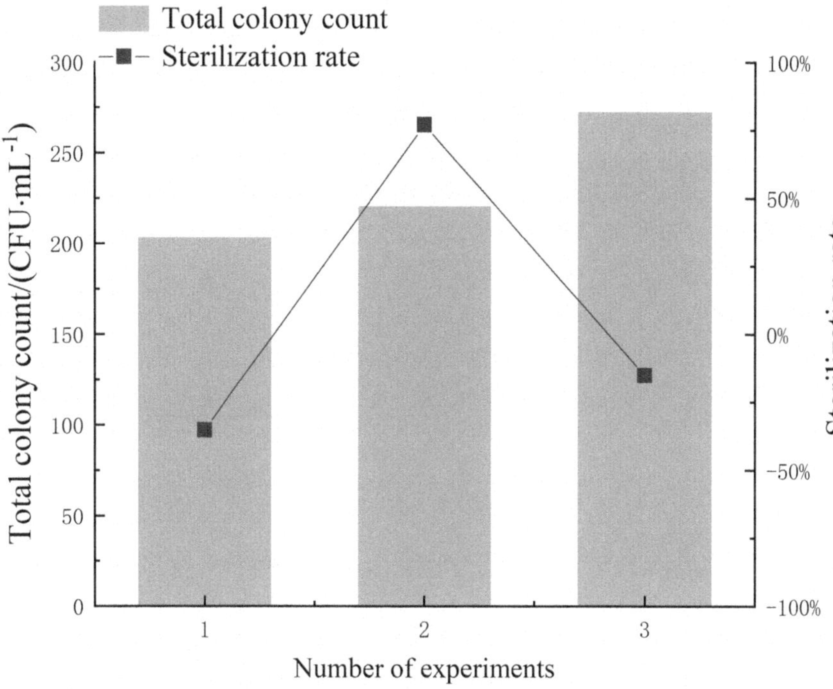

Fig. 17.4 Bactericidal effect of G600 when used alone

17.3.2 Second Phase of the Experiment

Since the first phase of the experiment did not achieve stable sterilization effects, switch to different models of equipment for the second phase of the experiment.

The disinfection effects of G800, G450-1, and G450-2 when used individually and in combination

Upon optimizing the first phase of the experimental equipment, devices with smaller diameters and higher power were tested for their disinfection efficacy. As indicated in Figs. 17.14 and 17.15, when only the G450-1 was activated, the instantaneous sterilization rate of G450-1 could reach up to 64%, displaying unstable disinfection effects with a lower sterilization rate. Moreover, the majority of the effluent water samples had a total colony count greater than 100 CFU/mL, and the removal rate of coliforms did not reach 100%, failing to meet the "Standards for Drinking Water Quality" (GB5749—2022). As shown in Figs. 17.16 and 17.17, when only G800 was activated, its disinfection performance was superior to G450, with sterilization rates ranging from 78 to 92%, and the total colony count in the effluent was consistently

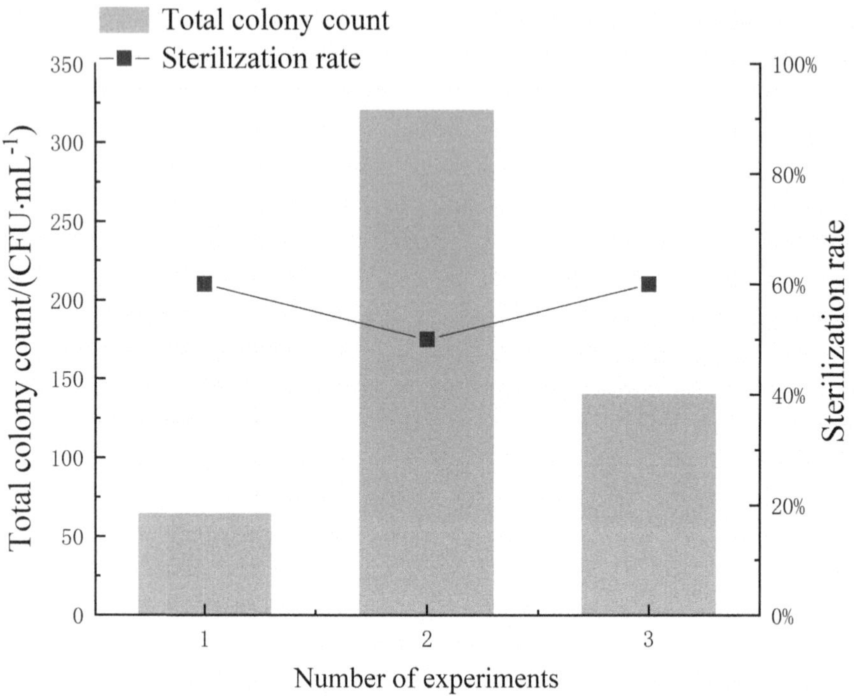

Fig. 17.5 Bactericidal effect of G600 and G680 when used in series

below 100 CFU/mL. However, the coliform removal rate ranged from 75 to 98%, indicating that coliforms were not completely removed, and the effluent did not meet the standards. Due to the high bacterial concentration in the source water and the large inflow volume, the insufficient UV radiation exposure resulted in a high bacterial count in the effluent, affecting the UV disinfection efficacy. The independent use of these two models of equipment is not suitable for the disinfection treatment of drinking water under these experimental conditions.

When G450-1 and G450-2 were activated simultaneously, multiple sets of experimental results, as shown in Fig. 17.18, indicated that their combined use typically achieved sterilization rates ranging between 70%–96%, with the total colony count in the effluent being less than 100 CFU/mL, and the coliform removal rate at 100%, meeting the "Standards for Drinking Water Quality" (GB5749—2022). When G800 was operated together with G450-1 and G450-2, the sterilization rates obtained in two sets of experiments were 99.2% and 96.6%, respectively, with the total colony counts in the effluent being 6 CFU/mL and 27 CFU/mL, significantly below the standards specified in the "Standards for Drinking Water Quality" (GB5749—2022) for effluent colony counts. The coliform removal rate was 100%, indicating the

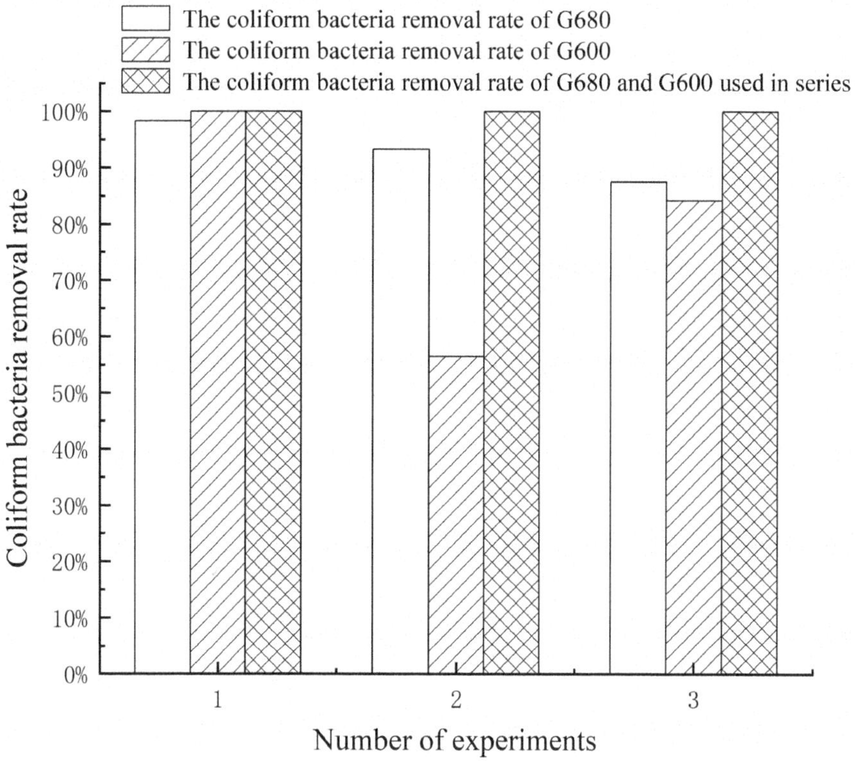

Fig. 17.6 Coliform removal rate of G600, G680 respectively and when used in series

highest sterilization rate and the best disinfection effect, suitable for the disinfection treatment of drinking water under these experimental conditions.

As shown in Fig. 17.19, when G800, G450-1, and G450-2 are opened simultaneously, the sterilization rate obtained from the analysis of water samples before and after disinfection of G450-1 and G450-2 is between 85 and 100%. This is significantly higher than the combined sterilization rate of G4501 and G4502 when G800 is not activated, indicating an enhanced disinfection effect. This suggests that the quantity of microorganisms in the raw water affect the disinfection efficiency of the equipment.

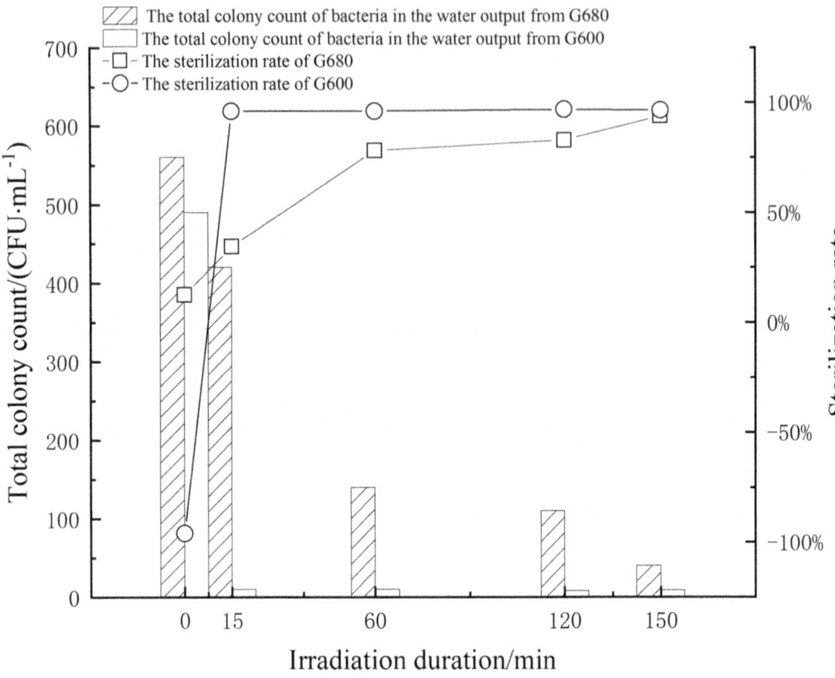

Fig. 17.7 Bactericidal effect of G680 and G600 when used separately

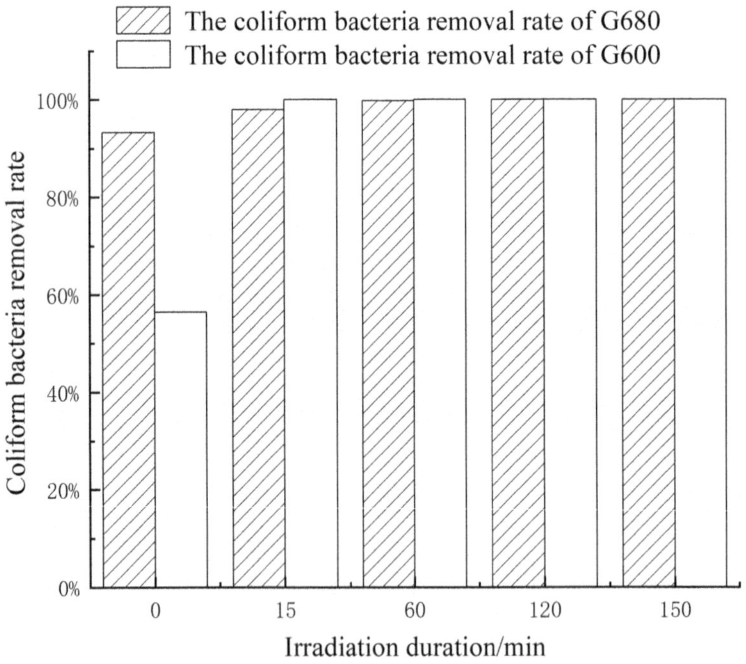

Fig. 17.8 Coliform removal rate of G680 and G600 when used separately

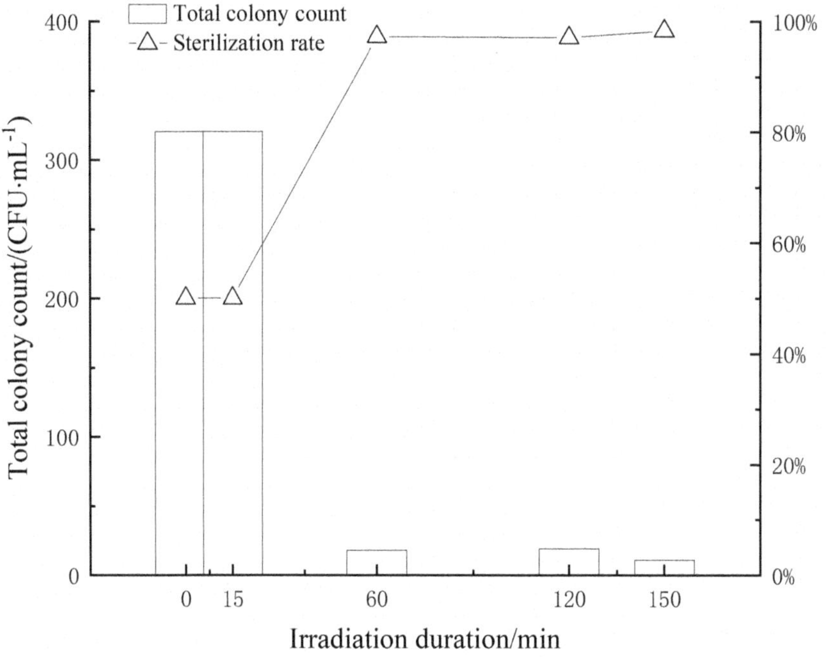

Fig. 17.9 Bactericidal effect of G600 and G680 when used in series with increasing irradiation duration

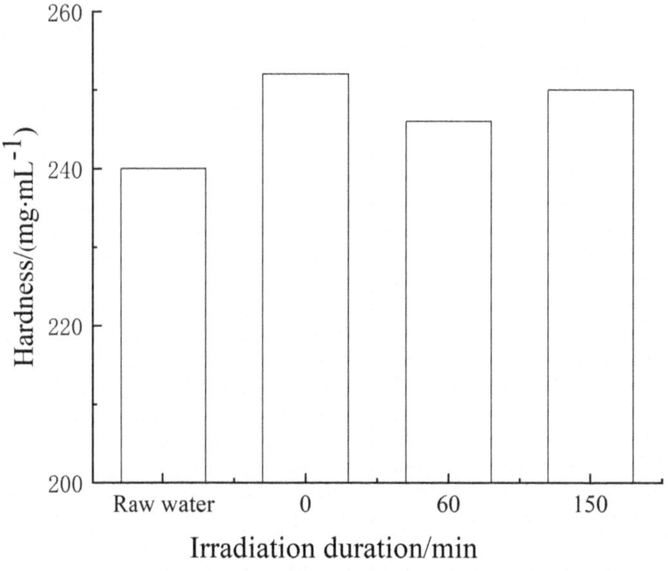

Fig. 17.10 Change of hardness after different duration of disinfection

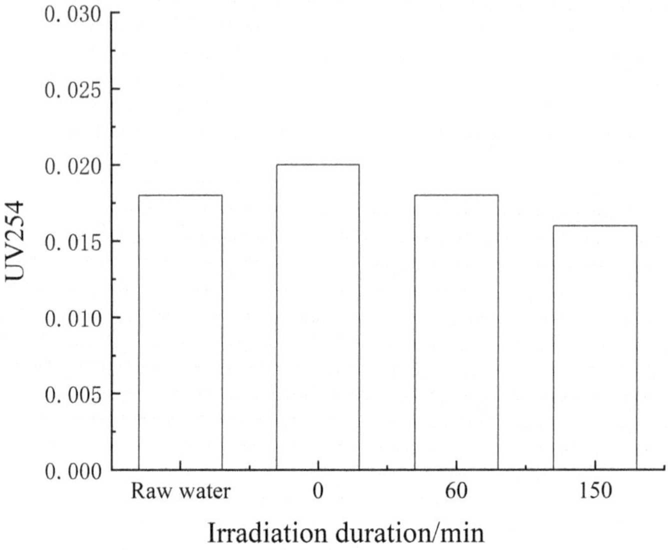

Fig. 17.11 Change of UV254 after different duration of disinfection

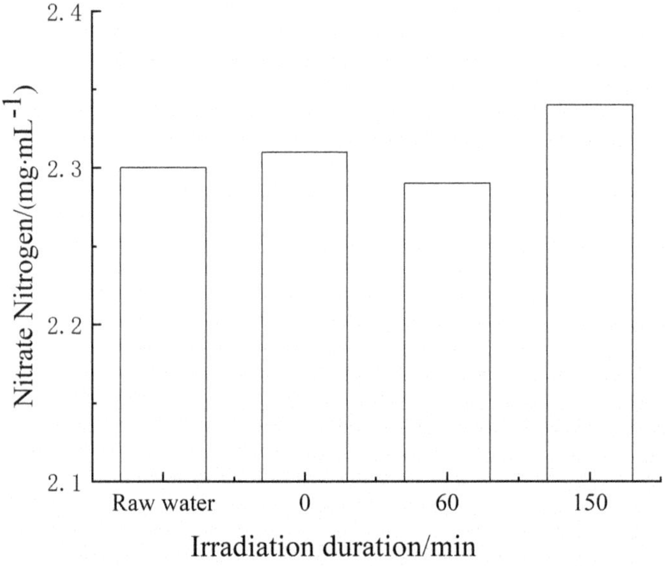

Fig. 17.12 Change of nitrate nitrogen after different duration of disinfection

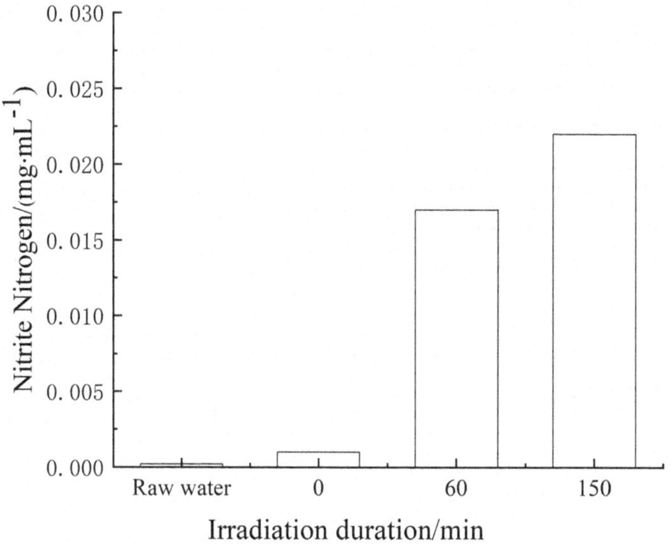

Fig. 17.13 Change of nitrite nitrogen after different duration of disinfection

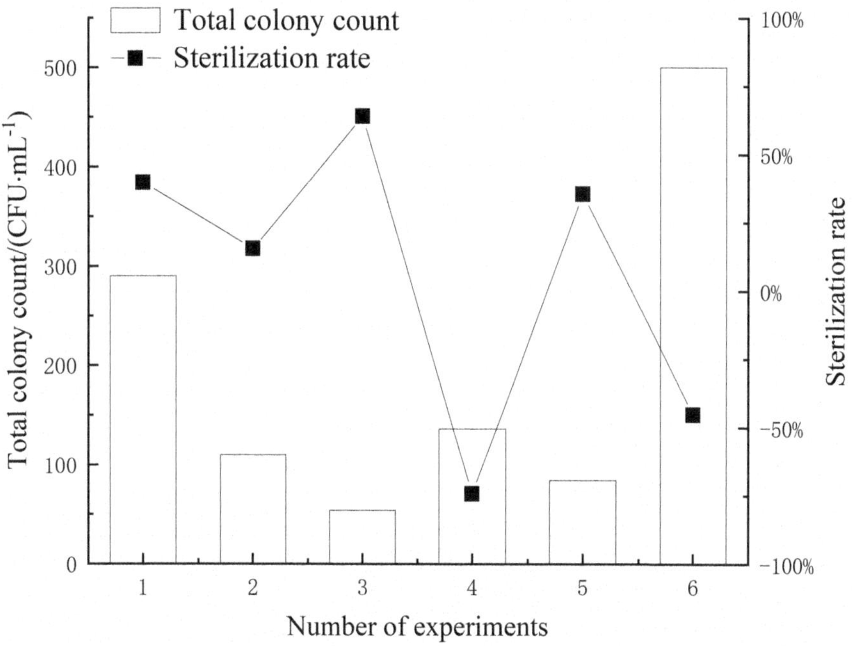

Fig. 17.14 Bactericidal effect of G450-1 when used alone

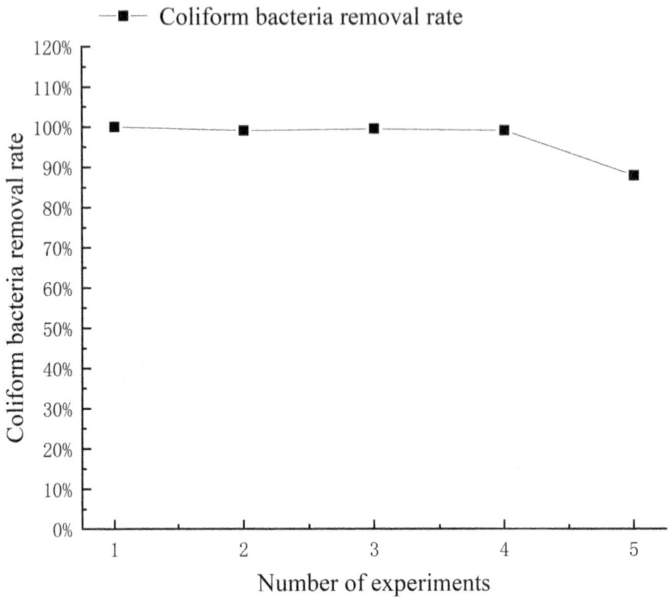

Fig. 17.15 Coliform removal rate of G450-1 when used

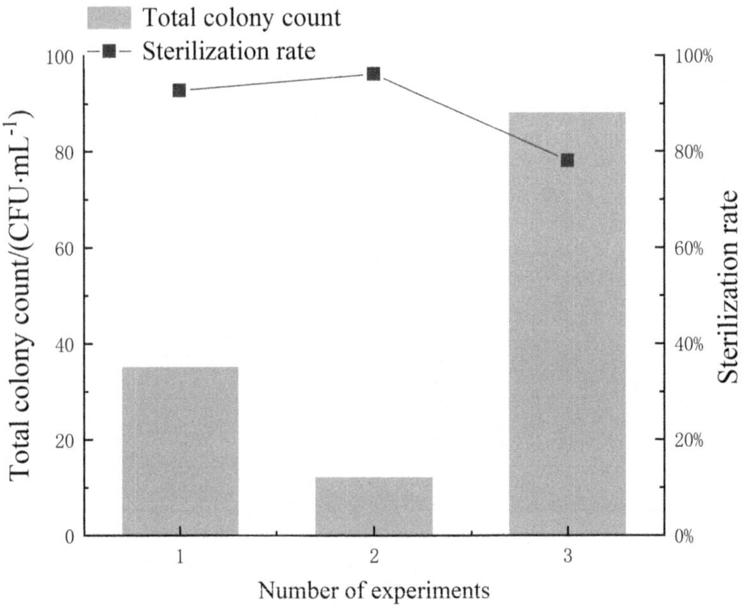

Fig. 17.16 Bactericidal effect of G800 when used alone

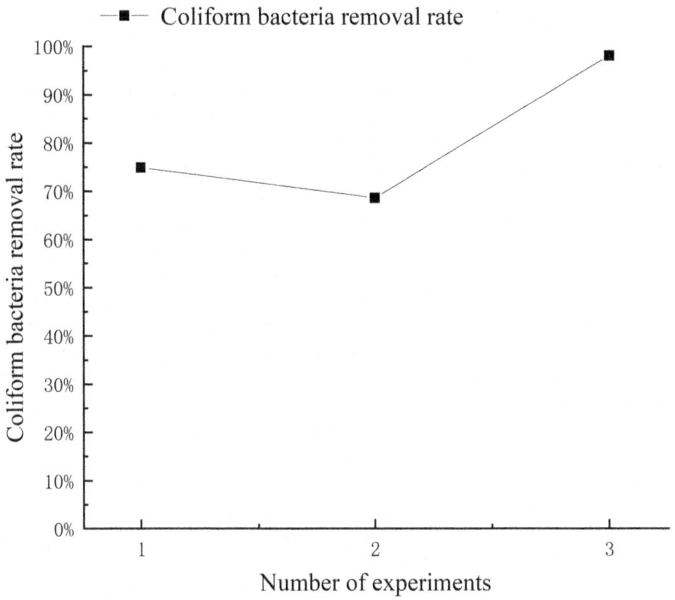

Fig. 17.17 Coliform removal rate of G800 when used alone

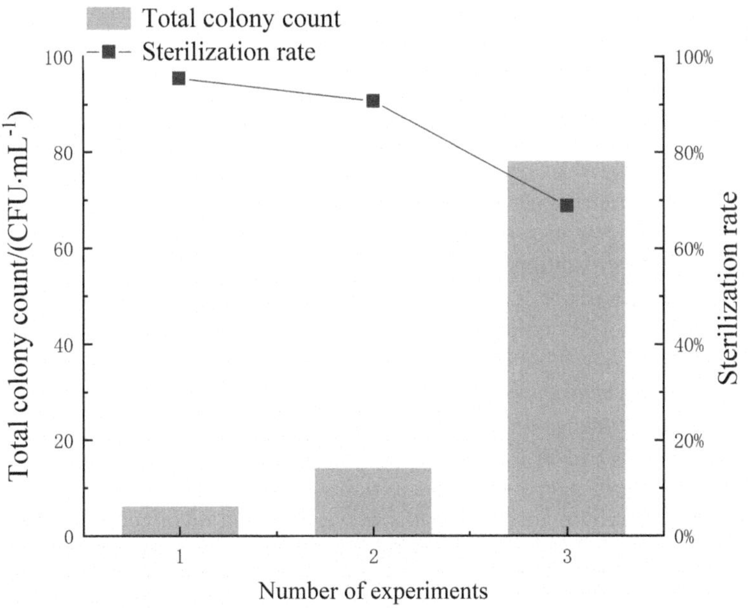

Fig. 17.18 Bactericidal effect of G450-1 and G450-2 when used in series

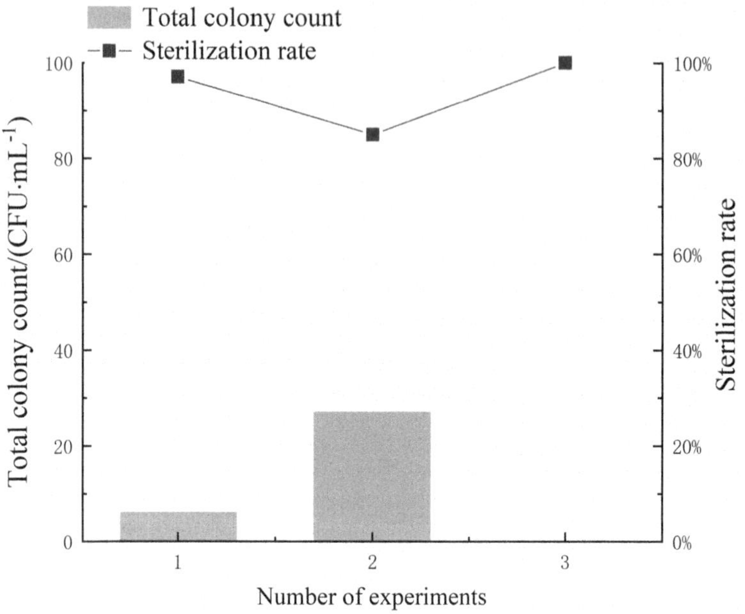

Fig. 17.19 Bactericidal effect of G450-1 and G450-2 when used in series after G800 is used

17.4 Conclusion

(1) The disinfection effect of the vertical microwave electrodeless UV equipment is closely related to the irradiation time. Under experimental conditions, the sterilization rate begins to increase with the duration of equipment disinfection but enters a slow growth phase or stabilizes after 60 min, with no significant increase in sterilization rate thereafter.

(2) Under the same conditions of lamp power and quantity, vertical microwave electrodeless UV equipment with smaller diameters irradiates thinner water layers, achieving more uniform sterilization and better disinfection effects.

(3) Microwave electrodeless UV disinfection only has a minor impact on the effluent's hardness, UV254, nitrate nitrogen, and nitrite nitrogen, essentially not altering the water's physicochemical properties.

(4) The quantity of microorganisms in raw water can affect the disinfection efficiency of the equipment. Due to the uneven distribution of microorganisms and significant differences in raw water microbiology, the disinfection efficiency of the same equipment is influenced by the microorganisms in the raw water. Therefore, ensuring that the water output from multiple groups of experiments by the equipment is stable and meets standards is essential before it can be put into use. Otherwise, there is an increased risk of the water output exceeding the standard.

(5) When multiple sets of vertical microwave electrodeless UV equipment are operated in series, sterilization rates can reach about 99%, achieving the desired treatment effect. Under experimental conditions, the effluent from two devices, each with a power of 1kw and a diameter of 450 mm operating in series, meets the "Standards for Drinking Water Quality" (GB5749—2022). This demonstrates the feasibility of using this type of vertical microwave electrodeless UV sterilization equipment in drinking water treatment. However, the use of vertical microwave electrodeless UV sterilization equipment still requires further consideration of cost among other factors.

Acknowledgements This research was supported by the Jinan Water Science and Technology Project (JNSWKJ202212) and Jinan Water Group Internal Project.

References

1. Wang XC (2022) Safe water reuse through a quasi-natural water cycle. J Water Reuse Desalin 12(3):366–372
2. Wallmann L, Krampe J, Lahnsteiner J et al (2021) Fate and persistence of antibiotic-resistant bacteria and genes through a multi-barrier treatment facility for direct potable reuse. J Water Reuse Desalin 11(3):373–390
3. Li X, Cai M, Wang L et al (2019) Evaluation survey of microbial disinfection methods in UV-LED water treatment systems. Sci Total Environ 659
4. Lin J, Lu J, Lin HZ et al (2014) Treatment of pesticide wastewater using Fenton reagent enhanced by a microwave electrodeless ultraviolet lamp. J Anhui Agric Sci 42(25):8696–8698
5. Wang XD (2023) Development of photon-photon coupling cold sterilization technology and its application in the field of microbial contamination control of raw milk. Inner Mongolia University
6. Zheng T, Zheng YL, Sun YL et al (2019) Research progress of microwave electrodeless ultraviolet lamp in the field of environmental pollution control. J Harbin Inst Technol 51(02):69–75
7. Xu B-J, Wu H, Su H-Z et al (2024) A novel ultraviolet light source of microwave discharge electrodeless ultraviolet lamp: luminescence mechanisms, reactor structures, and environmental applications. J Cleaner Prod 140706.
8. Yang B (2018) Experimental research and design of DR ultraviolet disinfector. Shenyang Jianzhu University
9. Sun Y (2017) Experimental studies of OMS type UV sterilizer for the drinking water treatment. Shenyang Jianzhu University
10. Yan YJ, Yang QX, Mo JC et al (2022) Detection and analysis of five different types of drinking water quality. Mod Food 28(01):188–191
11. Piao FS, Fu JX, Liu J (2005) Ultraviolet disinfection of drinking water. J Liaoning Tech Univ 03:440–443
12. Zhang LJ (2009) A pilot study on ultraviolet(UV) disinfection of a tertiary municipal effluent for reuse. Xi'an University of Architecture and Technology
13. Li LXY, Ma XJ (2006) Characteristics of hydroxide radical and its reaction mechanism in photochemistry oxidation process. Technol Dev Chem 08:27–29
14. Hong XH (2019) Effects on the concentration of nitrite and its content-change in the manufacture of drinking water. Zhejiang University

Chapter 18
Research on Scale and Corrosion Inhibition Performance of Circulating Water Treatment Agent

Zhan Liu, Mei-fang Yan, Yu-hua Gao, Na Li, Hai-hua Li, and Zhen-fa Liu

Abstract A compound product of green water treatment chemicals was studied, and its scale inhibition rate and corrosion rate on carbon steel were determined through experiments, which has good scale inhibition performance, and can improve the utilization rate of circulating water, and play a role in promoting the sustainable development of water resources.

Keywords Water treatment agent · Compounding · Scale inhibition · Corrosion rate

18.1 Introduction

Many industrial enterprises such as petrochemicals, pharmaceuticals, etc. are water-consuming households, and their water consumption is mainly concentrated in the circulating water [1, 2]. Now the vast majority of enterprises have adopted the circulating cooling water treatment system to realize the reuse of water. And with the strict control of the country on the amount of sewage discharged into the enterprise, the concentration of water times higher and higher, scaling and corrosion phenomenon is also more and more serious. This requires the use of scale and corrosion inhibitors with better performance, which can play a role in scale and corrosion inhibition [3–5]. At present, the vast majority of enterprises are still using phosphorus-containing water treatment agent formulations, agent usage, and scale and corrosion inhibition performance is poor [6–9]. Therefore, in the enterprise circulating cooling water system, the promotion and application of efficient, environmentally friendly green scale inhibitor should become the focus of development [10–12].

Z. Liu (✉) · M. Yan · Y. Gao · H. Li · Z. Liu
Institute of Energy Sources, Hebei Academy of Science, Shijiazhuang, Hebei, China
e-mail: liuzhan1216@126.com

Hebei Engineering Research Center for Water Saving in Industry, Shijiazhuang, Hebei, China

N. Li
Hebei Sun-Water Treatment Co., LTD., Shijiazhuang, Hebei, China

K. Y. Foo et al. (eds.), *Water Nexus Approach for Sustainable Development*, Environmental Science and Engineering, https://doi.org/10.1007/978-3-031-89238-7_18

In this article, we studied a compound product of green water treatment chemicals, and determined its scale inhibition rate and corrosion rate on carbon steel through experiments, which has good scale inhibition performance, and can improve the utilization rate of circulating water, and promote the sustainable development of water resources.

18.2 Experimental Part

18.2.1 Main Experimental Apparatus and Drugs

Maleic anhydride (MA), 2-acrylamide-2-methylpropane sulfonic acid (AMPS), ammonium persulfate, sodium bicarbonate, anhydrous calcium chloride, ferric chloride, sodium tetraborate, etc., and the purity of all the agents is analytically pure.

Electrothermal constant temperature water bath, vacuum drying oven, constant temperature magnetic stirrer, balance, syringe pump and so on.

18.2.2 Synthesis of Scale Inhibitor

A quantity of MA, AMPS and distilled water was added to a four-necked flask. The solids and solution were stirred to dissolve the solids, and the system was heated to 90 °C and then N_2 was poured in. After five minutes, the ammonium persulfate solution was injected into the system with a syringe pump. The system was held at this temperature for 6 h with stirring to obtain an amber colored liquid, which was the scale inhibitor product. The solution was washed with anhydrous ethanol and dried in a vacuum drying oven to obtain a solid product as the purified scale inhibitor.

18.2.3 Experimental Method

Evaluation of calcium carbonate scale inhibition performance.

The experiments were carried out using recycled cooling water from the enterprise. A vessel was filled with a test water solution containing a scale and corrosion inhibitor. It was kept in a water bath at 80 °C for 10 h. The container with the solution was removed and cooled in air. Titrate the concentration of calcium ions in the upper clear solution of the conical flask and calculate the scale inhibition rate η_1.

$$\eta_1 = \frac{V_1 - V_0}{V_2 - V_0} \times 100\%$$

where V_0 (ml), V_1 (ml), V_2 (ml) denote the volume of EDTA consumed by the blank solution after heating, the solution after dosing, and a certain solution in the original water, respectively.

18.2.4 Determination of Corrosion Rate and Corrosion Inhibition Rate

The water used for the experiment was enterprise circulating cooling water. The experiment referred to the national standard GB/T 18175–2014. rotary suspension method was used to study the corrosion rate of scale and corrosion inhibitor. Under the condition of 30 °C, the rotational speed is 75r/min, the experimental time is 72 h, the corrosion rate X (mm/a) is calculated as follows.

$$X = \frac{8760 \times 10 \times (W - W_0)}{A \times D \times T}$$

where W, W_0 indicate the weight of the specimen before and after the experiment, respectively, in grams; A is the surface area of the specimen, the value of which is 28 cm^2; D is the density of the specimen, the value of which is 7.850 g per square centimetre.

The corrosion inhibition rate X_2 was calculated by the formula:

$$X_2 = \frac{X_0 - X_1}{X_0} \times 100\%$$

where: x_0 is the corrosion rate of the specimen in the blank test; x_1 is the corrosion rate of the specimen in the test with the agent, and the unit is all millimeters per year (mm/a).

18.2.5 Biodegradability

Biodegradability was determined by biological shaker method, inoculum and aqueous solution containing agent were added to the experimental container, which was placed in a constant temperature oscillating shaker at 25 °C, and chemical oxygen demand, i.e., COD, of the solution was determined at 1, 7, 14, 21, and 28 days.Calculation was carried out to obtain the biodegradation rate of the agent by the following formula.

$$\text{Biodegradation rate} = \left(1 - \frac{C_t - C_{bt}}{C_0 - C_{b0}}\right) \times 100\%$$

where: C_{bt}, C_t, C_{b0}, C respectively represent COD in the inoculated reaction solution without agent, COD in the inoculated reaction solution with agent, COD in the inoculated reaction solution without agent, COD in the initial value of COD in the inoculated reaction solution without agent, and COD in the inoculated reaction solution with the addition of agent at the moment of t. The unit is mg/L.

18.3 Results and Discussion

18.3.1 Test Results of Scale Inhibition Performance of MA/AMPS

The scale inhibition and corrosion inhibition performance of different concentrations of MA/AMPS on calcium carbonate was determined, and the results are as follows. As can be seen from Fig. 18.1, its scale inhibition rate increases with the increase of concentration, and its scale inhibition rate reaches 100% when the dosage of the agent is 20 mg/L.

Figure 18.2 shows the corrosion rate. The corrosion rate of MA/AMPS is 0.1700 mm/a at a concentration of 25 mg/L, which does not comply with the national standard of less than 0.075 mm/a.

The experimental group also studied the biodegradability of the agent, and from the experimental data in Fig. 18.3, it can be understood that the biodegradability of the binary polymer reached 63% at 28 days on the ground, which belongs to a kind of agent with good biodegradability, and the long-term use of it will not cause negative impacts on the environment.

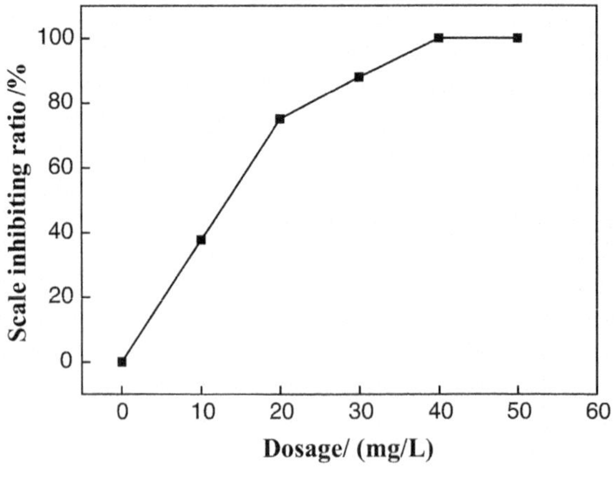

Fig. 18.1 The scale inhibition of MA/AMPS

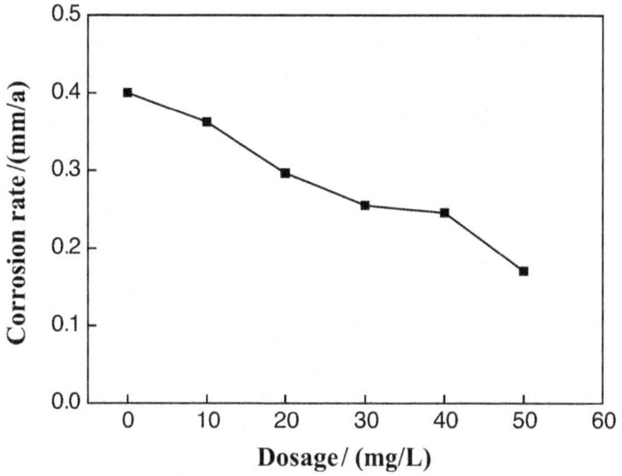

Fig. 18.2 The corrosion rate of MA/AMPS

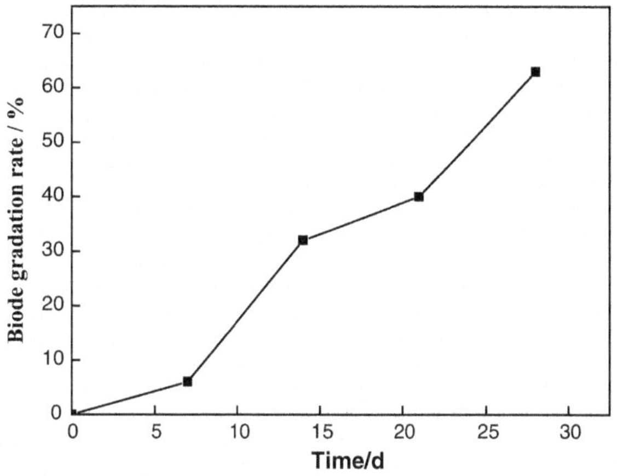

Fig. 18.3 The biodegradability of MA/AMPS

18.3.2 Performance Test Results of the Compounded Agent

As can be seen from the results in Fig. 18.2, the corrosion rate of 50mg/L MA/AMPS on carbon steel is 0.1700mm/a, which is not up to the national standard, indicating that the corrosion inhibition performance of the polymer needs to be improved. Through the study of polymer properties, types, etc., the agent and zinc salt compounding, compounding agent scale inhibition performance as shown in Fig. 18.4, corrosion inhibition performance as shown in Fig. 18.5. It can be seen that the addition of zinc salt has little effect on the scale inhibition performance. However, with the increase of zinc salt addition, the corrosion rate of the agent gradually decreased, that is, its corrosion inhibition performance gradually became better. When the addition of zinc salt is 2mg/L, the corrosion rate of the compound is 0.069mm/a, which is less than 0.075mm/a as required by the national standard.

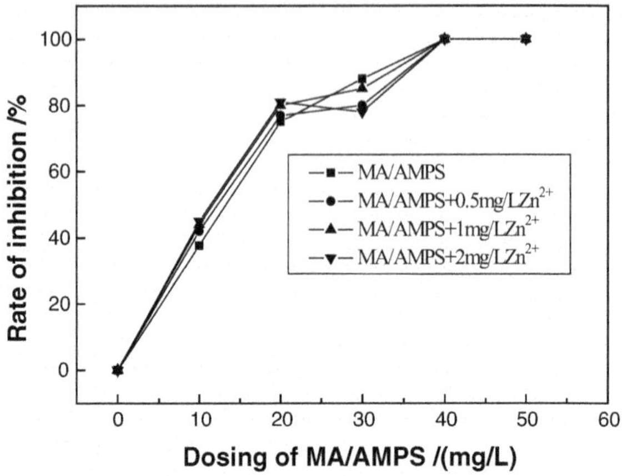

Fig. 18.4 Scale inhibition performance of compounding agents

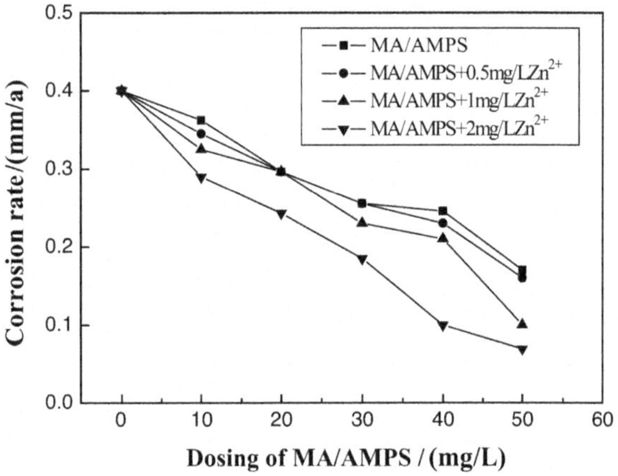

Fig. 18.5 Corrosion rate of compounded agents

18.4 Conclusion

(1) Synthesized agent MA/AMPS, and studied its scale inhibition and corrosion inhibition performance, which has excellent scale inhibition performance, but corrosion inhibition performance is poor, in a certain concentration failed to meet the national standard requirements.

(2) The synthesized environmentally friendly agent was compounded with other agents to determine its performance, and the results were obtained, under a certain concentration and proportion, the scale inhibition rate of the two compounded agents was 100%, and the corrosion rate was 0.0700 mm/a, which was better than the national standard "Design Code for Industrial Circulating Cooling Water Treatment" GB50050-2017 requirements.

Acknowledgements This work was financially supported by the Science and Technology Program of Hebei Academy of Sciences (22701, 23701, 24701)

References

1. Li H (2023) Application and development of chemical water treatment technology in power plants. Shandong Chem Ind 52(22):250–251+255
2. Teng X, Wang S, Wang X et al (2024) Analysis and treatment of evaporative condenser fouling. Shandong Chem Ind 53(01):179–186
3. Qiu J (2022) Engineering application of phosphorus-free water treatment chemicals in waste heat power generation boilers. Ind Boiler (05):56–58
4. Gao Y, Wang Z, Li H et al (2021) Research progress on the application of chitosan derivatives in industrial circulating cooling water treatment. Appl Chem Eng 50(12):3441–3445
5. Jia Y, Ping S, Xu X et al (2023) Research and application of new phosphorus-free furnace water treatment agent for industrial boilers. Petrochem Technol 30(10):4–6
6. Li H, Sun Z, Jing G (2024) Synthesis and performance study of Ly-CCQDs, a green scale inhibitor. Mod Chem Ind 44(03):212–217
7. Hao Y, Li G, Yan Q et al (2024) Analysis and improvement of scaling in external cold water system of regulator. Shandong Power Technol 51(02):79–84
8. Chen M, Sun B, Zhang Y et al (2024) Preparation of polymaleic anhydride-based green scale inhibitor and its application in cement production line. New Century Cement Herald 30(01):53–56
9. Tang H (2024) Application and optimization strategy of energy saving and consumption reduction technology in chemical enterprises. Chemical Management 02:49–51
10. Wang M-Q, Li Y-Q, Lin Y-P et al (2024) Preparation and mechanism of MA/AA/DETA terpolymer corrosion and scale inhibitor. Fine Petrochem 41(01):10–14
11. He F, Meng X, Wu Y et al (2023) Pilot comparison of variable-frequency electromagnetic and chemical inhibition of reverse osmosis membrane scale. Chem Eng 51(12):62–65
12. Zhang Q, Wang L (2023) Research on environmentally friendly low phosphorus circulating water treatment technology and its application. Cleaning World 39(11):47–49

Chapter 19
Research on Utilization of Sewage Sludge Resourceful Disposal Technology and Analysis of Green Ecological Application

Caixia Li, Junxi Bai, Min Dong, and Jinshan Li

Abstract With the expansion of economic scale and industrial development, the production of domestic sludge and industrial sludge is also increasing day by day. Sludge is highly susceptible to decay and odor, and contains a large amount of organic matter, thus reasonable disposal of industrial sludge can not only reduce its pollution of the environment, but also help the recovery of resources and energy. How to use resourceful disposal technology to deal with sewage sludge, to realize the virtuous cycle of solid waste economy, reduce costs and increase efficiency, is one of the keys to the green development of industry. This paper focuses on analyzing and analyzing the current situation of sewage sludge discharge and the degree of industrial sewage sludge resourceful disposal, to industrial solid waste economic cycle for green ecological application analysis and research, discussing industrial sewage sludge resourceful treatment process to reduce costs and increase efficiency methods, constructing a flow chart applicable to industrial solid waste economic cycle, promoting the economy to achieve green low-carbon rapid development, and constructing a good ecological harmonious coexistence of resources and the ecological environment The relationship between resources and the ecological environment will be constructed in a harmonious coexistence.

Keywords Sewage Sludge · Cost reduction and efficiency · Green ecology

C. Li (✉) · J. Bai · M. Dong · J. Li
School of Energy and Constructional Engineering, Shandong Huayu University of Technology, Dezhou, Shandong, China
e-mail: licaixia345440715@163.com

K. Y. Foo et al. (eds.), *Water Nexus Approach for Sustainable Development*,
Environmental Science and Engineering, https://doi.org/10.1007/978-3-031-89238-7_19

19.1 Introduction

In recent years, China has been actively carrying out important initiatives for energy conservation and emission reduction, pollution reduction and carbon reduction, and vigorously building a new pattern of ecological priority, green and low-carbon development, and rational resource-based disposal of industrial sludge, which not only reduces environmental pollution, but also contributes to resource recycling. The government should set up special funds to support the construction and operation of sludge resource disposal projects, and encourage the participation of social capital to form a diversified investment mechanism. With the acceleration of industrialization, its development has brought many employment opportunities and economic growth, but it is also accompanied by the generation and discharge of a large amount of industrial wastewater sludge. Although the government has taken a series of measures to solve the sewage problem, there are still some challenges. The coverage of sewage treatment plants is not high, which leads to a part of industrial wastewater can not be effectively treated, which may pollute the local water bodies and threaten the health of the ecosystem [1]. In addition, the sludge treatment technology is relatively backward, and a complete and efficient treatment technology system has not yet been formed, which limits the process of sludge resource disposal to a certain extent. Wastewater discharge standards are not strictly enforced and some enterprises have non-compliant discharge behavior. This exacerbates the sewage problem and negatively affects the living environment in the surrounding areas. In addition to the sewage problem, the treatment and management of sludge is also a key topic, and the sludge generated by the sewage treatment process needs to be properly disposed of, otherwise it may pose a threat to the environment and human health.

Research on industrial sewage sludge treatment has revealed that sludge treatment is faced with insufficient sewage sludge treatment capacity and limited sludge treatment facilities. There are relatively few sludge treatment facilities to meet the demand for treating large quantities of sludge. This results in some sludge being disposed of in an inefficient or non-compliant manner, which may pose a risk to the surrounding environment [2, 3]. Therefore, it is particularly important to explore more environmentally friendly and sustainable ways of resourceful sludge treatment [4].

Based on this, this paper investigates the status quo of sewage sludge production in the industrial production process, as well as the production and composition of industrial sewage sludge, analyzes the status quo of sewage sludge and the degree of industrial realization of sewage sludge resourcing, explores the status quo of industrial sewage sludge treatment and recycling as well as the utilization of resourcing disposal technology, compares and analyzes the industrial sewage sludge treatment technology and the standard of treated sludge, and constructs a flow chart applicable to the industrial solid waste economic, To build a flow chart for industrial solid waste economic cycle, to form a closed-loop economic circle of solid waste "no waste", and to deepen the analysis of sewage sludge resourceful disposal process to reduce costs and increase efficiency "win–win" method.

19.2 Study on the Utilization of Sewage Sludge Resourceful Disposal Technology

Sewage sludge is a sediment containing high concentrations of organic matter, inorganic matter, bacteria and viruses, and other harmful substances produced by the treatment of industrial wastewater. It is mainly produced with wastewater process with high wetness and viscosity. In recent years, the expansion of urban scale and the development of industry, a variety of industrial production and processing will produce a large number of sewage wastewater, mixed with many impurities, and contains many organic chemicals, bacteria, heavy metals and harmful substances. These waste-waters are treated by industrial wastewater treatment facilities, which produce corresponding sludge, leading to an increasing production of domestic and industrial sludge [5]. Heavy metals and toxic substances in sludge are difficult to degrade in nature, which in turn causes contamination of soil and groundwater, and hazardous substances in recycled sludge can cause various diseases, posing a serious threat to the environment and human health.

After the dewatering treatment of sludge, the common sludge disposal technologies in foreign countries mainly include sludge curing landfill, biocomposting treatment and incineration, gasification, pyrolysis technology, etc., but these disposal technologies have certain advantages and shortcomings [6].

For the common domestic and foreign sludge disposal technology mainly sludge curing landfill, biological composting technology, drying, incineration technology, etc., sludge incineration, sludge composting and sludge drying of the three feasible processes for comparison, its advantages and disadvantages are shown in Table 19.1.

Industrial sewage sludge treatment process is mainly the sewage sludge produced by industrial production after simple treatment, directly sent to the solid waste treatment plant to landfill, resulting in secondary pollution problems, causing more problems to the sewage sludge resource disposal process. For the sewage treatment industry at home and abroad, the main treatment process for the A2/O process (anaerobic—anoxic—aerobic method), in fact, is the use of the technology of the relevant chemicals in the sewage nitrogen removal of phosphorus, reduce the hazards of the sewage treatment process is the simplest, the application of the more mature, widely [5]. Through the analysis of the above treatment process, combined with solar energy related technology, the use of more mature A2/O process, the formation of solar energy sludge drying process, the dried sludge sent to the waste treatment plant incineration or landfill, can be used to operate the relevant equipment using convenient automatic control technology, to achieve the drying of sludge treatment.

Table 19.1 The comparison of sludge disposal processes

Project title	Sludge incineration	Bio-composting of sludge	Solar sludge drying
Dominance	Occupied area saving, good effect of reduction and harmlessness. Applicable to projects with high requirements for harmlessness, no conditions for landfill, land use, etc., and very tight land use. Mainly applicable to large-scale projects	The process is mature and stable, utilizing biomass energy to save investment and operation costs. The compost products can be land-utilized and can generate certain economic benefits. Suitable for medium-sized projects	Diversified use of energy, according to local conditions, can choose the investment and operating costs of energy as a heat source. The product can be used for building materials, organic fertilizer and fuel. Short drying cycle. Applicable to various scales
Drawbacks	High investment and operating costs; difficulty in completely solving pollution problems such as dioxins; uncertainty in the utilization of incineration residues	Large footprint; long composting time, usually about 20 days. Heavy metals in sludge can cause contamination of soil and plant systems	large footprint
Investment	20,000,000–30,000,000 yuan/ton	300,000–400,000 yuan/ton	300,000–400,000 yuan/ton
Running cost	200–3000 yuan/ton	80–120 yuan/ton	80–120 yuan/ton
Final disposition	Incineration residue building materials utilization; fly ash transported to the hazardous waste disposal site	Compost products can be used for agriculture and greening	The dried sludge can be used for fuel as well as building materials and ceramic granules

19.3 An Analytical Study of Green Ecologization Applications

19.3.1 Economic Recycling of Industrial Solid Waste

With the help of solar energy technology, the basic flow of the production process, which is based on new energy (solar) drying, is formed, and combined with the more mature technology of sewage sludge treatment A2/O process, which is used as a secondary treatment method, the core of which is the traditional activated sludge method [6]. The processes used in the treatment of sewage include A2/O sewage treatment process, A/O sewage treatment process, oxidation ditch treatment process and SBR treatment process. In this project, A2/O process was used for biochemical treatment of wastewater so that the wastewater could achieve the purpose of nitrogen and phosphorus removal.

The sewage sludge with 80% water content will be transported to the sludge storage tank, and then discharged into the dilution tank after precipitation, and the sludge water content will reach 92–95%, and then transported to the comprehensive conditioning tank by the pressurized pump, and the well-proportioned conditioner will be added to the tank and rapidly stirred, and then after the related oxidation–reduction reaction, it will increase the dewatering performance, and change the hydrophilicity to hydrophobicity, and reduce the leaching rate of the sewage sludge. The dewatered and dried sludge is conveyed to the next step of solar drying treatment system, and the dewatered water is discharged into the sludge tank for recycled water.

The sludge with 65% water content is crushed by plate and frame filter press and transferred to the paver silo by belt conveyor, the sludge is spread evenly after passing through the fabric screw to complete the longitudinal fabric, the sludge is panned horizontally after passing through the direction-change scraper, and the surface of the sludge is constantly refurbished after passing through the tumbling screw. At the same time, under the synergistic effect of active solar greenhouse system, air energy heat pump system and solar circulating hot water heating system, the water content of sludge is continuously reduced. When the water content of sludge is reduced to 40%, it is conveyed by belt conveyor to the dry sludge silo to be utilized. This sludge can be made into new building materials, thermal insulation materials, ceramic granules, filter media, etc. to achieve source utilization.

The main process flow cycle, as shown in Fig. 19.1.

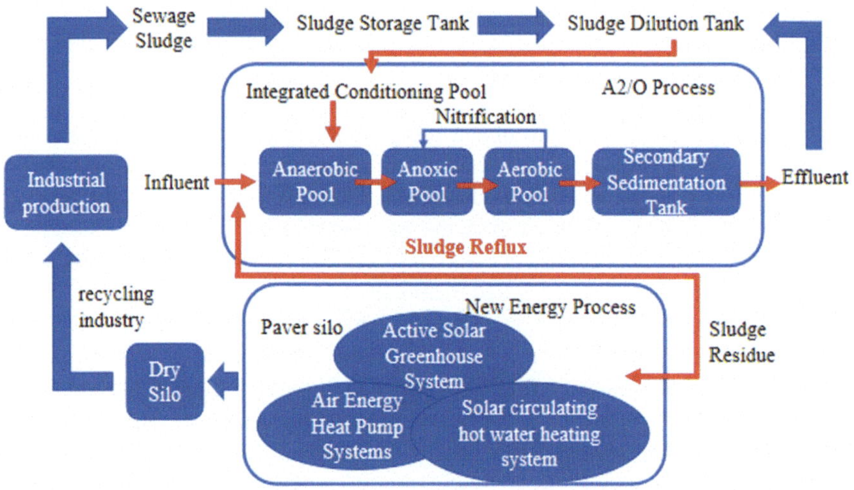

Fig. 19.1 Process flow cycle diagram

19.3.2 Industrial Sewage Sludge Resourceful Treatment Process Cost Reduction and Efficiency Improvement Methods

In recent years, the resourceful treatment of industrial sewage sludge is an important way to solve environmental problems and achieve sustainable development. Through the research and analysis of sewage sludge treatment plant, in practice, should be based on different industries, different processes and different nature of sewage sludge and other specific circumstances for comprehensive consideration, through the practice to improve the utilization efficiency. In order to reduce treatment costs and improve the efficiency of sewage sludge and other industrial solid waste resource utilization, the following four processes are more feasible for industrial sewage sludge resource treatment process to reduce costs and increase efficiency:

1. Wastewater pretreatment: before entering the sludge treatment stage, pretreatment of industrial wastewater can effectively reduce the sludge production and the difficulty of treatment. Common pretreatment techniques include screen filtration, physical precipitation, pH adjustment, etc., which can remove suspended matter, precipitate harmful substances, and minimize solid particles in sewage [7].
2. Biological pretreatment: biological pretreatment is in the sewage sludge process, through the use of microorganisms to degrade organic wastes, to achieve the purpose of purifying the water quality and reducing the volume of sludge. Selection of appropriate microbial flora and control of temperature, pH and other factors can improve the treatment efficiency and reduce waste generation.
3. Anaerobic digestion: Anaerobic digestion is a method of decomposing organic waste in an anaerobic environment using microorganisms (such as anaerobic bacteria and parthenogenic bacteria). Through this digestion technique, waste can be decomposed into resources such as CH_4 and CO_2 for recycling and to reduce odor and stench. In addition, the digested sludge can be used for farmland fertilization and horticulture cultivation to achieve recycling of the products of sludge wastewater resource-based treatment.
4. Sludge stabilization treatment: Stabilization treatment can reduce the water content of sludge, kill a variety of pathogens, and degrade organic matter to ensure safe treatment and reduce the impact on the environment. Common stabilization methods include composting and curing, which can reduce sludge volume and improve its treatment performance.

For the process, industrial sewage sludge resource treatment process cost reduction and efficiency method can be carried out by phases. In the pretreatment stage, sludge dewatering, crushing, screening and grinding are carried out to further reduce the particle size; in the resourceization stage, these sludge fuels can be utilized for incineration, bioelectricity generation, and brick making. These processes not only utilize the sludge but also reduce the environmental pollution. Through the emission reduction stage, sludge can be anaerobically digested, sterilized at high temperatures, and incinerated, which can effectively reduce the harmful substances in sludge, and also convert it into valuable energy or resources; in the cost reduction and efficiency stage, pretreatment, resourcing, and emission reduction can be integrated to improve the sludge digestibility and reduce the operating costs. In addition, it can also improve production efficiency and reduce operating costs by optimizing equipment selection and layout.

19.3.3 Feasibility Analysis of Resourceful Disposal Process of Industrial Sewage Sludge

Technically, the process technology combines A2/O process and new energy utilization, which is more mature, and the feed water source in the process adopts the recycled water of the sewage treatment plant, which can greatly save water and can effectively degrade pollutants, reduce the discharge of wastes, and obtain the reusable resources [8, 9].

In terms of economy, the construction of sludge disposal project solves the problem of secondary pollution of sludge, and more importantly, solves the problem of outlets for sludge. The implementation of the project protects the ecological environment and beautifies the urban environment. It has important environmental benefits for the local ecological environment and pollution prevention of Haihe River Basin. The implementation of sludge disposal resource recycling technology will certainly promote the construction of urban infrastructure and optimize the investment environment in order to promote the development of production and create favorable conditions for economic prosperity. After sludge disposal resource recovery, by reviewing the relevant laws and regulations to ensure that the treatment process meets the requirements of environmental protection, and at the same time comply with the local planning and management and land use policies, the raw sludge after treatment can be used as fuel or as landfill cover soil, turning waste into treasure.

19.4 Conclusion

In this paper, the sewage sludge-based solid waste economic cycle of industrial sewage sludge, the use of new energy (solar energy) drying-based production process basic flow, the use of solar energy and heat pumps (ground-source heat pumps, sewage-source heat pumps) as a heat source for low-temperature drying of sludge. And combined with the more mature technology of sewage sludge treatment A2 / O process, so as to achieve the purpose of nitrogen and phosphorus removal. The use of solar energy sludge drying investment and operating costs are low, the drying cycle is short, and it is conducive to promoting the industrial realization of building ecological priority, green and low-carbon development of a new pattern.

Acknowledgements This work was financially supported by the Green and low-carbon smart heating and cooling technology characteristic laboratory (PT2022TS01).

References

1. Zhu X et al (2019) Pyrolysis of pre-dried dewatered sewage sludge under different heating rates: Characteristics and kinetics study. Fuel 255:115591
2. Gao N et al (2020) Product property and environmental risk assessment of heavy metals during pyrolysis of oily sludge with fly ash additive. Fuel 266:117090
3. Oladejo J et al (2018) A Review of sludge-to-energy recovery methods. Energies 12(1):1–38
4. Gao N et al (2020) Modeling and simulation of coupled pyrolysis and gasification of oily sludge in a rotary kiln. Fuel 279:118152
5. Pan X (2021) Discussion on the deep dewatering technology of municipal sewage sludge. Energy Energy Conserv 000(012):74–76
6. Wang J (2022) Exploration on the development strategy of sludge treatment technology of urban sewage treatment plant. Ecol Environ Prot 5(2):4–6
7. Li S, Li C, Shao Z (2022) Microwave pyrolysis of sludge: a review. Sustain Environ Res 32(1):1–14
8. Lin K, Zeng JA, Chiang HU (2020) Microwave pyrolysis of sludge for potential use as land application and biofuel. J Chem Technol Biotechnol 95(4):975–984
9. Tran T, Giang LV, Loc HH et al (2022) The initial study on the application of sewage sludge for agriculture on the laboratory scale. Mater Sci Forum 1048

Chapter 20
Design and Implementation of Air Quality Prediction System Based on Federated Learning

Yuechun Feng

Abstract This study is based on federated learning technology and designs and implements a system for air quality prediction. The system consists of multiple regional terminals, each responsible for locally training air quality data within the region and generating regional models. By exchanging and integrating models between regions, global model updates and air quality predictions can be achieved. After testing, The system has achieved significant results in improving model training efficiency and prediction accuracy. This study provides a new approach and method for the development of air pollution monitoring systems. The system was developed using Python as the development tool and TensorFlow as the development platform.

Keywords Federation learning · Regional model · Air quality monitoring · Monitor system · TensorFlow

20.1 Introduction

In recent years, with the increasing attention paid to air pollution, more and more researchers have come up with various ways to monitor and predict air quality. For example, some researchers have developed portable sensing devices that can directly collect samples through physical methods and monitor air data, while others have used satellite shake sensing technology to directly infer upper air images. Of course, there are also more methods based on machine learning big data to model and analyze data, and then monitor and predict air quality data. Moreover, with the increase in data volume and the development of deep neural networks in the field of big data, models of time-series historical data and massive monitoring image data can be applied to air quality monitoring systems [1]. In addition, due to the low computing efficiency caused by massive data, distributed model training methods are also emerging. Among these methods, there are also many frameworks that can

Y. Feng (✉)
Ningxia Institute of Technology, Ningxia, China
e-mail: 67723847@qq.com

K. Y. Foo et al. (eds.), *Water Nexus Approach for Sustainable Development*,
Environmental Science and Engineering, https://doi.org/10.1007/978-3-031-89238-7_20

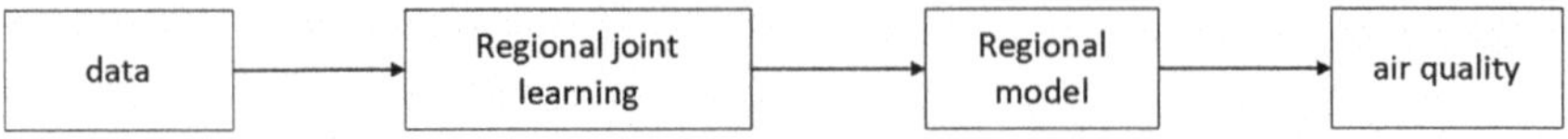

Fig. 20.1 Monitoring process

be applied to the Internet of Things data, edge computing and other fields. They are very suitable for the analysis of perceptual data [2].

20.2 Theoretical Basis

20.2.1 Overall System Design

This project aims to leverage the advantages of federated learning to study a new air pollution monitoring method based on Federated Learning Across Regions [3], which applies it to air pollution monitoring while ensuring the accuracy of existing distributed air pollution models. While introducing relevant technologies, an analysis was also conducted on the shortcomings of some existing technologies in model training efficiency, and it was pointed out that the quality of distributed training models often cannot surpass that of centralized training models. The overall idea of the system is to effectively and reasonably partition its edge devices, as well as air quality and image datasets, based on the framework of federated learning distributed algorithms. At the same time, during the training process, the focus is on the models trained by the edge devices representing each region, which we refer to as regional models, rather than just focusing on the shared models jointly trained by regional models. The overall monitoring process is shown in Fig. 20.1. Through this method, in a distributed framework, not only can the learning efficiency of the model be greatly improved, but the model can also better reflect the regional attributes (local attributes) of the region, thereby greatly improving the accuracy of the model; It is also possible to consider the common attributes (universal attributes) of the overall data, in order to avoid a decrease in model quality caused by a small sample size in a single region.

20.2.2 Regional Division Module Design

Adopting a suitable and efficient partitioning method can maximize the universality of partition data in each partition, thereby improving the performance of the partition model trained in the partition and facilitating better data management at the edge of the micro cloud in each partition [4]. In order to achieve more efficient partitioning, a weighted GN algorithm is adopted, which is suitable for clustering small and medium-sized communities and has good performance and effectiveness. Because data sampling points are usually calibrated by surrounding air monitoring stations, the number of sampling points is small, and not all air quality monitoring stations can work properly, meeting the requirements of small and medium-sized groups. In addition, we can also see that in some areas, the spacing between monitoring stations is too small, resulting in redundancy between monitoring stations, while in some areas, there is a lack of monitoring stations. This makes it more necessary to divide these stations to form an independent or common area, so as to make the data in the area closer and more complete.

Firstly, the weighted GN algorithm is used to partition the sample stations, that is, to partition the partitioned area and the nodes and edges within the area. In this project, the distribution map of air pollution monitoring stations is treated as a graph G to be segmented, and the λ Consider each site as a node group V, and then connect these nodes with an edge to form an edge group E. E_{ij} refers to node i and node j (i, j Σ [1,2, λ]) An edge between nodes, where $E_{ij} = 0$ represents non connectivity, $E_{ij} = 1$ represents connectivity, and the edge weight is set to A, A_{ij} represents the edge weight between node i and node j. To express the relationship between any node, the system changes the correlation rij based on the distance between node i and node j, as well as the air quality between nodes, and sets a weight value A. Here, the correlation degree rij of air quality changes between two nodes is represented as the Pearson correlation degree, as shown in equation:

$$r_{ij} = \frac{\operatorname{cov}(v_{i,} v_j)}{\sqrt{\operatorname{var}[v_i]\operatorname{var}\left[v_j\right]}} \tag{20.1}$$

In the formula, V_i represents the daily average air quality dataset of node i, and cov (v_i, v_j) represents the covariance of two quantities taken in parentheses. It is obvious that the correlation coefficient r_{ij} should be set to be directly proportional to the weighting, while the distance d_{ij} should be inversely proportional to the weighting. Therefore, a weighted A_{ij} should be defined between node i and node j, as shown in equation:

$$A_{ij} = \frac{r_{ij}}{d_{ij}} \tag{20.2}$$

Next, we can continue to define the core function in the GN algorithm, which represents the clustering degree Q function, as shown in Eq. (20.3):

$$Q = \frac{1}{2M}\sum\nolimits_{j,j}\left[\left(A_{ij} - \frac{k_i k_j}{2M}\right)\delta\left(\sigma_i - \sigma_j\right)\right] \tag{20.3}$$

20.2.3 Model Union Module Design

The model integration module is generally deployed on a central server, and its function is not limited to integrating the data uploaded by various sub terminals. In the design of the model joint module, on the one hand, it is responsible for collaborative training of all regional models to ensure that the system can perform distributed model training normally; On the other hand, it is responsible for regulating the overall situation, including parameter changes, instruction issuance, data allocation, and some settings during transmission. The most core task of the partition mode training module is to use the data from each partition to train the partition mode for the micro clouds in each partition. The module needs to have corresponding interfaces for accessing and setting global parameters sent by the central server, and training the model according to the parameter requirements. Next, it is necessary to select and design the neural network required to train the region model according to the requirements. The neural network mainly consists of two parts: a time series model used to analyze historical weather and air quality data, and a convolutional neural network used to analyze sky photo data. In addition, this module also needs to implement the client program in the regional federated learning algorithm. The client's job is to download the global model sent from the server, train the downloaded global model with local data, update the two neural networks above, and finally obtain a new model. It is then uploaded to the server to predict the PM2.5 two-layer LSTM recurrent neural network as shown in Fig. 20.2 [5, 6]. In addition, the client program also needs to have the ability to collect and process experimental data, which can be achieved through embedded preprocessing modules. After training the new model, its parameters and accuracy are recorded and uploaded, allowing the central server to make quick judgments, thereby improving the efficiency of the system.

The model adopts a multi-level structure, which makes it highly adaptable, hierarchical, and has strong learning ability. This system can only monitor the air quality within Ningxia at present, but if this system is expanded nationwide, it may improve training efficiency and save a lot of data transmission costs.

20.3 Implementation of Air Quality Prediction System Based on Federated Learning

20.3.1 Technology Roadmap

We jointly developed an air quality prediction platform using Python and TensorFlow, and built an efficient and accurate air quality prediction system through Python's flexible programming capabilities and TensorFlow's powerful machine learning capabilities. We will utilize Python's rich libraries and tools to process data, build system frameworks, and combine TensorFlow's deep learning technology for air quality model training and optimization, in order to achieve accurate prediction of air quality.

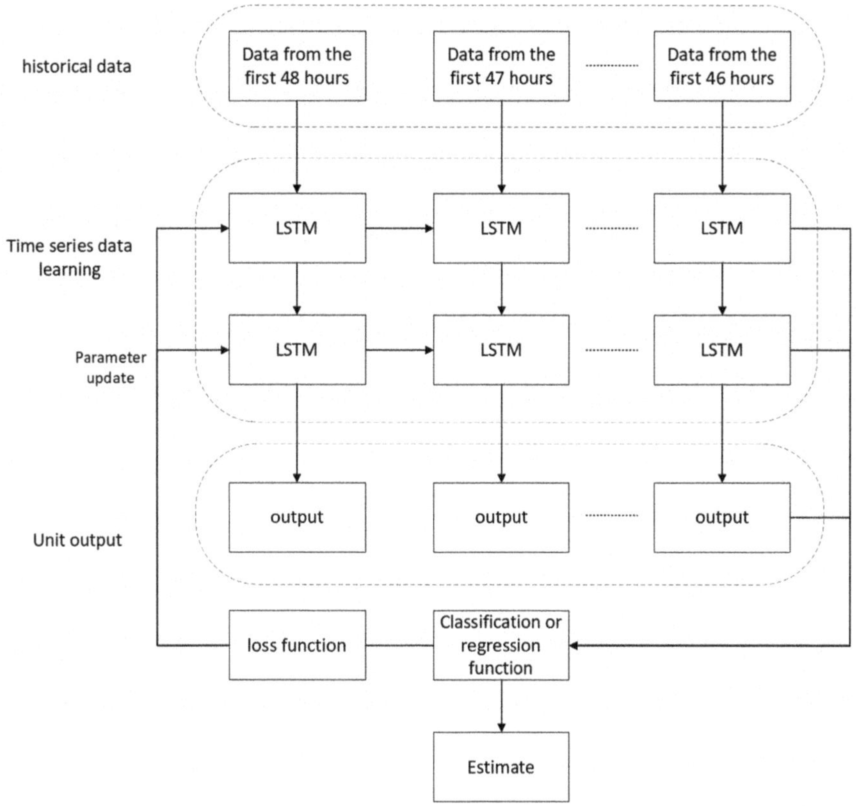

Fig. 20.2 Two layer LSTM recurrent neural network

Write a data collection program using Python to obtain environmental data such as air quality monitoring data, meteorological data, etc. Clean, transform, and integrate data using Python's data processing libraries (such as Pandas and NumPy) to prepare for subsequent modeling [7]. Using Python for feature engineering, including feature selection, feature extraction, and feature transformation. By analyzing the importance of data features and extracting features related to air quality, effective features are provided for model training. Build machine learning or deep learning models using TensorFlow. Select appropriate models based on the characteristics of the data, and train and optimize the models to achieve accurate prediction of air quality. Use Python for model evaluation, use cross validation and other methods to evaluate model performance, and optimize the model to improve prediction accuracy and generalization ability.

20.3.2 Implementation of Data Preprocessing Module

Data preprocessing plays a crucial role in air quality prediction platforms, and can be implemented through Python programming language and related libraries to remove non-standard data such as missing values and outliers.

```
import pandas as pd
data = pd.read_csv('air_quality_data.csv')
data.dropna(inplace = True)
data = data[(data['PM2.5'] > = 0)& (data['PM2.5'] < = 500)]
features = ['Temperature', 'Humidity', 'Wind_Speed', 'PM2.5']
data_selected = data[features]
from sklearn.preprocessing import MinMaxScaler
scaler = MinMaxScaler()
data_scaled = scaler.fit_transform(data_selected)
from sklearn.model_selection import train_test_split
X = data_scaled[:,:-1]
y = data_scaled[:, -1]
X_train, X_test, y_train, y_test = train_test_split(X, y, test_size = 0.2, random_
state = 42)
```

Through the above steps, the data preprocessing module is implemented to provide clean and standardized data for subsequent model training.

20.3.3 Implementation of Regional Division Module

The region segmentation module mainly completes the segmentation of existing samples using the weighted GN algorithm, in order to obtain the segmented adjacency matrix P_{ij}. In implementation, edges are only established when the correlation coefficient $r_{ij} > 0.95$ and the distance $d_{ij} < 10$ between two nodes i and j. Use the weighted GN algorithm to perform region segmentation on samples, load a dataset containing sample data, which includes the feature information and position information of each sample, and calculate the similarity matrix based on the similarity between samples. Using the weighted GN algorithm to segment the similarity matrix, samples with higher similarity are grouped into the same region. The process and code of using weighted GN algorithm for region segmentation of samples are as follows:

```
import numpy as np
from sklearn.metrics.pairwise import euclidean_distances
def weighted_gn_algorithm(similarity_matrix, threshold):
  n = similarity_matrix.shape[0]
  labels = np.arange(n)
  while True:
```

```
max_similarity = np.max(similarity_matrix)
if max_similarity < threshold:
  break
    i, j = np.unravel_index(np.argmax(similarity_matrix), similarity_
matrix.shape)
labels[labels = = labels[j]] = labels[i]
  similarity_matrix[i,:] = np.maximum(similarity_matrix[i,:], similarity_
matrix[j,:])
similarity_matrix[:,i] = np.maximum(similarity_matrix[:,i], similarity_matrix[:,
j])
similarity_matrix[i, i] = 0
  return labels.
data = np.array([[0, 0], [1, 1], [2, 2], [5, 5] [6, 6], [7, 7]])
similarity_matrix = euclidean_distances(data)
threshold = 2.0
labels = weighted_gn_algorithm(similarity_matrix, threshold)
print(labels)
```

20.3.4 Implementation of Model Training Module

The basic architecture of the model training module is relatively simple. It is mainly responsible for training it based on the global hyperparameters received from the central server. The machine learning training model is shown in Fig. 20.3 [9].

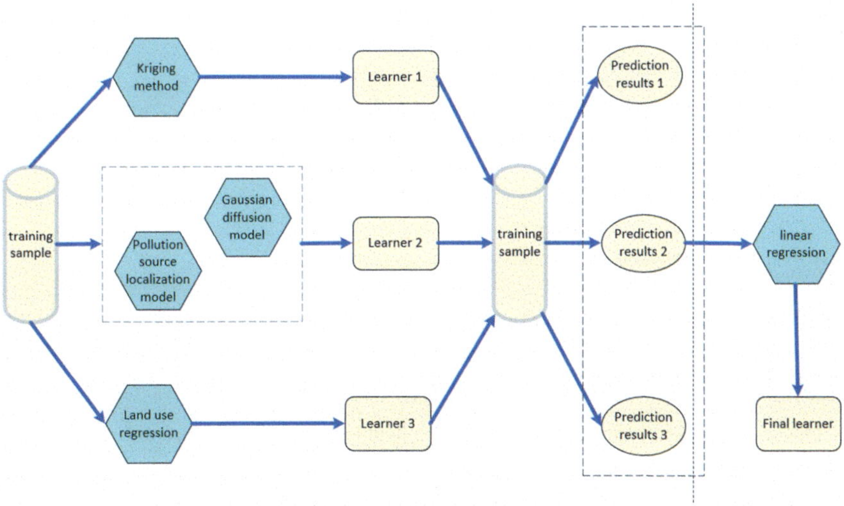

Fig. 20.3 Federated Learning Training Model

We plan to use a Stacking learner that integrates pollution source localization model, Gaussian diffusion model, Kriging method, land use regression, and linear regression [8]. The Stacking learner will use three primary learning algorithms, namely Kriging method, land use regression, and a joint model that combines pollution source localization model and Gaussian diffusion model. These three primary learning algorithms will make predictions on the training samples in different ways and obtain different prediction results as inputs for the secondary learning algorithm. Among them, the joint model first uses a pollution source localization model to locate the pollution source, and then uses a Gaussian diffusion model to predict the concentration of pollutants at any spatial point. The Stacking learner will use linear regression as the secondary learning algorithm to train the final learner based on the prediction results of the primary learning algorithm.

In the specific implementation, the high-level API of TensorFlow platform TFLearn was used to provide a detailed description of the designed network [10]. In the TFLearn library, there are already mature basic neural network layers such as input layer, output layer, fully connected layer, etc., and standard LSTM layer interfaces are directly provided. When in use, it only needs to be adjusted. Among them, the most important one is TFlearn. The key parameters of the LSTM interface are described in Table 20.1. When setting them, due to the limited data dimensions, the number of neurons was set to a commonly used 128. Moreover, since no changes were made to the internal structure of the LSTM, the activation function was left as default. In addition, since this method is a two-layer least squares method, all data of this method must be passed back to the next layer in the first layer, so return_seq is set to true. The global hyperparameters E, B, and C of the system are set in Table 20.2.

Table 20.1 TFlearn.LSTM Parameter details

Parameter name	Parameter function description	data type
incoming	Input the 4-dimensional tensor of this layer	Ternsor
nb_blocks	The number of blocks in this layer	int
bottleneck_size	Convolutional kernels	int
out_channels	Number of output channels	int
downsample	Whether to downsample	bool
downsample_stride	Downsampling step size	int

Table 20.2 Global hyperparameter settings

Parameter name	Parameter function description	data type
E	Epoch during training model at the control region sub end	int
B	Mini batch for training models at the sub end of the control region	int
C	Control how many ratios of region endpoints will be trained in each round	float
Buffer size	The size of data cached each time during transmission	int
IPAddr; Port	Store the IP addresses and port numbers of all machines	String

20.4 Conclusion

The focus of this article is to design and implement an air pollution monitoring system based on federated learning. The line graph of AQI and PM2.5 data predicted using Python is shown in Fig. 20.4. It can be seen that the predicted data shows that there will be significant fluctuations in the values of AQI and PM2.5 in the future. Through analysis, we conclude that this is related to the complex environment and perceived factors in the northwest region. But the overall PM2.5 value is 75 μ G/m^3 Below, the air quality is relatively good, and the PM2.5 value is occasionally around 75 μ G/m^3 Above, there will be slight pollution.

Adopting a distributed model training method based on regional federated learning to solve the problem of low training efficiency in current centralized training methods. The system divides the data from different sampling points into different regions, configures distributed training sub terminals in each region, and predicts the air quality in that region through the regional model of the regional terminals, thereby

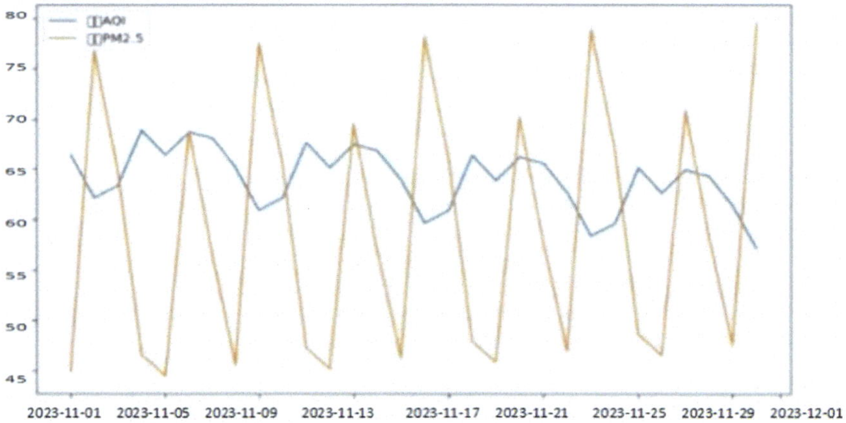

Fig. 20.4 AQI and PM2.5 prediction

improving the training efficiency and accuracy of the model. Users can use this system to real-time understand the air quality status of various monitoring stations in Yinchuan City, analyze the trend of air quality changes, and provide strong data support for environmental protection and healthy living.

Acknowledgements Fund Project: Ningxia Natural Science Foundation Project.

Project Name: Research on Mobile Group Intelligence Perception Air Quality Analysis Model and Prediction System.

Project number: 2022AACO3346.

References

1. Li J (2024) Research and practice on deep learning model for air quality prediction. Intell Comp Appl 7
2. Fang W, Li F (2021) Design and development of remote indoor air monitoring system based on Yeelink. Comp Program Skills Maintenance (10):26–27
3. Lei T (2020) Design and implementation of air quality prediction system based on federated transfer learning. Beijing University of Posts and Telecommunications, vol 4, pp 5–20
4. Hu B (2019) Design and implementation of air quality monitoring system based on federated learning, vol 5, pp 10–25
5. Bai S, Shen X (2019) PM2.5 prediction based on LSTM recurrent neural network. Comp Appl Softw 1
6. Wang J, Song Y, Wang T (2023) Prediction model based on LSTM network and error compensation. Comp Technol Dev
7. Liu Y, Cao Y (2022) Air quality visualization platform integrating CNN-LSTM prediction model. Inform Technol Informatization 4
8. Shi J, Dou Y (2023) Air quality index prediction in urumqi based on stacking fusion model. J Agric Catastrophol 4
9. Wang X, Yu X, Wang T (2021) Prediction of the impact of air pollution in chemical industrial parks based on ensemble learning strategy. Oper Res Manag Sci 11
10. Yang C, Wang Y (2021) The application of LSTM model based on TensorFlow in the prediction of air quality index. Digit Technol Appl 3

Chapter 21
Experimental Study on the Acidification Purification Technology of a Certain Cyanide Containing Wastewater

Yanbo Chen, Guangsheng Li, Xingfu Zhu, and Junjie Wang

Abstract For a certain cyanide containing wastewater, acidification method was used for purification treatment. pH, aeration rate, reaction time, and reaction temperature conditions were tested successively to determine the optimal reaction conditions. The concentrations of Cu, Zn, and Fe heavy metal elements in the cyanide containing wastewater were reduced to below 14.22 mg/L, 1.12 mg/L, and 326.40 mg/L, respectively, achieving the expected purification goals.

Keywords Cyanide containing wastewater · Acidification method · Purification treatment

21.1 Introduction

Cyanide containing wastewater refers to industrial wastewater containing cyanide, which is produced in gold smelting, coke oven steelmaking, electroplating, and other industries. Cyanide is a highly toxic substance that can be fatal if absorbed by the human body at a concentration of 0.4 mg/kg to 3.6 mg/kg. During the cyanide leaching process of gold bearing ores, the accompanying compounds such as sulfur, iron, copper, zinc, or their decomposition products in the ore will directly or indirectly react with CN^-, O_2, OH^- in the wastewater, generating various impurity anions, mainly including free cyanide CN^-, copper cyanide complex $Cu(CN)_3^{2-}$, zinc cyanide complex $Zn(CN)_4^{2-}$, iron cyanide complex $Fe(CN)_6^{4-}$, and thiocyanate SCN^-. These reactions not only consume a large amount of cyanide, leading to a significant increase in the cost of gold leaching, reducing the leaching rate of gold, but also making the composition of cyanide containing wastewater more complex, which is not conducive to comprehensive treatment. The concentrations of Cu, Zn, and Fe in a certain cyanide containing wastewater reached 2600 mg/L,600 mg/L, and

Y. Chen (✉) · G. Li · X. Zhu · J. Wang
Selection and Metallurgy Laboratory of Shandong Gold Mining Technology Co. Ltd, Yantai, Shandong, China
e-mail: 595907901@163.com

K. Y. Foo et al. (eds.), *Water Nexus Approach for Sustainable Development*, Environmental Science and Engineering, https://doi.org/10.1007/978-3-031-89238-7_21

Table 21.1 Analysis results of cyanide containing wastewater

Assay elements	Cu	Zn	Fe
Element content (%)	1780.0	870.5	1441.0

1200 mg/L, respectively, and the concentration of thiocyanate ions reached 20 g/L. At present, no measures have been taken to purify the cyanide solution, which has caused a large amount of valuable metal accumulation in production, affecting the economic benefits of the enterprise; The second is to increase the zinc consumption in zinc powder replacement operations, resulting in high levels of impurity metals such as copper, zinc, and iron in the gold mud, which increases the difficulty and cost of subsequent smelting [1–3].

21.2 Sample Analysis

The analysis results of cyanide containing wastewater samples are shown in Table 21.1.

21.3 Experimental Research

The experiment uses dilute sulfuric acid to adjust the pH, and the system conducts pH conditions, inflation rate conditions, temperature conditions, and time conditions tests [4].

21.3.1 pH Condition Test

Examine the treatment effect of acidification method under pH conditions of 2, 3, 4, 5, and 6, and determine the optimal pH [5].The experimental process and conditions are shown in Fig. 21.1, and the experimental results are shown in Table 21.2 and Fig. 21.2.

According to Table 21.2, under non inflating conditions, within the pH range of 2–5, the Zn concentration is below 2 mg/L, achieving the expected target (reducing to below 10 mg/L); The concentrations of Cu and Fe increase with the increase of pH, and although the expected goals were not achieved at pH 2 (Cu dropped below 20 mg/L and Fe dropped below 200 mg/L), the concentrations were the lowest, and the pH value was determined to be 2.

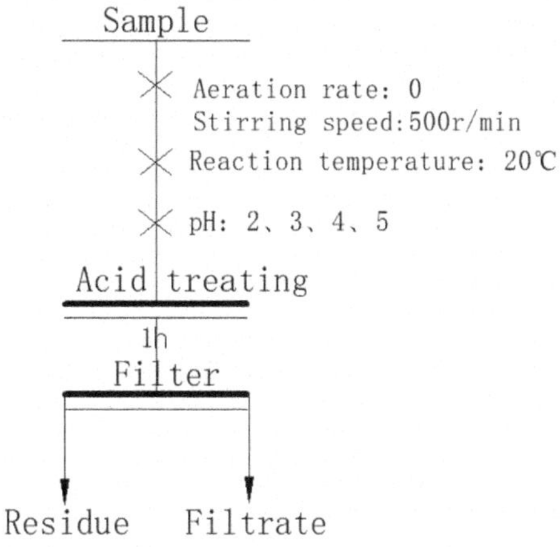

Fig. 21.1 pH condition test process of acidification method

Table 21.2 Results of pH condition test using acidification method

pH	Assay elements (mg/L)		
	Cu	Zn	Fe
1	52.82	1.58	418.7
2	126.87	1.34	510.5
3	1035.38	1.25	754.9
4	1345.07	1.82	1012

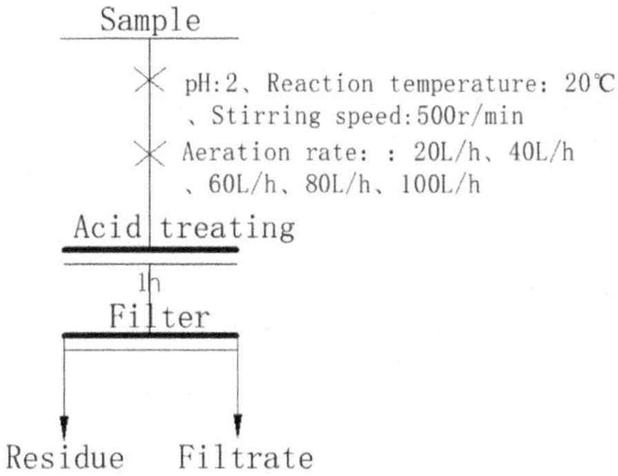

Fig. 21.2 Aeration rate condition test process

21.3.2 Aeration Rate Condition Test

Examine the acidification treatment effect under the conditions of Aeration rates of 20 L/h, 40 L/h, 60 L/h, 80 L/h, and 100 L/h, and determine the optimal inflation rate [6]. The experimental process and conditions are shown in Fig. 21.2, and the experimental results are shown in Table 21.3.

According to Table 21.3, within the range of 0–100 L/h per liter of water, the concentrations of Cu and Fe gradually decrease as the inflation rate increases. When

Table 21.3 Aeration rate condition test results

Aeration rate (L/h)	Assay elements (mg/L)		
	Cu	Zn	Fe
0	52.82	1.58	418.7
20	34.49	1.8	376.1
40	17.16	1.01	289.4
60	16.78	1.24	225.7
80	14.42	1.12	176
100	11.22	0.75	154.2

the inflation rate is 40L/h per liter of water, the Cu concentration drops to below 20 mg/L, achieving the expected target. The Fe concentration is 289.40 mg/L, which is close to the expected target of 200 mg/L. Taking into account the inflation cost comprehensively, the inflation rate is determined to be 40L/h per liter of water.

21.3.3 Reaction Temperature Condition Test

The boiling point of hydrogen cyanide is 26 °C. Set a temperature above one boiling point (30 °C) and compare the treatment effect of acidification method at room temperature (20 °C) [7]. The experimental process and conditions are shown in Fig. 21.3, and the experimental results are shown in Table 21.4.

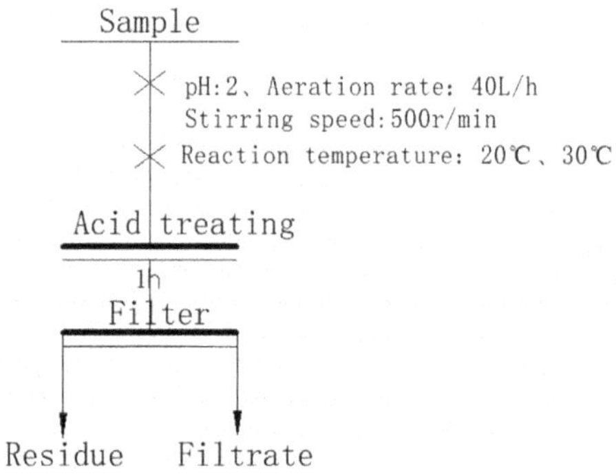

Fig. 21.3 Temperature condition test process

Table 21.4 Temperature condition test results

Temperature (°C)	Assay elements (mg/L)		
	Cu	Zn	Fe
20	17.16	1.01	289.40
30	12.82	1.58	268.70

According to Table 21.4, compared to 20 °C, under the condition of 30 °C reaction temperature, the concentrations of Cu and Fe did not significantly decrease. Considering the increase in treatment cost due to heating, the reaction temperature was determined to be 20 °C.

21.3.4 Reaction Time Condition Test

Examine the effect of reaction time on the treatment effect of acidification method and determine the optimal reaction time. The experimental process and conditions are shown in Fig. 21.4, and the experimental results are shown in Table 21.5.

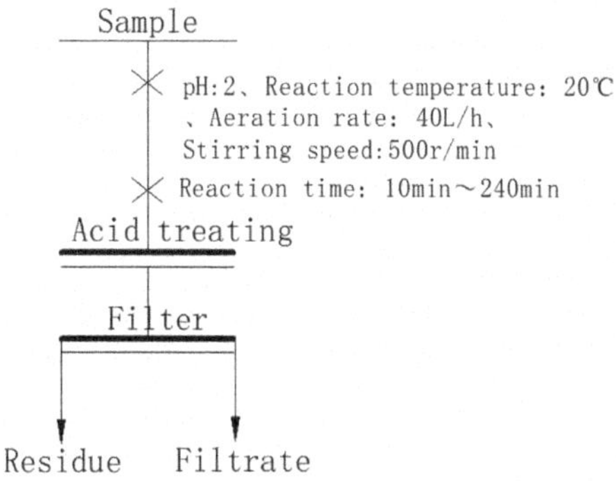

Fig. 21.4 Reaction time condition test process

Table 21.5 Reaction time condition test results

Reaction time (min)	Assay elements (mg/L)			
	Cu	Zn	Fe	FCN
Sample	1780.00	870.50	1441.00	3204.51
10	18.18	0.94	543.60	3505.58
20	15.37	0.62	524.50	2853.38
30	14.04	0.10	552.30	2337.05
40	13.18	0.34	500.00	2038.13
50	12.35	0.60	457.10	1793.55
60	10.97	0.53	433.80	1467.45
80	9.13	0.85	377.60	913.08
100	13.31	2.22	345.40	597.85
120	11.24	2.46	324.00	524.20
140	8.09	2.15	301.40	452.20
180	8.20	0.39	185.10	369.58
240	6.95	0.46	161.80	244.58

According to Table 21.5, within the reaction time range of 10–240 min and with a reaction time of 10 min, the concentrations of Cu and Zn decrease to below 20 mg/L and 10 mg/L, respectively. As the reaction time prolongs, the Fe concentration gradually decreases, and the FCN concentration first slightly increases (at the beginning of the experiment, some chelated cyanide dissociates into free cyanide, leading to an increase in free cyanide concentration), and then gradually decreases, and the decrease amplitude gradually decreases. Due to the limited effect of long reaction time on reducing Fe and FCN concentrations and increasing electricity costs, it is advisable to choose a reaction time of 120 min.

21.3.5 Comprehensive Condition Test

Conduct comprehensive condition experiments according to the determined optimal parameters to examine the purification effect of lean liquid. The experimental process and conditions are shown in Fig. 21.5, and the experimental results are shown in Table 21.6.

According to Table 21.6, the optimal conditions for the acidification stage of a certain cyanide lean solution are determined as follows: pH 2, inflation rate of 40 L/h per liter of water, reaction temperature of 20 °C, and reaction time of 120 min. Under optimal conditions, the concentrations of Cu, Zn, and Fe in a certain cyanide lean solution were reduced from 1780.0 mg/L, 870.5 mg/L, and 1441.0 mg/L to 14.22 mg/L, 1.12 mg/L, and 326.40 mg/L, respectively.

Fig. 21.5 Comprehensive condition test process

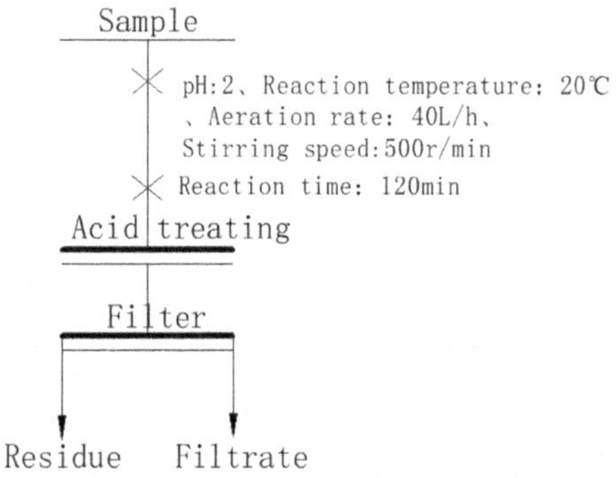

Table 21.6 Comprehensive condition test results

Assay elements	Cu	Zn	Fe
Element content (%)	14.22	1.12	326.40

21.4 Conclusion

The optimal conditions for the "acidification method" treatment stage of a certain cyanide lean solution are determined to be: pH 2, inflation rate of 40 L/h per liter of water, reaction temperature of 20 °C, and reaction time of 120 min. Under optimal conditions, the concentrations of Cu, Zn, and Fe in a certain cyanide lean solution were reduced from 1780.0 mg/L, 870.5 mg/L, and 1441.0 mg/L to 14.22 mg/L, 1.12 mg/L, and 326.40 mg/L, respectively.

References

1. Jiang K et al (2023) Optimization and application of acidification process for cyanide containing wastewater. Gold 44(9):123–127
2. Li X et al (2023) Research and application of advanced treatment technology for cyanide containing wastewater in gold mines. Gold 44(7):117–120
3. Ebel M, Evangelou MWH, Schaeffer A (2007) Cyanide phytoremediation by water hyacinths (*Eichhornia crassipes*). Chemosphere 66(5):816–823
4. Estay H, Ortiz M, Romero J (2013) A novel process based on gas filled membrane absorption to recover cyanide in gold mining. Hydrometallurgy 134–135:166–176
5. Feng X et al (2023) Research progress on purification and recovery process of cyanide containing wastewater in gold mines. Gold 44(4):80–85
6. Shan Z et al (2022) Experimental study on copper salt precipitation method for treating high concentration cyanide containing wastewater. Gold 43(7):84–87
7. Alvarez R, Ordonez A, Loredo J et al (2013) Wetland-based passive treatment systems for gold ore processing effluents containing residual cyanide, metals and nitrogen specie. Environ Sci Process Impact 15:2115–2124

Chapter 22
Experimental Study on the Adsorption of Cr(VI) in Wastewater by Self-Made Black Tea Residue Activated Carbon

Jinyong Huang

Abstract Tea residue activated carbon was prepared using discarded black tea leaf residue and zinc chloride activation. Analytical methods such as infrared spectroscopy and X-ray diffraction were utilized, combined with theories of adsorption kinetics and thermodynamics, to study and discuss the characteristics of Cr(VI) adsorption by black tea residue activated carbon from wastewater, as well as the effects of oscillation time, dosage, pH, and other factors on the adsorption process. The results showed that the optimal conditions for Cr(VI) removal by black tea residue activated carbon are: an oscillation time of 60 min, an activated carbon dosage of 0.4 g, a solution temperature of 40 °C, a solution pH of 2, and an initial Cr(VI) concentration of 50 mg/L.

Keywords Black tea residue activated carbon (black tea residue AC) · Cr(VI) · Adsorption

22.1 Introduction

Chromium (Cr) is one of the most widely used heavy metals in the industrial field, and its usage has gradually increased with the expansion of industrial production scales. Due to the previous lax environmental management, a significant amount of Cr has entered the soil and groundwater, posing serious threats to environmental safety and human health [1]. Chromium in the natural environment mainly exists in two valence states: Cr(III) and Cr(VI). Among them, Cr(III) has weaker mobility and lower toxicity, causing less harm in the environment [2], while Cr(VI) is a confirmed high-risk carcinogen [3].

With the development of industry and the enhancement of environmental awareness, water treatment issues have increasingly become problems that need to be directly faced and resolved. The treatment of heavy metal wastewater and the hazards

J. Huang (✉)
College of Chemical and Materials Engineering, Beijing Institute of Technology, Zhuhai, China
e-mail: 294244481@qq.com

K. Y. Foo et al. (eds.), *Water Nexus Approach for Sustainable Development*, Environmental Science and Engineering, https://doi.org/10.1007/978-3-031-89238-7_22

of eutrophication to the environment and human health are undeniable, and how to better handle water pollution issues has always been a focus of research for scholars at home and abroad. According to the data from the first national pollution source census, chromium pollution ranks second among heavy metal pollutions, with a production volume (in 2010) of 4906.012 tons and an emission volume of 94.987 tons [4]. Among them, hexavalent chromium is one of the most toxic pollutants, which can cause cellular mutations in the human body, thereby endangering human health. How to efficiently and reasonably handle chromium-containing wastewater is a hot topic of research among scholars both domestically and internationally [5].

This article uses the optimal combination of prepared black tea residue activated carbon(black tea residue AC), and by changing conditions such as solution pH value, initial solution concentration, and the dosage of activated carbon, studies the impact of black tea residue activated carbon on the Purification rate of Cr(VI) and total phosphorus of water. In addition, it compares the removal rate of Cr(VI) by commercial activated carbon(commercial AC) adsorption.

22.2 Materials and Methods

22.2.1 Raw Materials and Reagents

The activated carbon prepared from black tea residue (prepared under optimal conditions with zinc chloride method and sieved through a 200-mesh screen), commercial activated carbon, potassium dichromate (AR), sodium hydroxide (AR), sulfuric acid (AR), phosphoric acid (AR), diphenylcarbazide (AR); sodium thiosulfate (AR), ascorbic acid (AR), ammonium molybdate (AR), potassium sodium tartrate (AR), sulfuric acid (AR), sodium hydroxide (AR).

22.2.2 Instrumentations

The UV757CRT type UV-Vis spectrophotometer, CP214 electronic balance, SHA-C type constant temperature oscillator, and GZX-9240MBE type electric heating constant temperature air circulation drying oven.

22.3 Results and Discussion

22.3.1 *The Effect of Adsorption Time on the Adsorption of Cr(VI)*

Take 0.3 g of black tea residue activated carbon and 0.3 g of commercial activated carbon, respectively, and place them into 250 mL conical flasks. Add 30 mL of Cr(VI) solution of 100 mg/L to each flask, then place them into a constant temperature oscillator. Adjust the temperature to 25 °C and set the oscillation speed to 175r/min. The oscillation times are 30 min, 60 min, 90 min, 120 min, 150 min, 180 min, 210 min, and 240 min. After filtration, measure the absorbance of the Cr(VI) solution and use the Cr(VI) standard curve to determine the Content of Cr(VI) in the filtrate. Then, calculate the purification rate of Cr(VI). The results are shown in Fig. 22.1.

From Fig. 22.1, it can be observed that the removal rate of Cr(VI) increases with the extension of oscillation time. This is because when the time is less than 60 min, the adsorption of Cr(VI) by the activated carbon has not reached saturation. When the time reaches 60 min, the removal rate of Cr(VI) by the black tea residue activated carbon reaches its maximum. If the oscillation time is increased further, the Cr(VI) adsorbed inside the activated carbon may undergo desorption, leading to a decrease in the removal rate. Therefore, an oscillation time of 60 min is appropriate. The optimal oscillation time for commercial activated carbon is also 60 min.

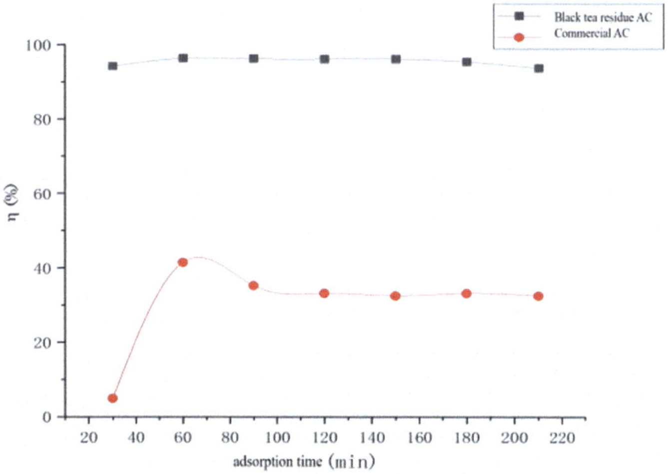

Fig. 22.1 The effect of adsorption time

22.3.2 The Impact of Different Dosages on the Adsorption of Cr(VI) by Black Tea Residue Activated Carbon

Weigh out 0.1 g, 0.2 g, 0.3 g, 0.4 g, 0.5 g, and 1.0 g of black tea residue activated carbon and commercial activated carbon, respectively, and add them to 250 mL conical flasks. Then, using a pipette, take up 30 mL of a 100 mg/L Cr(VI) solution and add it to the flasks. Place the flasks into a constant temperature oscillator, set the temperature to 25 °C, and adjust the oscillation speed to 175r/min. Oscillate for 60 min. Afterward, filter the sample and measure the absorbance. Use the Cr(VI) standard curve to determine the concentration. Calculate the purification effect of Cr(VI) according to Eqs. 1–2. The results are shown in Fig. 22.2.

From Fig. 22.2, it can be seen that the impact of the dosage of black tea residue activated carbon on the removal rate of Cr(VI) is less than that of commercial activated carbon. When the dosage of the former reaches 0.4 g and the latter reaches 0.3 g, the removal rates of Cr(VI) reach their maximum values, which are 99.86% and 37.83%, respectively. The removal rate of Cr(VI) by black tea residue activated carbon does not change significantly when the dosage exceeds 0.3 g. The reasons for this could be the agglomeration or aggregation between activated carbon particles, or the electrostatic interactions between the adsorbate and the binding sites on the adsorbent [6].

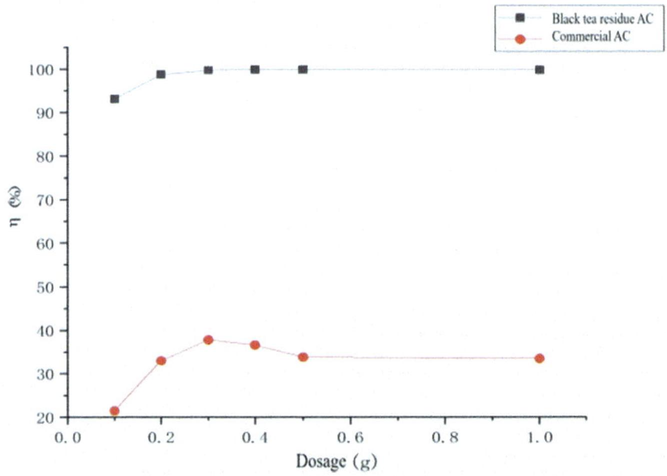

Fig. 22.2 The impact of different dosages

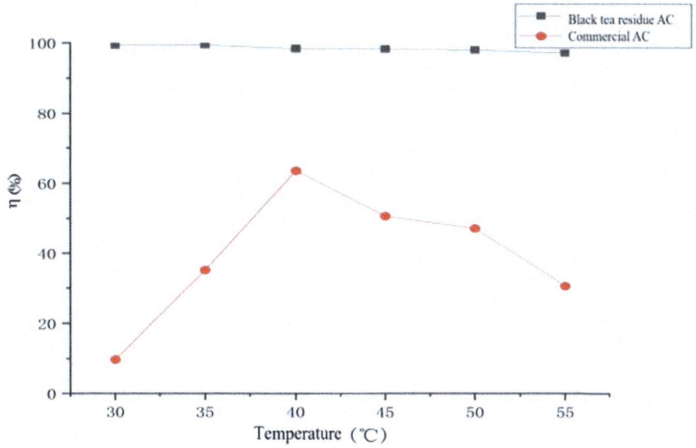

Fig. 22.3 The effect of temperature on the removal rate

22.3.3 The Effect of Temperature on the Removal Rate of Adsorbed Cr(VI)

Weigh out 0.4 g of black tea residue activated carbon and 0.3 g of commercial activated carbon, and add them to a 250 mL conical flask. Use a pipette to take up 30 mL of a 100 mg/L Cr(VI) solution and pour it into the conical flask. Then place the flask into a constant temperature oscillator, set the temperatures to 30 °C, 35 °C, 40 °C, 45 °C, and 50 °C respectively, and adjust the oscillation speed to 175r/min with an oscillation time of 60 min. Afterward, filter the sample and measure the absorbance. Use the Cr(VI) standard curve to determine the content of Cr(VI). Then, calculate the purification effect of Cr(VI) according to Eqs. 1–2. The results are shown in Fig. 22.3.

From Fig. 22.3, it can be observed that the change in temperature does not significantly affect the removal rate of Cr(VI) by black tea residue activated carbon. In contrast, for commercial activated carbon, the removal rate of Cr(VI) initially increases and then decreases with the rise in the temperature of the wastewater. This could be due to the less developed pore structure of commercial activated carbon compared to black tea residue activated carbon, hence the temperature has a slight influence on the removal rate. Both types of activated carbon perform best at a temperature of 40 °C.

22.3.4 The Effect of pH on the Adsorption of Cr(VI)

Weigh 0.4 g of black tea residue activated carbon and 0.3 g of commercial activated carbon and add them to a 250 mL conical flask. Then, using a pipette, take up

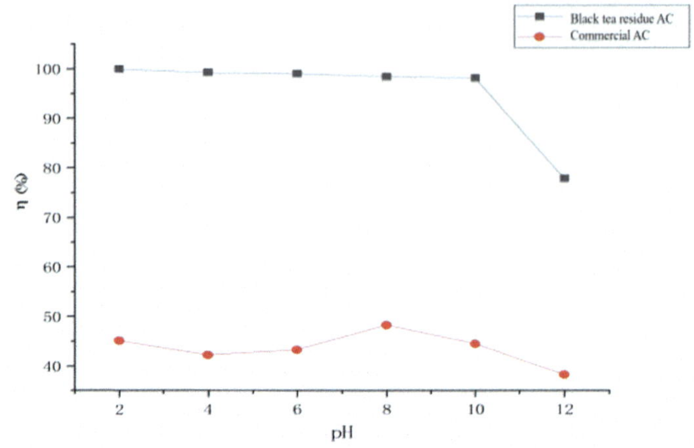

Fig. 22.4 The effect of pH on the adsorption of Cr(VI)

30 mL of a 100 mg/L Cr(VI) solution and add it to the flask. Adjust the pH of the solution to 2, 4, 6, 8, 10, 12, and 14, respectively. Subsequently, place the flask into a constant temperature oscillator, set the temperature to 40 °C, and maintain the oscillation speed at 175r/min with an oscillation time of 60 min. After filtration, measure the absorbance of the Cr(VI) solution obtained. Use the Cr(VI) standard curve to determine the quantity of Cr^{+6} in the filtrate. Calculate the removal rate of Cr(VI) according to Eqs. 1–2. The results are shown in Fig. 22.4.

From Fig. 22.4, it can be seen that as the pH value increases, the removal rate of Cr(VI) by black tea residue activated carbon gradually decreases, reaching its maximum at a pH of 2.0. The pH value primarily affects the form in which Cr(VI) exists as well as the surface charge of the adsorbent. At different pH values, the hexavalent chromium ions exist in different forms, such as H_2CrO_4 (aq), $Cr_2O_7^{2-}$, $HCrO_4^-$, $KCrO_4^-$, CrO_4^{2-}. When the pH is less than 6.51, the main form of Cr(VI) is $HCrO_4^-$; when the pH is greater than 6.51, the main form of Cr(VI) is CrO_4^{2-} [6, 7]. The above results indicate that Cr(VI) mainly exists in the form of anions.

When black tea residue activated carbon is added to the solution, Cr(VI), which exists in the form of negative charges, is attracted to the positively charged surface of the black tea residue activated carbon due to electrostatic attraction. This force is strengthened as the pH value decreases. The increase in solution pH also increases the electronegativity of the surface of the black tea residue activated carbon, thereby weakening the adsorption of hexavalent chromium by the activated carbon. When the pH value reaches 7.0, the OH^- in the solution begins to show a significant competitive adsorption with CrO_4^{2-} for the activated carbon. Therefore, the increase in solution pH value has an inhibitory effect on the ability of black tea residue activated carbon to adsorb Cr(VI) [8].

22.3.5 The Effect of Initial Concentration on the Adsorption of Cr(VI)

Weigh out 0.4 g of black tea residue activated carbon and add it to a 250 mL conical flask. Also, weigh 0.3 g of commercial activated carbon and add it to a separate 250 mL conical flask. Then, using a pipette, take up 30 mL of Cr(VI) solutions with concentrations of 50, 75, 100, 125, 150, and 175 mg/L, respectively, and add them to the corresponding flasks. Place the flasks into a constant temperature oscillator, set the temperature to 40 °C, and maintain the oscillation speed at 175r/min. Adjust the pH to 2 and oscillate for 60 min. Afterward, filter the solutions and measure the absorbance of the Cr(VI) solutions obtained. Use the Cr(VI) standard curve to determine the content of Cr^{+6} in the filtrates. Calculate the removal rate of Cr(VI) according to Eqs. 1–2. The results are shown in Fig. 22.5.

From Fig. 22.5, it can be observed that when the initial concentration of Cr(VI) is 50 mg/L, both types of activated carbon achieve the maximum removal rates for Cr(VI), which are 99.84% and 69.71%, respectively. As the initial concentration increases, the removal rates of Cr(VI) by both activated carbons gradually decrease. The reason for this is that the adsorption capacity of activated carbon is limited. Once a certain amount of Cr(VI) is adsorbed, saturation occurs. An excessively high initial concentration can affect the adsorption by the activated carbon, leading to a slower increase in the removal rate. Therefore, it is advisable to have an initial concentration of Cr(VI) in the wastewater of 50 mg/L.

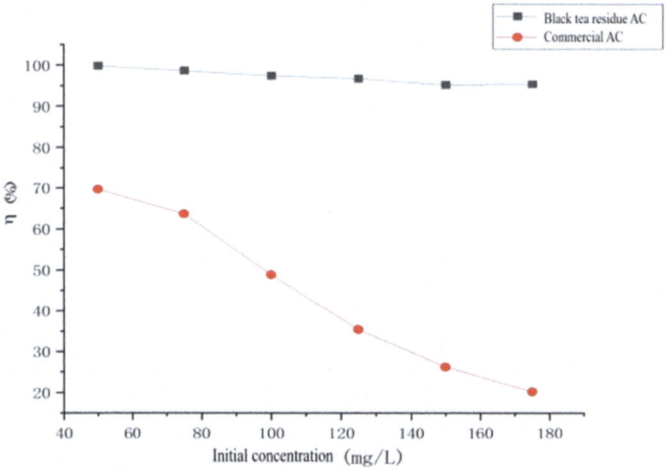

Fig. 22.5 The effect of initial concentration on the adsorption of Cr(VI)

Table 22.1 Kinetic equation for the adsorption of Cr(VI) onto black tea residue AC

Kinetic model	Parameter	Black tea residue AC (pH = 2)
Pseudo-first-order kinetics model	$q_{e,exp}$(mg/g)	0.95040
	$q_{e,cal}$(mg/g)	0.89035
	K_1(min^{-1})	0.29571
	R_1^2	0.54423
Pseudo-second-order kinetics model	$q_{e,exp}$(mg/g)	0.95040
	$q_{e,cal}$(mg/g)	0.96586
	K_2(min^{-1})	0.32334
	R_2^2	0.99964

22.4 Fitting Effect of Kinetic and Thermodynamic Models

22.4.1 Fitting Effect of Kinetic

The kinetic fitting results for the adsorption of Cr(VI) by activated carbon derived from black tea waste are presented in Table 22.1. The results indicate that the R_1^2 value for the pseudo-first-order kinetic equation is 0.54423, while the R_2^2 value for the pseudo-second-order kinetic equation is 0.99964, with the latter being significantly higher. This suggests that the adsorption of Cr^{+6} by AC derived from black tea waste is more accurately described by the pseudo-second-order kinetic model. Consequently, the role of chemical adsorption and the surface functional groups of the activated carbon is more pronounced during the purification process of Cr^{+6} by the black tea waste-derived activated carbon.

22.4.2 Fitting Effect of Thermodynamic Models

The isothermal adsorption fitting parameters and correlation coefficients for the adsorption of Cr(VI) by activated carbon derived from black tea waste are presented in Table 22.2. The results show that the R^2 value for the Langmuir model is 0.96245, while the R^2 value for the Freundlich model is 0.97919, with the latter being higher. This suggests that the thermodynamic process of adsorption by the activated carbon is more consistent with the Freundlich model. The adsorption process involves porous and multilayer molecular adsorption, along with the occurrence of chemical reactions.

Table 22.2 Thermodynamic equation for the adsorption of Cr^{+6} onto black tea residue AC

Adsorption model	Parameter	Black tea residue AC (pH = 2)
Langmuir model	Q_m(mg/g)	47.53093
	K_L(L/mg)	0.00684
	R^2	0.96245
Freundich model	K_F(mg/g/(L/mg)$^{1/n}$)	0.52928
	1/n	0.80012
	R^2	0.97919

22.5 Conclusion

The optimal factors for the removal rate of Cr(VI) by black tea residue activated carbon are as follows: the oscillation time is 60 min, the amount of activated carbon used is 0.4 g, the solution temperature is 40 °C, the pH is 2, and the initial content of Cr^{+6} is 50 mg/L. The increase in the pH value of the Cr(VI) solution has a significant impact on the adsorption capacity of the activated carbon for Cr(VI). When the pH is between 2 and 8, the change in the removal rate of Cr(VI) is not obvious. When the pH is between 10 and 14, the removal rate of Cr(VI) decreases significantly.

The adsorption of Cr(VI) by activated carbon derived from black tea waste is more consistent with pseudo-second-order kinetics, indicating that chemical adsorption and the role of surface functional groups on the black tea activated carbon are more pronounced in the adsorption of Cr(VI). The thermodynamic process of adsorption is more in line with the Freundlich model, suggesting that the adsorption process involves porous and multilayer molecular adsorption, along with the occurrence of chemical reactions.

Acknowledgements This work was financially supported by Outstanding On-Campus Course of Beijing Institute of Technology, Zhuhai.

References

1. Wu H (1986) Investigation on the health hazards of residents due to hexavalent chromium contamination in drinking water wells. J Environ Health (06):9
2. Zhu H (2018) Prevention and control of hexavalent chromium pollution in the electroplating industry. Shanxi Metall 41(04):59–61
3. Wang D, Wang S, Zhao X (2018) Current status and development of hexavalent chromium inhibitors in cement. J Chin Ceram Soc 37(05):1625–1631
4. Xu H (2022) Research on the remediation effect of organic materials on hexavalent chromium in soil and the application of landscaped soil after remediation. Beijing Academy of Environmental Sciences
5. Jin X (2018) Preparation of cassava sludge activated carbon and its adsorption study on hexavalent chromium. Anhui University of Technology

6. Huangg Q, Hu Y (2010) Current status and prospects of electroplating wastewater treatment technology. Guangdong Chem Ind 37(04):128–130
7. Sun J, Li J, Han L et al (2021) The magical family of activated carbon. Univ Chem 10:1–7
8. Duo Z (2018) Preparation of buckwheat hull-based activated carbon and its adsorption performance for hexavalent chromium. Beijing University of Chemical Technology, Beijing

Part III
Agriculture Water and Sustainability

Chapter 23
A Habitat Level Analysis Method for Port Shoreline Utilization Planning

Yi Hu, Min Dong, Chen Shen, Dachuan Wang, and Wenyuan Wang

Abstract It is very important to consider the impact of port shoreline utilization on the surrounding habit environment during port planning. Taking planning of Ningbo-Zhoushan Port as an example, this paper proposes a habitat level analysis method for port shoreline utilization planning based on the existing air quality monitoring, population distribution and other data. Factors of impact on the surrounding habit environment are considered during port planning. The research results can provide decision support for port shoreline utilization selection.

Keywords Habit environment · Port shoreline utilization planning · Analysis method · Ningbo-Zhoushan port

23.1 Introduction

In the new period, continuously satisfying people's aspirations for a better life, realizing people-oriented development and achieving development in harmony with nature have become the important starting and ending points for economic and social development, as well as the important prerequisites for the development of various industries.

The development and utilization of port shoreline may bring certain problems such as air and noise pollution, which will have impacts on the human environment of the surrounding area. In the past, port shoreline utilization planning mainly considered factors such as transportation demand, port construction conditions, collection and evacuation conditions, etc. [1–3], and lacked quantitative consideration of the impact on the human environment in the port area. The existing evaluation of the level

Y. Hu (✉) · M. Dong · C. Shen · D. Wang
Transport Planning and Research Institute, Ministry of Transport, Chaoyang, Beijing, People's Republic of China
e-mail: huyi@tpri.org.cn

W. Wang
State Key Laboratory of Coastal and Offshore Engineering, Dalian University of Technology, Ganjingzi, Dalian, Liaoning, People's Republic of China

K. Y. Foo et al. (eds.), *Water Nexus Approach for Sustainable Development*, Environmental Science and Engineering, https://doi.org/10.1007/978-3-031-89238-7_23

of habitat environment is also mostly based on the evaluation object of the city, focusing on the consideration of topography, climate, vegetation, etc. [4–6], which is difficult to guide the layout of port shoreline utilization. Thus this paper proposes a method of analyzing the level of habitat environment for port shoreline utilization planning, in order to solve the problem of lack of consideration of impacts on habitat environment during the port planning stage. Applying this method, the level of human environment in the port area is analyzed at the beginning of the port shoreline site selection. The research results can provide decision-making support for the selection of the port shoreline utilization, avoiding the problem of over-reliance on the later environmental protection measures to make up for the problem.

23.2 Theory and Method

23.2.1 Technical Route

The method proposed in this paper includes following steps. The flow chart is shown in Fig. 23.1:

1. Collect relevant data for the study area, including air quality monitoring data from air quality monitoring stations over a long time series (not less than five years), historical data on population distribution (not less than five years), data on the distribution of scenic spots, meteorological data, and topographical data.

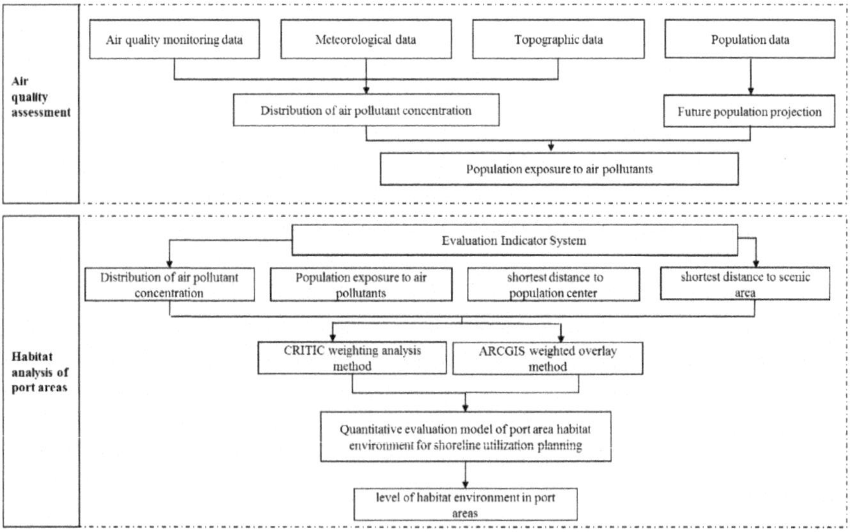

Fig. 23.1 Technology roadmap

2. The shoreline of the research area is divided into multiple analysis units by equidistant segmentation.
3. Based on the long time series monitoring data of the air quality monitoring stations, the annual average concentrations of PM_{10} and $PM_{2.5}$ are obtained from each air quality monitoring station. Combined with meteorological and topographic data, the spatial distribution of PM_{10} and $PM_{2.5}$ are obtained using ArcGIS spatial interpolation analysis method.
4. Based on the historical population distribution data of the study area, combined with the national spatial planning, a gray GM (1, 1) prediction model [7–9] was used to predict the future population distribution of the study area. Spatial differences of population exposure to air pollutants in the port area are analyzed by the application of Population Exposure Assessment Model.
5. The analysis factor values of each analysis unit are obtained. Based on the magnitude of the impact of shoreline development on the habit environment, the analysis factors include: air pollutant concentration, population exposure to air pollutants, shortest distance to population center, and shortest distance to scenic area.
6. Based on the analysis factor values of each analysis unit obtained in Step 5, CRITIC weight analysis method [10, 11] is applied to determine the weights of the four analysis factors of each analysis unit. ArcGIS weighted overlay analysis method is applied to comprehensively analyze the habitat level of each analysis unit.
7. Recommendations for port shoreline utilization planning are proposed based on the results of the habitat level analysis.

23.2.2 Main Algorithms

Models for assessing population exposure to air pollutants.

In step 4, the formula for calculating population exposure to air pollutants is as follows:

$$Exposure = \frac{Concentration_j * Pop_j * IR}{Area_j} \tag{23.1}$$

where Exposure is the population exposure to air pollutants, $Concentration_j$ is the concentration of air pollutants in Population Colony j, Pop_j is the number of people in Population Colony j; IR is the long-term average daily respiration of people of different genders and ages; and $Area_j$ is the land area of Population Colony j.

CRITIC weighting analysis method.

CRITIC weighting analysis method is a better objective weighting method compared to the entropy weighting and standard deviation methods. It combines objective weights of indicators based on analyzing the comparative strength of the indicators and the conflict between them. Considering the size of the variability of the indicators while taking into account the correlation between the indicators, it is not the case that the larger the number means the more important it is, but the objective attributes of the data itself are fully utilized for scientific analysis. The calculation steps of the CRITIC weighting analysis method are as follows:

Assuming a total of m samples and n indicators, the x_{ij} denotes the value of the jth indicator for the ith sample, the analysis matrix can be expressed as follows:

$$X = \begin{pmatrix} x_{11} & \dots & x_{1n} \\ \vdots & \ddots & \vdots \\ x_{m1} & \cdots & x_{mn} \end{pmatrix} \tag{23.2}$$

(1) Normalization of indicators. For the larger and better indicators, there are

$$y_{ij} = \frac{x_{ij} - \min(x_j)}{\max(x_j) - \min(x_j)} \tag{23.3}$$

For smaller and better indicators, there are

$$y_{ij} = \frac{\max(x_j) - x_{ij}}{\max(x_j) - \min(x_j)} \tag{23.4}$$

y_{ij} is the normalized form of the x_{ij} normalized form of the indicator, from which the normalization matrix is obtained.

(2) Calculation of the mean $\overline{x}_j$ and standard deviation S_j

$$\overline{x}_j = \frac{1}{m}\sum_{i=1}^{m} x_{ij} \tag{23.5}$$

$$S_j = \sqrt{\frac{\sum_{i=1}^{m}(x_{ij} - \overline{x}_j)^2}{m}} \tag{23.6}$$

(3) Calculation of the coefficient of variation v_j

$$v_j = \frac{S_j}{\overline{x}_j} \tag{23.7}$$

(4) Calculation of the matrix of correlation coefficient ρ_{ij}

$$\rho_{ij} = \frac{\mathrm{cov}(y_k, y_l)}{S_k * S_j}, \quad k = 1, 2, \dots, n; \quad l = 1, 2, \dots, n \tag{23.8}$$

in the formula, ρ_{ij} is the correlation coefficient between the jth indicator of the ith sample. $\mathrm{cov}(y_k, y_l)$ denotes the covariance between the kth indicator and the l th indicator.

(5) Calculation of the amount of information contained in the indicator:

$$\eta_j = v_j \sum_{k=1}^{n} (1 - \rho_{ij}), j = 1, 2, \ldots, n \tag{23.9}$$

(6) Determination of objective weights:

$$\beta_{ij} = \frac{\eta_j}{\sum_{j=1}^{n} \eta_j} \tag{23.10}$$

23.3 Case Study

Ningbo-Zhoushan Port is the largest port in China. Taking Ningbo-Zhoushan Port as an example, this paper analyzes the level of habitat environment in the port area and puts forward suggestions for port shoreline utilization planning.

First, air quality monitoring data from 2015–2020, population distribution data from 2010–2020, distribution of scenic spots, meteorological data, and topographical data were collected in the study area.

Then the segmentation tool in ArcGIS was used to segment the shoreline at 1000 m spacing, forming 7141 analysis units.

Based on the long time series monitoring data of air quality stations, the annual average concentrations of PM_{10} and $PM_{2.5}$ were obtained from each air quality station. The spatial distribution of PM_{10} and $PM_{2.5}$ was obtained by using ArcGIS spatial interpolation analysis. The analysis results were compared with the concentration limits stipulated in the Ambient Air Quality Standards (GB3095-2012), and the annual average concentrations of PM_{10} and $PM_{2.5}$ were classified into four levels: relatively highest, relatively high, relatively low and relatively lowest, as shown in Figs. 23.2 and 23.3.

Long time series population data of Ningbo and Zhoushan City were input to the GM (1, 1) forecasting model to construct the Ningbo-Zhoushan population forecasting model. Since the overall natural growth rate of the population of Ningbo and Zhoushan City has been relatively stable from 2010 to 2020, the forecast structure mainly showed a steady increase. According to the model prediction results, combined with the functional orientation of different districts in the territorial spatial planning, the population distribution of Ningbo and Zhoushan city was predicted in the year 2030, as shown in Fig. 23.4.

Based on the population predict results, combined with the statistical results of the Seventh National Population Census Main Data Bulletin of Zhoushan City in 2020 and the Seventh National Population Census Main Data Bulletin of Ningbo City on the characteristics of the population distribution of Ningbo and Zhoushan City in terms of gender and age groups, and the long-term respiratory volume of population in China published by the Ministry of Environmental Protection, the population

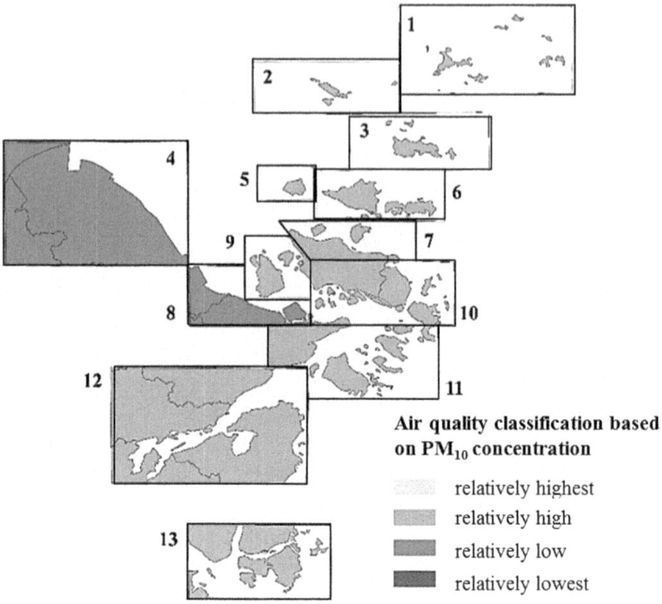

Fig. 23.2 Air quality classification based on PM_{10} concentration

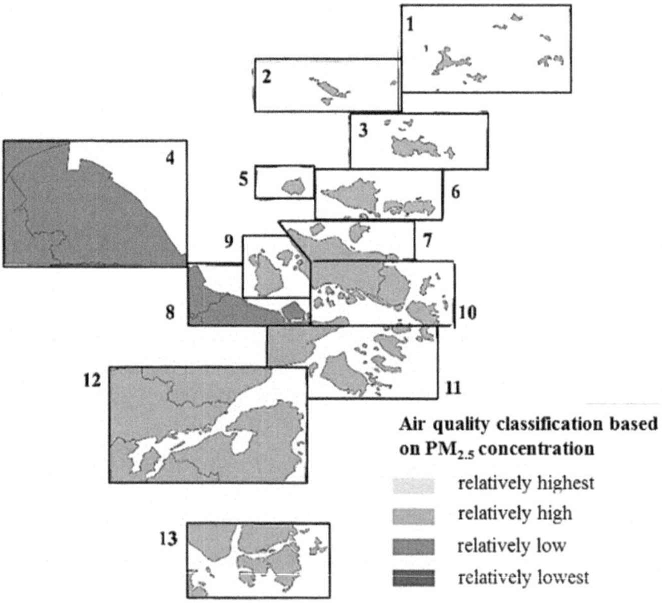

Fig. 23.3 Air quality classification based on $PM_{2.5}$ concentration

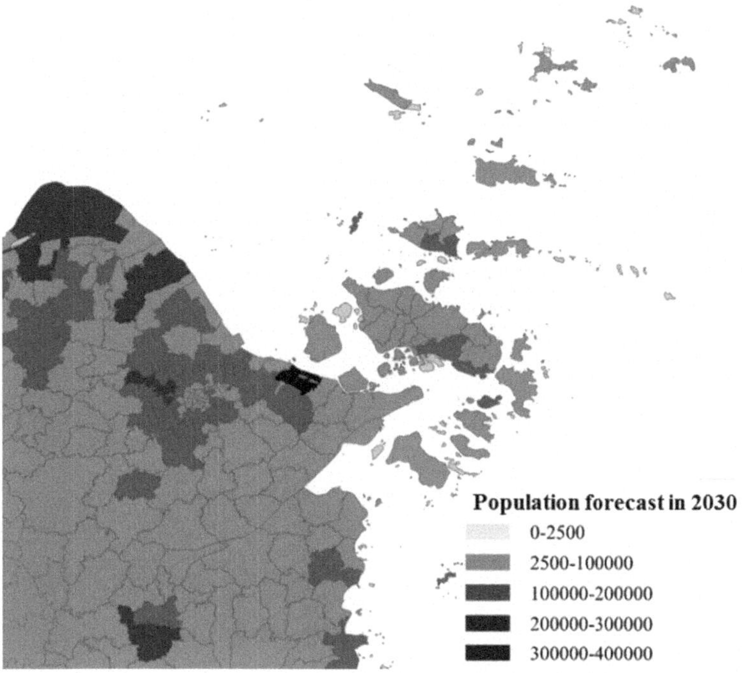

Fig. 23.4 Population prediction for the study area in the year 2030

exposure to air pollutants was calculated to analyze the population exposure of air pollutants. The results were categorized into four classes: relatively highest, relatively high, relatively low and relatively lowest, as shown in Figs. 23.5 and 23.6.

The nearest neighbor analysis tool in ArcGIS was used to calculate four indicators of each analysis unit: air pollutant concentration, population exposure to air pollutants, shortest distance to population catchment areas, and shortest distance to scenic spot.

Based on the CRITIC weight analysis method, the weights of the four analysis factors of each analysis unit were determined. ARCGIS weighted overlay method was used to comprehensively analyze the level of habitat environment of each analysis unit. According to the thresholds of ≥ 8, 6–8, 4–6, 2–4, < 2, the level of habitat environment in the port area was divided into four grades: relatively highest, relatively high, relatively low and relatively lowest, as shown in Fig. 23.7.

The results show that most of the shorelines in Ningbo and Zhoushan are suitable for development, and appropriate shoreline development will not have a significant impact on the surrounding habitat. However, during the development of some shorelines in Beilun District, Zhenhai District and Cixi City, attention should be paid to the control of air pollutant emissions, restricting the emission of air pollutants from ships, transport vehicles and shipping enterprises in the port area, so as to minimize the impacts on the habitat environment near the ports.

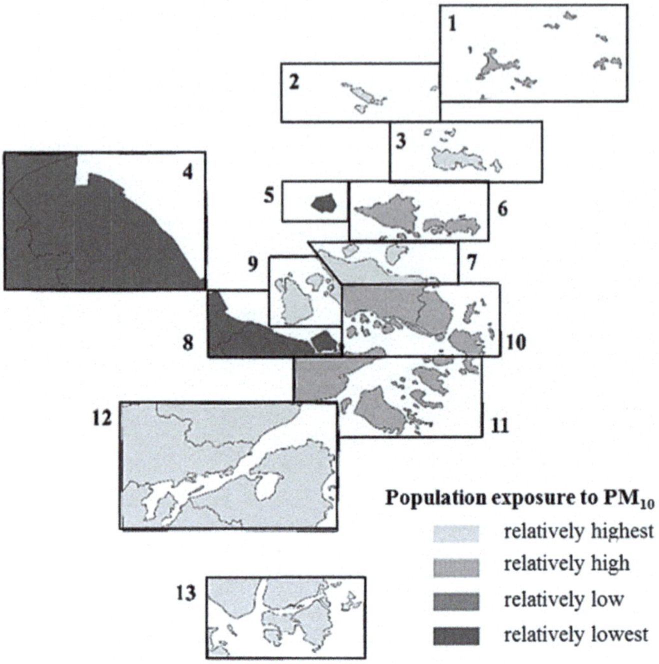

Fig. 23.5 Population exposure to PM_{10}

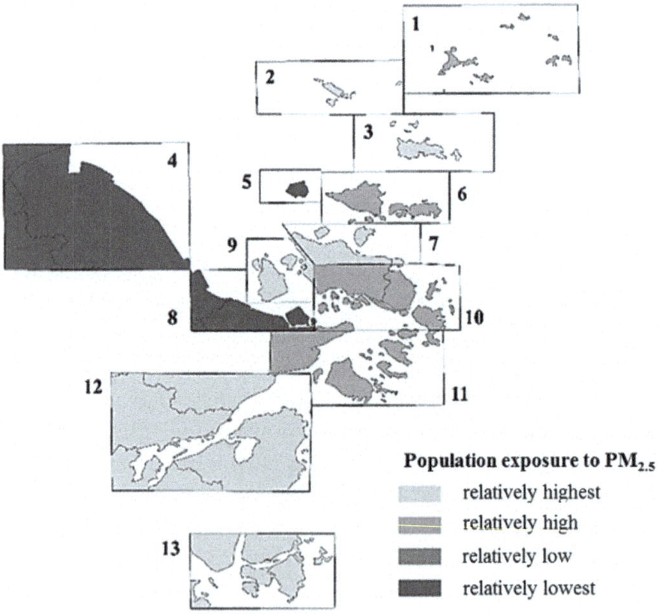

Fig. 23.6 Population exposure to $PM_{2.5}$

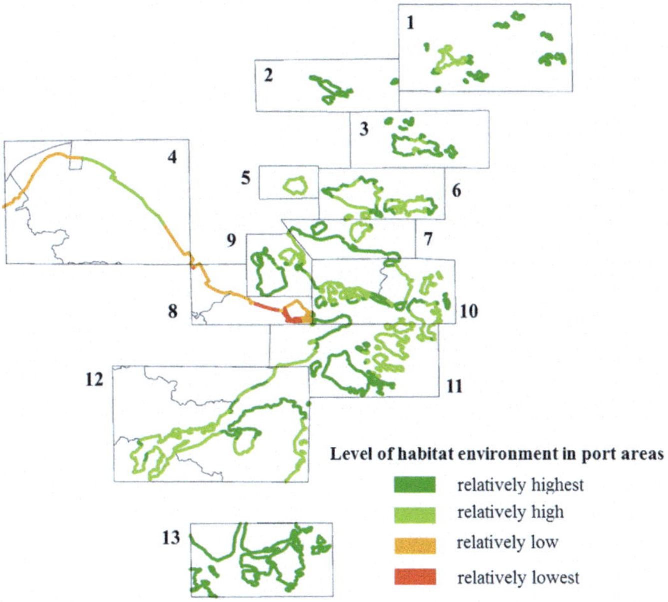

Fig. 23.7 Level of habitat environment in port areas

23.4 Conclusion

The development and utilization of port shoreline may bring certain problems such as air and noise pollution, which will have certain impacts on the habitat environment of the surrounding area. The method proposed in this paper can quantitatively analyze the level of habitat environment in the port area through the existing air quality monitoring, population distribution and other data, and the results of the analysis can provide decision support for the port shoreline utilization planning. Long-term environmental impacts of port shoreline development would be further analyzed in depth during the environmental impact assessment (EIA) of planning and project stages, thus have not been considered in this study. Research results show that impacts on the surrounding habitat environment should be fully considered during the utilization planning of port shoreline, and the planning should be carried out in the area with relatively high level of habitat environment as far as possible. When planning port shorelines in areas with a relatively low level of habit environment, relevant environmental protection measures should be taken to minimize the impact on the surrounding habit environment.

References

1. PIANC (2014) Masterplans for the development of existing ports. The world association for Waterborne transport infrastructure
2. Chen Y (2007) Research on spatial decision support system framework of port location based on GIS. Shanghai Maritime University
3. Dai ZJ, Yun CX, Han Z (2006) Application of satellite remote sensing combined with digital topographic information in port engineering: a case study of Maoming deep-water port site selection. Ocean Eng 24:115–121
4. Liu JG, Zhang WZ (2014) Review of habitat environment assessment method. Urban Dev Stud 21:46–52
5. Lee MH, Rho HS, Lee HG, Park CH, Kim CH (2020) Analysis of seabottom and habitat environment characteristics based on detailed bathymetry in the northern shore of the East Sea (Gyeongpo Beach, Gangneung). Econ Environ Geol 53:729–742
6. Xu W, Sun T (2022) Evaluation of rural habitat environment in under-developed areas of Western China: a case study of Northern Shaanxi. Environ Dev Sustain 24:10503–10539
7. Zhao YC, Cao SY (2020) Research on forecast of coordinated development of economy and environment in coastal economic belt based on GM model. J Coast Res 115:338–344
8. Bao KY, He G, Lan J, Yang JW, Zhou QT (2020) Study on sustainable development of mining cities by the method of relative resources carrying capacity and GM (1,1) model. Pol J Environ Stud 29:3983–3995
9. Manickam A, Indrakala S, Kumar P (2023) A novel mathematical study on the predictions of volatile price of gold using grey models. Contemp Math 4:270–285
10. Zhang QJ, Liu CJ, Guo S, Wang WT, Luo HM, Jiang YH (2023) Evaluation of the rock burst intensity of a cloud model based on the critic method and the order relation analysis method. Min Metall Explor 40:1849–1863
11. Thi KLN, Hoang NL, Van HN, Bang AH (2020) Critic method and grey system theory in the study of global electric cars. World Electr Veh J 11:1–15

Chapter 24
Effects of Rural Domestic Reclaimed Water Irrigation on Heavy Metals Changes in Soil-Crop System

Lei Wang, Luyao Zhou, Jie Yang, Yanshu Chen, Jiafang Cai, and Peipei Kong

Abstract Reclaimed water reuse is a topic of growing interest in the field of reclaimed water management. This study took the paddy, seedling, vegetable fields in south China as the research object, aimed at the safety of irrigation regulation of reclaimed water, and was based on the field experiment, to research the effect of paddy irrigation regulation (low water level W1, medium water level W2 and high water level W3) and vegetable、seedling conventional irrigation of different water sources (rural domestic sewage primary treatment water R1, secondary treatment water R2, ecological pond purified water R3 and river water CK/R4) on the content of typical heavy metals (Cd, Pb, Cr, Cu, Zn) in soil and crop. The study examines the levels of typical, harmful heavy metals (Cd, Pb, Cr, Cu, Zn) in soil and crops. Our results show that the concentration of heavy metals in the rice plant's stem is greater than that of the grains and leaves, with Zn being the highest concentration, followed by Cr, Pb, and Cd to decrease concentration. Compared to the irrigation water source, it was also observed that the concentration of harmful elements in the grains did not show a significant rise when using recycled water for irrigation. In order to adhere to the specified limits on impurities found in rice grains, it is essential to follow strict guidelines. As for the germination of seeds, it is crucial to ensure compliance with

L. Wang (✉) · J. Cai
Zhejiang Institute of Hydraulics and Estuary (Zhejiang Institute of Marine Planning and Design), Hangzhou, Zhejiang, China
e-mail: wanglei0704@zjwater.gov.cn

L. Zhou
Zhejiang Water Resources and Hydropower Management Center (Zhejiang Water and Soil Conservation Monitoring Center), Hangzhou, Zhejiang, China

J. Yang
Hangzhou Agricultural and Rural Bureau of Qiantang District, Hangzhou, Zhejiang, China

Y. Chen
Hangzhou River and Rural Water, Conservancy Management Service Center, Hangzhou, Zhejiang, China

P. Kong
Hohai University, Nanjing, Jiangsu, China

K. Y. Foo et al. (eds.), *Water Nexus Approach for Sustainable Development*, Environmental Science and Engineering, https://doi.org/10.1007/978-3-031-89238-7_24

the established standards. For seedlings, Cd and Cu content increased in the 20 cm soil layer. For vegetable planting, Cd and Pb content increased in the 20 cm soil layer, 40 cm soil layer, and 60 cm soil layer. Overall, the fluctuations in heavy metal levels in soils irrigated by R3 and CK/R4 water sources are largely similar. In general, the use of reclaimed water for irrigation did not result in significant contamination of agricultural fields. However, the potential for a sustained increase in the risk of heavy metal pollution due to long-term reclaimed water irrigation requires further investigation.

Keywords Rural domestic reclaimed water · Paddy water level regulation · Heavy metals · Ecological risk analysis

24.1 Introduction

In view of the increasing scarcity of global water resources, it is crucial to use sewage resources for agricultural production to replace some fresh water irrigation, thereby conserving water resources to some extent and providing a prominent strategy to alleviate water supply demand and local water scarcity contradictions [1–3]. At present, about 10% of the cultivated land in China uses reclaimed water for agricultural irrigation, and good utilization results have been achieved [4]. Rural domestic sewage typically has a higher N and P content than urban sewage, low toxicological indexes such as heavy metal pollutants and good biodegradability, so it can be used as a kind of traditional water and fertilizer resources. Utilizing treated irrigation water can effectively mitigate the release of contaminants into the environment while conserving precious freshwater resources. This sustainable practice serves as a viable strategy to advance economic and social progress towards higher quality standards [5, 6]. The challenge of heavy metal pollution stemming from the utilization of reclaimed water for irrigation purposes presents a significant obstacle in both the study and implementation of reclaimed water reutilization. Soil contamination caused by heavy metals leads to a range of intricate reactions and processes involving plant growth. This contamination encompasses various chemical, physical, and biological interactions that affect the soil environment and impact crop development [7, 8]. The heavy metals commonly found in untreated wastewater from rural areas typically consist of Cu, Cr, Pb, Cd, and Zn, with concentrations usually below 5 ug/L. Among these metals, Cr and Cd are identified as posing the greatest ecological risks [9, 10]. Zhang et al. [11] found that the amount of Cd in the above-ground part of alfalfa irrigated with reclaimed water increased by 98.6% compared with that under clean water irrigation. The heavy metal Cd content showed a certain degree of accumulation, but the content was lower than the national health standard limit [11]. The same results were also reflected in the studies on alfalfa, soybean and corn [12, 13]. Wu et al. [14] examined the impact of using reclaimed water for irrigation on the distribution of heavy metals in sugarcane. The results showed that the roots accumulated a higher number of heavy metals compared to the stems and leaves of the plant [14]. Jung et al. [15] found that there was no significant difference in the heavy metal content of rice under domestic sewage irrigation and that under groundwater irrigation [15]. Li et al. [16] also had similar results in their study on alfalfa [16]. Al-lahham et al. [17] tested the effects of different mixing ratios of clean water and recycled water on tomato quality and showed that tomato pulp would not be contaminated by microorganisms in recycled water [17]. Xin et al. [18] reported no significant changes in heavy metal content in seeds during the reclaimed water irrigation process. However, when measured under regular irrigation conditions, there was a distinct pattern of heavy metal accumulation in root > stem > seed crops. Notably, the heavy metal content in seeds was found to be within the acceptable range of food hygiene standards (GB 2715-2005) and food contaminant limit standards (GB 2762-2005) [18]. Therefore, it is of scientific research value and practical significance to study the effect of reclaimed water irrigation on soil heavy metal content and evaluate the risk of soil heavy metals in order to control the environmental pollution of rural cultivated land and maintain the soil safety of farmland in China [19, 20]. The focus of this study was on examining the variations in heavy metal levels in paddy fields and crop plants when irrigated with reclaimed water from rural areas. Additionally, an investigation into the correlation between heavy metal concentrations in paddy fields and the overall environmental quality of the soil was conducted. The impact of utilizing recycled water for watering paddy fields was examined in relation to the potential presence of harmful substances like heavy metals in the soil.

24.2 Materials and methods

24.2.1 Experimental site

The experiment took place at Jinhua Water-saving pilot Base of Zhejiang Province from May to November in 2020 and 2021. The specific research site was situated in Zhoushan Town, Yongkang City, Jinhua City, the topography is characterized by gentle slopes and high precipitation levels, leading to an uneven distribution of rainfall across the region with an average yearly rainfall of 1787 millimeters, the research area saw the establishment of a total of 36 experimental plots for testing rice standards. These plots, measuring 20 meters by 5 meters each, were carefully set up to conduct the study which were supplied with water through a network of underground conduits. Within the confines of the experimental plot, a new wastewater treatment facility has been constructed in a rural area, serving a population of 4500 people with a daily capacity of 400 m^3.The study focused on identifying the wastewater origin. In terms of treatment, a secondary treatment approach was implemented following a traditional primary treatment process. The A2O process was improved for secondary treatment, resulting in the effluent meeting the highest standard prescribed by the Urban Sewage Treatment Plant Pollutant Discharge Standard (GB18918-2002). The

Table 24.1 Physical and chemical properties of topsoil

Soil depth	PH	EC (mS/m)	TN (%)	TP (%)	OM (g/kg)	NH_4^+-N (g/kg)	NO_3^--N (g/kg)
0–20 cm	5.6	2.6	0.12	0.07	17.7	8.24	2.84
20–40 cm	5.9	2.9	0.12	0.07	17.7	8.24	2.84
40–60 cm	6.2	2.8	0.09	0.03	14.8	5.75	2.69
60–80 cm	6.6	2.8	0.07	0.03	12.9	4.71	2.5

test site consists of a combination of sandy and clay soil types. The properties of various soil layers are detailed in the data provided in Table 24.1.

24.2.2 *Experimental design*

Our irrigation water resources are varied, including primary treated water from the RDRW (initially designated as R1), secondary treated water from the same source (initially designated as R2), water from an advanced treatment ecological pond (initially designated as R3) (the effluent from the secondary treatment unit is stored in this ecological pond with a storage capacity of 6000 m^3), and river water (initially designated as CK/R4). We use a simple, yet efficient submersible pump to distribute this water for irrigation. In 2020, the limited availability of water sources was a result of the development of new ecological ponds in 2021. Only R1, R2, and CK water sources were accessible during that time. In 2021, there are four types of irrigation water sources: R1, R2, R3 and CKR4. All irrigation water quality indexes meet the irrigation water quality standard (GB5084-2021). The experimental rice field was divided into three water level management treatments, with each treatment being replicated three times. Each stage of growth was meticulously managed to control water levels in the field, ensuring prompt replenishment of water when levels dropped below the designated lower limit, and effective drainage of excess water during periods of heavy rainfall that exceeded the field's capacity for rain storage. The regulations regarding the water levels in the field can be found in Table 24.2, while the arrangement of the experimental plots is displayed in a visual representation Figure 24.1. The tested rice variety was Jiayou Zhongke 13-1, and the fertilization method was based on fertilizer + topdressing according to the local fertilizer habit. The crops planted in the seedling test area in 2020 and 2021 are podocarpus. During the period of September to October in 2020 and June to August in 2021, the vegetable test area was used for radish planting and hollow cabbage planting experiments respectively. Vegetables and seedlings were conventional irrigation mode. The on-site planting of rice, seedlings and vegetables in the test area is shown in Figure 24.2.

Table 24.2 Paddy field water level control standard

Water control	Upper and lower limit	Turning green	Early tillering	Later tillering	Jointing booting	Heading flowering	Milk ripening
W1	Upper limit of sewage	0	3-5 d exposing field	1-2 d exposing field	1-2 d exposing field	1-2 d exposing field	3-5 d exposing field
	Lower limit of sewage	30	30	Exposing field	40	40	30
	Upper limit of sewage (rain) storage	50	70		80	80	60
W2	Upper limit of sewage	0	10	10	10	10	10
	Lower limit of sewage	30	50	Exposing field	50	50	50
	Upper limit of sewage (rain) storage	50	70		100	100	100
W3	Upper limit of sewage	0	40	40	40	40	10
	Lower limit of sewage	30	60	Exposing field	60	60	60
	Upper limit of sewage (rain) storage	50	100		150	150	100

24.2.3 Indicators and measurements

The upper layers of soil, ranging from 0–20 cm to 60–80 cm depth, were sampled both prior to and following the planting process, in order to analyze the levels of five common heavy metals

(Cd, Pb, Cr, Cu, Zn) in soil. After the rice crop was collected, samples were taken from the plants to analyze the presence of various metals in different parts such as

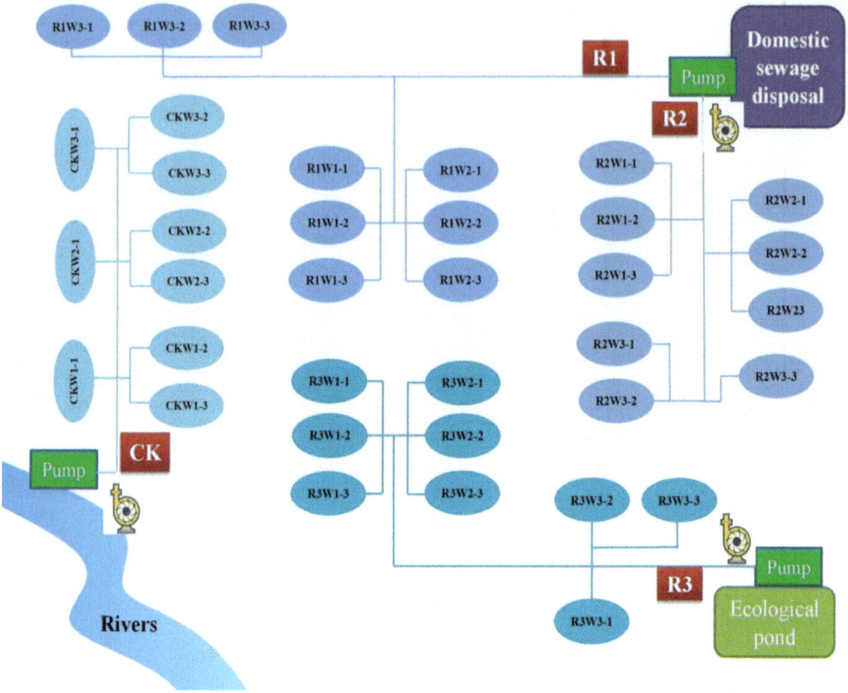

Fig. 24.1 Paddy field layout of the experimental area

stems, leaves, and grains. The process of heavy metal analysis can be found detailed in the corresponding reference materials (Table 24.3).

24.3 Results and analysis

24.3.1 Changes of heavy metal content in paddy soil

Figure 24.3 illustrates the variations in the concentrations of heavy metals across diverse soil depths in a rice paddy field, under various irrigation types and water levels. The elevated presence of Cd in the topsoil compared to the subsoil is attributed to Cd's limited movement in soil and its tendency to be immobilized in the upper soil layer. The Cd content of all soil layers under R1 water source irrigation increased the most, followed by R2 and R3 water source increased the least, indicating that the Cd content of soil under primary and secondary reclaimed water irrigation increased significantly. The Pb concentration in the various soil layers of the paddy field exhibited a general increase as the level of the water in the field rose. The level of lead in the paddy field showed significant variations depending on different sources of

[a] A and B are rice planting, C is seedling planting and D is vegetable planting.

Fig. 24.2 On-site planting of rice, seedlings and vegetables in the test area

Table 24.3 Detection method of heavy metals in soil and plants

Sample	Test item	Detection method	Standard number
Paddy soil	Cd	Graphite furnace atomic absorption spectrophotometry	GB/T 17141-1997
	Pb, Cr, Cu, Zn	Flame atomic absorption spectrophotometry	HJ 491-2019
Rice plant	Cu, Zn	Flame atomic absorption spectrometry	GB 5009.13-2017
	Pb, Cr, Cd	Inductively coupled plasma mass spectrometry	GB 5009.268-2016

water. The Cr content in most treated soils showed a decrease in 2021, despite the exceptions of increased Cr content in W1R2 and W3R2 treated soils compared to 2020. Overall, the change in Cr content exceeded the background value in 2021. Under the influence of decreased water levels, a minor reduction was observed in the presence of Cr in the soil. As the duration of irrigation increased, there was a notable shift in the levels of Cr detected in the soil. In rural villages during the years 2020 and 2021, there was a noticeable trend of decreasing Cu content in soil as a result of implementing reclaimed water irrigation practices. This trend was observed across all treatment groups, indicating the effectiveness of this method in controlling Cu levels in the soil. Under the control of middle and low water level, the Cu levels in rice fields are lower than usual during periods of reduced water levels, as compared to the normal background levels The field saw a significant decline, with a noticeable decrease in the upper end The level of water is currently low, and there seems to be a correlation with the decrease in Cu levels, as the amount of Zn in the soil also appears to be on a downward trend. The outcomes concerning the analysis of differences can be found in Table 24.4. The results suggest that variations in Cd content in soil layers of 0–20 cm and 20–40 cm are influenced significantly by different water level regulations and irrigation water sources ($P < 0.01$). Furthermore, the presence of irrigation water sources has been found to have a significant impact on the levels of Cr content in the soil layer between 20–40 cm. ($P < 0.01$) and a highly significant impact on the change in Cu content in the soil layer at a depth of 40–60 cm ($P < 0.01$).

24.3.2 *Changes of heavy metal content in rice plants*

Figure 24.4 illustrates the cumulative variations in heavy metal content within plant organs resulting from distinct irrigation water sources and varying water levels. For the stems of rice, the contents of Pb, Zn, Cd and Cr decreased with the change of irrigation water source from inferior to superior. The total content of heavy metals is R1 > R2 > CK/R4 (2020), R1 > R2 > R3≈ CK/R4 (2021), indicating that the total amount of heavy metals in the stems of rice irrigated by R3 and CK/R4 water sources is similar. W1 ≈ W2 > W3 (2020) and W1 ≈ W2 ≈ W3 (2021) were the changes of the total heavy metal content under different water levels, and the influence of the controlled water level on the heavy metal content in stems was small. The Zn, Pb, Cd, and Cr contents in rice leaves decreased as the irrigation water source changed from lower quality to higher quality, and the total heavy metal content followed the order R1 > R2 > CK/R4 (2020) and R1 > CK/R4 > R2 ≈ R3 (2021). Compared with the stem, the influence of irrigation water source on the heavy metal content in leaves was weakened. Under the control of different water levels, the total heavy metal content is W1 > W3 ≈ W2 (2020), W3 > W1 > W2 (2021). The content of heavy metals in rice grains varied under different irrigation water sources, with R2 ≈ R1 > CK/R4 in 2020 and R1 > R2 > R3 ≈ CK/R4 in 2021. Under the control of different water levels, the total heavy metal content showed W3 ≈ W1 ≈ W2 (2021), W3 > W2 ≈ W1 (2021). After conducting extensive research on whether grains irrigated with reclaimed water showed any increase in heavy metal levels, the findings were not able to provide conclusive evidence. However, the levels of heavy metals in the grains met the safety standards for rice outlined in the National Food Safety Standard GB 2762-2017 (Table 24.5). The heavy metal content of each part of rice showed Stem > Grain ≈ Leaf, and heavy metal composition showed Zn > Cr > Pb > Cd. The impact of the source of irrigation water on the concentration of heavy metals in the stem, leaf, and grain of rice gradually diminished. The outcomes of the analysis

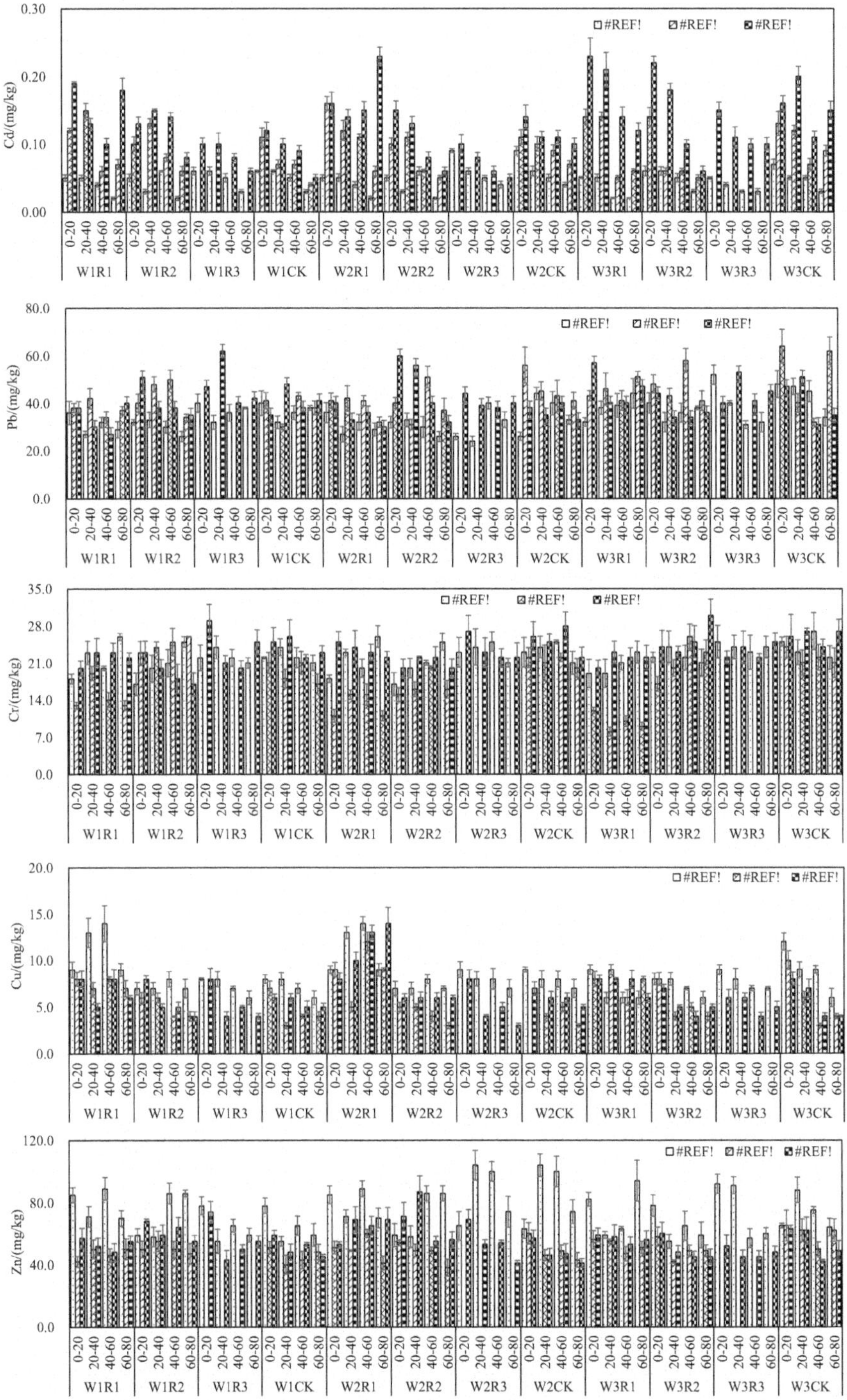

Fig. 24.3 Changes of heavy metals content in paddy field under different irrigation water sources and different water level regulations

Table 24.4 Two factor analysis of variance of heavy metals content in soil

Soil depth(cm)	Cd		Pb		Cr		Cu		Zn	
	W	R	W	R	W	R	W	R	W	R
0–20	**	**	NS	NS	NS	NS	NS	NS	NS	NS
20–40	*	*	NS	NS	NS	**	NS	NS	NS	NS
40–60	NS	NS	NS	NS	NS	NS	NS	**	NS	NS
60–80	NS	NS	NS	NS	NS	NS	NS	NS	NS	NS

W stands for water level control, R stands for irrigation water source
** means $P < 0.01$, * means $P < 0.05$, NS shows no significant difference

of variations can be seen in the detailed results provided in Table 24.6. A notable change in the Zn content of stems and leaves in 2020 can be linked to the source of irrigation water. Similarly, the source of irrigation water has a highly significant impact on the Cr content of leaves in 2020, influencing the Pb content of stems in 2021, and the Cu content of leaves and seeds in 2021. Furthermore, the origin of the irrigation water exerts a notable influence on the fluctuations in Pb levels in seeds. In the year 2021, the management of water levels caused notable changes in the Cd content present in the stems and leaves. Furthermore, there was a notable impact on the changes observed in the levels of Cr in the foliage during both the years 2020 and 2021.

24.3.3 Distribution and accumulation of heavy metals in vegetable planting soil

The changes in heavy metal levels in the soil used for vegetable cultivation under different irrigation conditions are depicted in Figure 24.5. As the depth of the soil increased, there was a noticeable increase in heavy metal concentrations across various soil layers. The distribution of soil metal concentrations varied across different ranges, with the highest values observed in the 60 ~ 80 cm range, followed by the 40 ~ 60 cm, 20 ~ 40 cm, and 0 ~ 20 cm ranges. The heavy metals with the highest soil concentrations were Zn, Pb, Cr, Cu, and Cd. In the layer of soil measuring 20 centimeters deep, the concentration of Cd and Pb increased. The increase in Cd concentration was R1 (0.1 mg/kg) > R2 (0.08 mg/kg) ≈ R3 (0.05 mg/kg) > CK/R4 (0.04 mg/kg), and the increase in Pb concentration was R1 (11 mg/kg) > R2 (9 mg/kg) > R3 (5 mg/kg) > CK/R4 (1 mg/kg). The contents of Cr, Cu, and Zn exhibited a downward trend. Compared to the baseline levels, the concentrations of soil Cr in the experimental site decreased by 0.03%, 8.3%, 8.3%, and 4.2%, respectively. In the layer of soil between 20 and 40 centimeters deep, the concentration of Cd showed the following trend: R1 (0.03 mg/kg) > R2 (0.02 mg/kg) ≈ CK/R4 (0.02 mg/kg) > R3 (0.01 mg/kg). Meanwhile, the quantities of other toxic metals displayed a decreasing trend. Compared with the background values, under water source irrigation of R1,

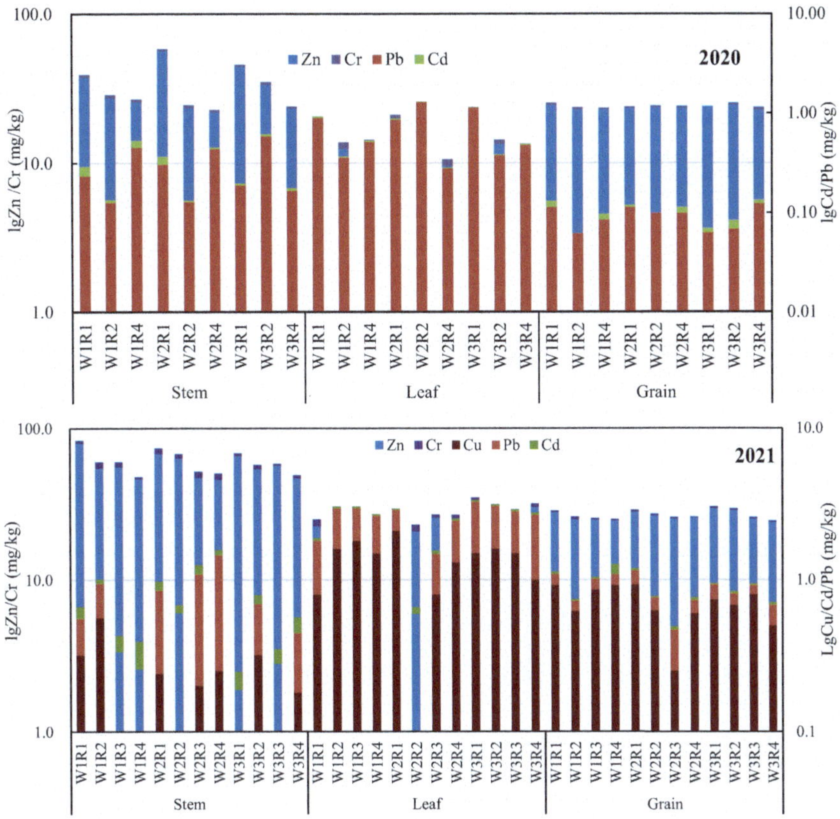

Fig. 24.4 Changes of heavy metals content in plant organ under different irrigation water sources and different water level regulations

Table 24.5 Limit requirements for heavy metal content in rice grains

Heavy metal category	Cd	Cr	Pb
Limited quantity(mg/kg)	0.2	1.0	0.2

R2, R3 and CK/R4, the content of Pb in the soil of the test area decreased by 5.7%, 2.9%, 5.7% and 8.6%, respectively, by 13.3%, 10.0%, 20.0% and 26.7%, respectively, by 50.0%, 60.0%, 40.0% and 50.0%, respectively. The content of Zn decreased by 51.1%, 39.8%, 54.5% and 60.2%, respectively. In the layer of soil measuring between 40 and 60 centimeters deep, similar to the soil layer found at a depth of 0–20 cm, the contents of Cd and Pb increased, Cd content increased as R1 (0.03 mg/kg) > R2 (0.01 mg/kg) > R3 (0.0 mg/kg) ≈ CK/R4 (0.0 mg/kg), Pb content increased as R1 (9 mg/kg) > R2 (9 mg/kg) > R3 (1 mg/kg) ≈ CK/R4 (1 mg/kg). In contrast, other heavy metal contents showed a declining trend compared with the background values. For example, the soil Cr content decreased by 12.5%, 31.3%, 28.1%, and

Table 24.6 Two factor analysis of variance of heavy metals content in plant organ

Year	Organ	Zn		Pb		Cd		Cr		Cu	
		W	R	W	R	W	R	W	R	W	R
2020	Stem	NS	*	NS	NS	NS	NS	NS	NS	–	–
	Leaf	NS	*	NS	NS	NS	NS	*	**	–	–
	Grain	NS	NS	NS	NS	NS	NS	NS	NS	–	–
2021	Stem	NS	NS	NS	**	**	NS	NS	NS	NS	NS
	Leaf	NS	NS	NS	NS	**	NS	*	NS	NS	**
	Grain	NS	NS	NS	*	NS	NS	NS	NS	NS	**

*Indicates that there is no significant difference between the two groups ($P < 0.05$), and **indicates that there is no significant difference between the two groups ($P < 0.01$).

31.3% for water source irrigation with R1, R2, R3, and CK/R4, respectively, and decreased by 55.6%, 55.6%, 55.6%, 55.6%, and 66.7%, respectively. Similarly, the soil contents of Cu, Zn, and Pb decreased by 40.0%, 31.4%, 40.0%, and 44.3%. In 60–80 cm soil layer, the contents of Cd and Pb increased, Cd content increased as R1 (0.02 mg/kg) ≈ R2 (0.02 mg/kg) > R3 (0.01 mg/kg) ≈ CK/R4 (0.01 mg/kg), and Pb content increased as R1 (12 mg/kg) > R2 (8 mg/kg) > CK/R4 (7 mg/kg) > R3(6 mg/kg). The heavy metal contents show a decreasing trend. The Cr content decreased by 7.4%, 0.0%, 7.4%, and 11.1% in R1, R2, R3, and CK/R4, and by 44.4%, 55.6%, 55.6%, and 55.6% in the soil irrigated with water source. From the soil layer point of view, the content of heavy metals in Zn > Pb > Cr > Cu > Cd, 0 ~ 20 cm, 40–60 cm, 60–80 cm soil layers decreased as soil depth increased. During the study, it was observed that the levels of Cd and Pb showed an increase specifically in the soil layer between 20–40 cm. In general, this pattern was consistent throughout the research period. The similarity in the trends of heavy metal concentrations between the soil irrigated with water from R3 and CK/R4 sources was evident.

24.3.4 Distribution and accumulation of heavy metals under seedling planting

Figure 24.6 illustrates the fluctuations in heavy metal content in seedling soil under various water irrigation conditions. Figure 24.6 depicts the variations in heavy metal concentrations found in the soil of seedling beds under different water irrigation regimens. The levels of heavy metals were observed to be highest in the soil layers ranging from 0 to 20 cm, followed by the layers from 40 to 60 cm, then the layers from 20 to 40 cm, and approximately equal in the layers from 60 to 80 cm. The order of heavy metal content in the soil is Zn followed by Pb, Cr, Cu, and Cd. There was a significant increase in the levels of Cd and Cu by R1 (0.11 mg/kg) > CK/R4 (0.068 mg/kg) ≈ R2 (0.06 mg/kg) > R3 (0.04 mg/kg) for Cd and R1 (5 mg/kg) > R2(3 mg/kg) > R3(2 mg/kg) > CK/R4(1 mg/kg) for Cu. The contents of Cr, Pb and Zn

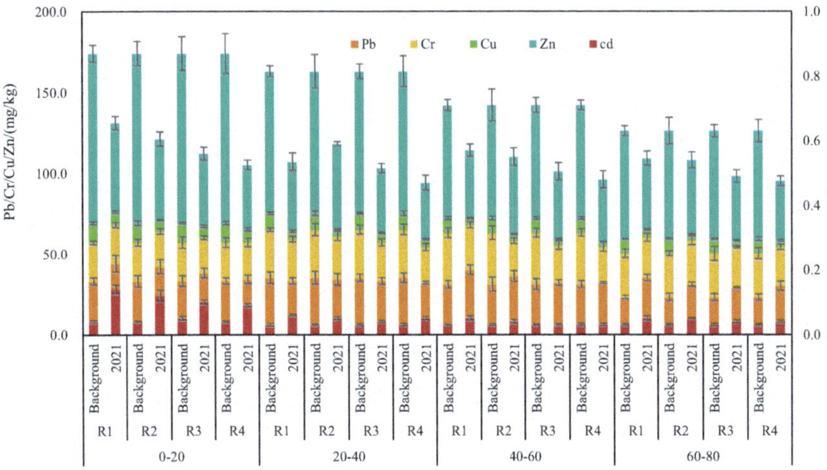

Fig. 24.5 Changes of heavy metal content in vegetable growing soil under different water source irrigation conditions

decreased, compared with the background values, R1, R2, R3, CK/R4 water source irrigation, soil Cr decreased by 17.1%, 14.6%, 4.9%, 7.3% respectively. The content of Zn in the test area decreased by 12.0%, 16.0%, 36.0% and 32.0%. Respectively, within the soil layer ranging from 20 to 40 centimeters, the Cd content in the soil followed the order of R1 (0.07 mg/kg) > R2 (0.03 mg/kg) ≈ CK/R4 (0.03 mg/kg) > R3 (0.01 mg/kg). The heavy metal levels in the soil exhibited a decreasing trend over time. With the exception of Zn, which remained steady. In comparison to the background values, the Pb content in the soil decreased by 0.0%, 17.5%, 12.5%, and 17.5% under R1, R2, R3, and CK/R4, respectively, while the Cu content decreased by 16.0%, 20.0%, 12.0%, and 8.0%, respectively. The Zn content decreased by 46.8%, 39.0%, 38.7% and 36.4%, respectively. 40–60 cm soil layer, similar to 20 ~ 40 cm soil layer, the content of Cd in soil increased in the order of R1(0.04 mg/kg) > R2(0.03 mg/kg) > R3(0.02 mg/kg) ≈ CK/R4(0.02 mg/kg). The contents of other heavy metals showed a downward trend. Compared with the background values, under water source irrigation of R1, R2, R3 and CK/R4, the content of soil Pb in the test area decreased by 39.5%, 5.3%, 5.4% and 13.2%, respectively. The content of Cr decreased by 59.6%, 63.5%, 56.9% and 55.8%, respectively. The content of Zn decreased by 46.3%, 26.8%, 35.0% and 41.5%, respectively. In the soil layer of 60–80 cm, the content of Cd increased in the order of R1 (0.07 mg/kg) > R2(0.05 mg/kg) > CK/R4(0.05 mg/kg) > R3(0.04 mg/kg). The contents of other heavy metals showed a downward trend. The levels of other high-density metals demonstrated a decline. In comparison to the baseline measurements, the irrigation of R1, R2, R3, and CK/R4 resulted in decreases of 8.1%, 8.1%, 19.4%, and 27.0% in the content of Pb in the tested area, respectively. The content of Cu decreased by 24.0%, 16.0%, 16.7% and 20.0%, respectively. The content of Zn decreased by 46.3%, 31.3%, 43.8% and 47.5% respectively.

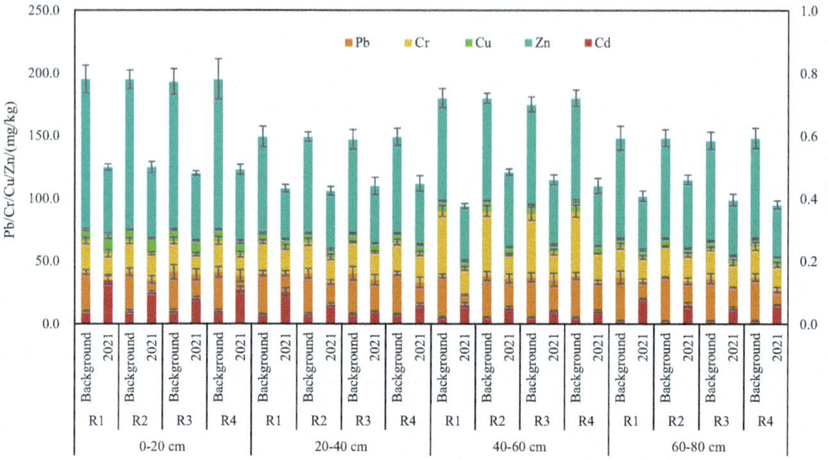

Fig. 24.6 Changes of soil heavy metal content in seedling test area under different water source irrigation conditions

The variation in heavy metal content differs across soil layers, with the highest levels found in the 0 ~ 20 cm layer, followed by the 40 ~ 60 cm layer, and similar levels in the 20 ~ 40 cm and 60 ~ 80 cm layers. Zn, Pb, Cr, Cu, and Cd were the heavy metals detected, with the highest concentrations of Cd and Cu in the 0 ~ 20 cm layer, and an increase in Cd levels observed from the 20–80 cm layer. Overall, the heavy metal content changes were similar in soils irrigated with R3 and CK/R4 water sources.

24.4 Conclusion

In this paper, we studied the change law of heavy metals in soil—crop system under the influence of rural domestic reclaimed water irrigation. Key findings are as follows:

1. The research findings indicated that there was a slight increase in the levels of certain contaminants, namely Cd and Pb, in rural domestic sewage utilized for irrigation in paddy fields. On the other hand, there was a decrease observed in the concentrations of Cr, Cu, and Zn. The concentration of Cd in the top 20 cm of soil was found to be notably influenced by varying water levels and treatments to the water source ($P < 0.01$), causing a significant impact on the concentration of Cd in the soil layer between 20–40 cm ($P < 0.05$). The presence of certain levels of trace elements in different soil layers was noticeably influenced by the type of water used for irrigation. This impact was particularly significant in the 20–40 cm soil layer for chromium levels and in the 40–60 cm soil layer for copper concentrations. This suggests a strong relationship between irrigation water quality and the distribution of these elements within the soil ($P < 0.01$).

2. The content of heavy metals in different parts of rice was as follows: stem>grain ≈ leaf. In addition, the composition of heavy metals in rice was considered to be: Zn > Cr > Pb > Cd. Conversely, the influence of irrigation water sources on the accumulation of heavy metals in rice stems, leaves and grain gradually weakened with time, while the regulation of water level had only a marginal effect on the overall content of heavy metals in all parts of the rice plant. Reclaimed water irrigation did not significantly affect the content of heavy metals in grains, meeting the stringent limits of contaminants in rice as per the National Standard for Food Safety GB 2762-2017.
3. There was an observed rise in the levels of Cd and Cu in the 20 cm soil layer as seedlings were planted, accompanied by an increase specifically in the Cd levels within this layer. However, the levels of Cadmium (Cd) and Lead (Pb) increased in the soil layers at depths of 20 cm, 40 cm, and 60 cm when it came to growing vegetables. Specifically, there was a notable rise in Cd concentration within the top 20 cm layer. The similarities in the variation of heavy metal levels in soils that are irrigated with R3 and CK/R4 water sources are quite noticeable.

Acknowledgments Our study received financial support from the Water Conservancy Science and Technology in Zhejiang Province (RB2410), the President's Fund of Zhejiang Institute of Hydraulics and Estuary (Zhejiang Institute of Marine Planning and Design) (ZIHE23Z003), the Digitalization Project of the President's Fund of Zhejiang Institute of Hydraulics and Estuary (Zhejiang Institute of Marine Planning and Design) (ZIHESZH23005, ZIHESZH23001), and the Provincial Research Institute Special Project (Scientific Research Category) (ZIHEYS23001).

References

1. Wang Z, Li JS, Li YF (2017) Using reclaimed water for agricultural and landscape irrigation in China: a review. Irrig Drain 66:672–686
2. Chopyk J, Nasko DJ, Allard S et al (2020) Comparative metagenomic analysis of microbial taxonomic and functional variations in untreated surface and reclaimed waters used in irrigation applications. WaterResearch 169:115250
3. Wang SY, Wu WY, Fei L et al (2018) Assessment of human health risks of heavy metals in the typical sewage irrigation areas. China Environ Sci 38:1550–1560
4. Cui BJ, Gao F, Hu C et al (2019) The use of brackish and reclaimed waste water in agriculture: a review. J. Irrig Drain 38(7):60–68
5. Martinez-Perez S, Bienes R, Garcia-Diaz A, Sastre-Merlin A (2019) Salt accumulation in soils and plants under reclaimed water irrigation in urban parks of Madrid (Spain). Agric Water Manag 213:468–476
6. Poustie A, Yang Y, Verburg P et al (2020) Reclaimed wastewater as a viable water source for agricultural irrigation: a review of food crop growth inhibition and promotion in the context of environmental change. Sci Total Environ 739(5):139756
7. Han Y, Qi XB, Li P et al (2018) Effects of reclaimed water irrigation on crop and soil security: research progress. Chin Agric Sci Bull 34(20):96–100
8. Christou A, Eliadou E, Michael C et al (2014) Assessment of long-term wastewater irrigation impacts on the soil geochemical properties and the bioaccumulation of heavy metals to the agricultural products. Environ Monit Assess 186(8):4857–4870

9. Ugulu I, Khan Z I, Sheik Z et al (2021) Effect of wastewater irrigation as an alternative irrigation resource on heavy metal accumulation in ginger (Zingiber officinale Rosc.) and human health risk from consumption. Arab J Geosci 14(8):702
10. Chen W, Lu S, Chi P et al (2013) Accumulation of Cd in agricultural soil under long-term reclaimed water irrigation. Environ Pollut 178:294–299
11. Zhang ZH, Chen WF, Shi YF et al (2009) Effects of irrigation with regenerated water on the growth and quality of alfalfa. Chin J Appl Ecol 20(11):2659–2664
12. Cao YJ, Tian JC, Shen H et al (2021) The effects of reclaimed water irrigation on growth and quality of alfalfa in Ningxia. J Irrig Drain 40(1):55–61
13. Huang ZB, Miao ZX, Hou LW et al (2007) Effect of irrigation time and mode with reclaimed water on growth and quality of crops. J Agro-Environ Sci 06:2257–2261
14. Wu WX, He LZ, Shao JH et al (2016) Effect of fresh water and sewage irrigation on yield and quality of sugarcane. Water Sav Irrig 09:74–78
15. Jung K, Jang T, Jeong H et al (2014) Assessment of growth and yield components of rice irrigated with reclaimed wastewater. Agric Water Manag 138:17–25
16. Li XN, Wu JY, Sun WY et al (2011) Influence of reclaimed water irrigation on growth and quality of alfalfa and white clover. Acta Agrestia Sinica 19(3):463–467
17. Al-Lahham O, El-Assinm NM, Fayyad M (2003) Impact of treated wastewater irrigation on quality attributes and contamination of tomato fruit. Agric Water Manag 61:51–62
18. Xin HJ, Liu QY, Bu QW et al (2012) Study on yield, quality and heavy metal content of corn irrigated by reclaimed water. J Irrig Drain 31(6):98–102
19. Njuguna SM, Makokha VA, Yan X et al (2019) Health risk assessment by consumption of vegetables irrigated with reclaimed waste water: a case study in Thika (Kenya). Environ Manage 231:576–581
20. Cui BJ, Gao F, Hu C et al (2019) Effect of different reclaimed water irrigation methods on bacterial community diversity and pathogen abundance in the soil-pepper ecosystem. Environ Sci 38(07):5151–5163
21. Ju C, Li K, Jiang S et al (2011) Heavy metal Pb and Cd distribution in winter-wheat with reclaimed water irrigation. J Agro-Environ Sci 30(01):78–83

Chapter 25
Antibiotics in the Coastal Waters and Sediments of China: A Brief Review and Perspective

Yiwen Luo, Xiaodan Chen, Cheng Cheng, Wei Li, Gong Cheng, and Guowei Liao

Abstract Substantial amounts of antibiotics have been consumed in recent years. Coastal ecosystem is highly sensitive to human activities, therefore, antibiotics in coastal waters and sediments should be given higher concern. This study reviewed the presence of antibiotics within the marine environment along the coastlines of China's Bohai Sea, Yellow Sea, East China Sea, and South China Sea regions. According to related research, the results suggest that in coastal waters, the concentration of antibiotics was observed to be maximum in the Yellow Sea, subsequently higher in the East China Sea, then in the South China Sea, and the least in the Bohai Sea. SAs, QNs and MLs are commonly detected in coastal waters. The pollution in the Bohai Sea and the East China Sea may be related to intensive husbandry and aquaculture, with main contaminants were TCs, SAs and QNs. The main contaminants in the South China Sea and the Yellow Sea were SAs, QNs, and MLs, indicating that humans, animal husbandry, and aquaculture all contribute to environmental risks. The concentrations of antibiotics found in the coastal sediments within the East China Sea exhibited a notably higher level compared to those in other areas. To control and prevent antibiotics pollution, management of administration, safe alternatives and comprehensive monitoring should be enhanced and restricted further.

Keywords Antibiotics · China · Coastal waters · Review · Sediments

---

Yiwen Luo, Xiaodan Chen—These authors contributed equally to this work.

Y. Luo · X. Chen · W. Li (✉) · G. Cheng · G. Liao
Shenzhen Academy of Environmental Sciences, Shenzhen, Guangdong, China
e-mail: liwei@meeb.sz.gov.cn

C. Cheng
Shenzhen Baoan District Environmental Protection Association, Shenzhen, Guangdong, China

K. Y. Foo et al. (eds.), *Water Nexus Approach for Sustainable Development*,
Environmental Science and Engineering, https://doi.org/10.1007/978-3-031-89238-7_25

25.1 Introduction

Antibiotics are antimicrobial drugs used in the clinical treatment to prevent bacterial infections, as they can kill or inhibit the growth of micro-organisms [1]. Over half a century, the excessive and inappropriate utilization of antibiotics has dramatically fueled the rise of antibiotic resistance, posing a critical worldwide health crisis. According to the 2020 report on Global Antimicrobial Resistance and Usage Surveillance by the WHO, there has been a substantial hike in the number of nations grappling with antibiotic resistance [2]. Moreover, the COVID-19 pandemic has added an unanticipated dimension to this issue [3]. Projections indicate that without intervention, antibiotic resistance could lead to an estimated 10 million fatalities and a $100 trillion economic loss by 2050.[4]. Since 2009, China remained the greatest antibiotics production and export country. In 2020, China witnessed a 6.06% rise in the overall consumption of veterinary antibiotics, reaching a total of 32,776 tons compared to the previous year's figures [5]. In terms of antibiotic usage, China registers a notably higher intake compared to developed countries. The country's Defined Daily Dosage (DDD) stands at 157 g for every 1000 inhabitants. By comparison, the DDD in the United Kingdom, the United States, Canada, and Europe registers at 27.4, 28.8, 20.4, and 20.1 g, respectively [6].

Coastal areas interact frequently with the biosphere, lithosphere, atmosphere, and hydrosphere and are greatly influenced by human development [7]. During 2005–2016, 45 types of antibiotics were detected in the coastal waters of China's Bohai Sea, Yellow Sea, East China Sea, and South China Sea regions [8]. In 2017, a substantial presence of antibiotics was observed in the Yellow Sea, reaching an overall concentration as high as 1349.2 ng/L [6]. Sediments with adsorbed antibiotics exhibit resilience, primarily owing to the sluggish sediment mobility and limited oxygen availability. These antibiotics have the potential to re-release into the aquatic environment under certain conditions. It emphasizes the need for heightened attention towards antibiotics present in coastal water and sediment environments.

Studying the spatial distribution characteristics and ecological risk assessment of antibiotics is important for administrative departments to formulate management policies and improve public and environmental health. In recent years, there were many literatures on water and sediment in each coastal area separately, but few on nationwide comparison especially for the sediments. Acknowledging these apprehensions, this research aims to examine the prevalence of antibiotics across China's coastal waters and sediment, with a view to offering scientific direction for mitigating antibiotic pollution in the impending future.

25.2 Antibiotics in Coastal Waters

This review mainly focuses on the commonly prescribed antibiotics. Based on the chemical structure and function, antibiotics are commonly classified into seven categories as follows: tetracyclines (TCs), sulfonamides (SAs), quinolones (QNs), macrolides (MLs), chloramphenicols (CAPs), β-lactams (β-Ls), and various others fall into this category. Some antibiotics can inhibit the synthesis of proteins, such as TCs which are commonly used as growth-promoting agents and veterinary medicines; MLs which are widely used to treat humans; CAPs which are widely used in aquaculture [6]. Some antibiotics can inhibit the synthesis of nucleic acids, such as SAs and QNs which are commonly used in animal husbandry and aquaculture [6]. Some antibiotics can inhibit the synthesis of the bacterial cell wall, such as β-lactams and others which are widely used for treating human ailments.

In this study, 27 antibiotics which commonly identified in coastal water samples from China were chosen for comparative analysis, including four TCs, nine SAs, five QNs, four MLs, three CAPs, and two β-Ls (Table 25.1).

The most prevalent antibiotic category detected in the coastal waters of the Bohai Bay in 2018 was TCs [9]. OTC was the dominant pollutant at a maximum concentration of 200.9 ng/L [11, 12]. The Bohai Sea stands as China's shallowest semi-enclosed inland body of water, characterized by an exceptionally sluggish water turnover rate and a restricted ability to assimilate and detoxify contaminants. The pollution in the Bohai Sea has seriously threatened the livelihoods and health of the local inhabitants via the food chain [15].

The most prevalent antibiotic category detected in the Yellow Sea was QNs, both in Jiaozhou Bay and the Southern Yellow Sea. AMX and OFL were the dominant antibiotics, recorded within the range of 7.81 to 90.97 ng/L [10], and even non-detectable to 497.6 ng/L in certain cases. Overall, the antibiotic pollution in the Southern Yellow Sea in 2017 was more severe than in Jiaozhou Bay in 2019 [10]. The shores of the Yellow Sea boast substantial population density and have witnessed a meteoric rise in the aquaculture sector. Major rivers and ports, tourist areas, and the aquaculture industry discharge significant quantities of antibiotics into the coastal area every year. This ecosystem is severely threatened, as evidenced by the vast shrinkage of its landmass, pervasive pollution, persistent algal blooms, and significant decline in both insect and animal populations [16].

The main antibiotics in the East China Sea differed between Hangzhou Bay and the Yangtze Estuary. The main category in Hangzhou Bay was QNs, and the most concentrated antibiotic was OFL with a concentration range of ND–50.8 ng/L [7]. The primary group of antibiotics identified in the Yangtze Estuary comprised of SAs, among which SPD held the highest concentration, varying from non-detectable levels to 219 ng/L [11]. In 2013, the antibiotic contamination in the Yangtze Estuary exhibited a higher severity compared to the state of Hangzhou Bay in 2020 [7, 11]. The East China Sea region is home to numerous producers of antibiotics, such as QNs, MLs, and SAs, that discharge these substances into the receiving ecosystems through solid waste and wastewater discharges [6]. Furthermore, the East China Sea

Table 25.1 Concentrations (ng/L) of antibiotics from China's coastal waters

Compound		The Bohai Sea	The Yellow Sea		The East China Sea		The South China Sea		
		Bohai Bay (2018, n[a] = 10)[9]	Jiaozhou Bay (2019, n = 25)[10]	Southern Yellow Sea (2017, n = 30)[6]	Hangzhou Bay (2020, n = 5)[7]	Yangtze Estuary (2013, n = 7)[11]	Beibu Gulf (2018, n = 31)[12]	Pearl River Estuary (2013, n = 28)[13]	Shenzhen Bay (2018, n = 5)[14]
TCs	DC				0.2–1.2				
	TC	3–11.8				ND-2.37		< LOD[b]–13.1	
	OTC	ND[c]–200.9				ND-22.5			
	CTC				0.3–0.9	ND-3.5			
SAs	SDZ		< LOD		0.2–21.9	0.28–71.8	< LOD–0.87	< LOD–18	23 ± 12
	SMD					0.53–89.1		< LOD–218	31 ± 20
	SMX	1.9–17.5		ND–48.1	0.8–16.7	0.30–56.8	0.09–7.43	< LOD–37.6	195 ± 53
	SPD				ND–4.5	ND-219	< LOD		
	SDM			ND–9.3			< LOD–0.24		
	TMP	ND	ND–12.99	1.4–95.8	ND–1.7		0.02–1.13		
	SMM		1.20–3.88		ND–3.1		< LOD–0.53		
	SMR					ND	< LOD		
	STZ					0.03–5.23	< LOD–0.10		
QNs	CIP	ND	ND–21.48			ND-2.27	< LOD–2.76		
	ENR	ND	5.74–14.06	ND–121.2	ND–17.7	ND-4.77	< LOD–1.10	ND	
	ENX			ND–100.2	ND–1.6		< LOD–2.95		
	NOR		4.98–48.31	ND–21.1	ND–32.2	ND-14.2	0.43–6.17	< LOD–136	ND
	OFL	4.3–15.3	ND–30.79	ND–497.6	ND–50.8	ND-12.4	< LOD–0.46	< LOD–15.8	ND

(continued)

Table 25.1 (continued)

Compound		The Bohai Sea	The Yellow Sea		The East China Sea		The South China Sea		
		Bohai Bay (2018, n[a] = 10)[9]	Jiaozhou Bay (2019, n = 25)[10]	Southern Yellow Sea (2017, n = 30)[6]	Hangzhou Bay (2020, n = 5)[7]	Yangtze Estuary (2013, n = 7)[11]	Beibu Gulf (2018, n = 31)[12]	Pearl River Estuary (2013, n = 28)[13]	Shenzhen Bay (2018, n = 5)[14]
MLs	ATM		< LOD	ND–138.9	ND–1.5		< LOD		
	CTM			ND–89			< LOD–0.27		
	ETM–H2O		ND	ND–1.7			< LOD–3.43	< LOD–121	165 ± 98
	RTM		ND–30.62	ND–77.1	2.4–34.8	0.05–8.2	< LOD–0.58	ND	8 ± 5
CAPs	CAP		< LOD	ND–73.2		ND-8.63	< LOD–0.24		123 ± 100
	FF		ND–25.7	ND–42		0.45–89.5	< LOD–2.41		
	TAP			ND–77.5		0.91–110			
β-Ls	PNC	ND–5.1		ND–11.8					
	AMX	ND–5.5	7.81–90.97						ND

[a] Numbers of samples
[b] Limit of detection
[c] Not detected

receives an inflow of water and sediments from the Yangtze River. From 2006 to 2018, a total of 61 distinct antibiotics were identified in the water of the Yangtze River, while 43 different types were found in its sediments, as per statistical analysis [8].

In the Beibu Gulf region of the South China Sea, antibiotic levels were observed to be generally modest. Among them, the maximum presence was recorded for SMX, which exhibited a concentration span varying from 0.09 to 7.43 ng/L [12]. In the Pearl River Delta, the primary pollutants were found to be SAs, MLs, and QNs, whereas for Shenzhen Bay, the major contaminants were SAs and MLs. In the Pearl River Estuary, SMD was identified as the most prevalent antibiotic, with a peak concentration recorded at 218 ng/L [13], while in Shenzhen Bay, ETM-H_2O held the highest concentration at 263 ng/L [14]. The South China Sea has a lucrative pond pisciculture, which is crucial for the food security of millions of people in Southeast Asia. In 2018, a total of 13 antibiotic substances were identified within the surface waters originating from the coastal and distant shelf areas surrounding the South China Sea's coral reef ecosystems [17].

Among China's four principal marine regions, the highest antibiotic concentration was detected within the Yellow Sea, followed by a relatively elevated level in the East China Sea, then the South China Sea, with the Bohai Sea registering the lowest levels. Overall, antibiotics like SAs, QNs, and MLs were frequently observed in coastal water bodies. Their presence and variation can largely be attributed to the degree of regional economic development, alongside the methods used for treating illnesses in humans, livestock, and aquaculture. Possible origins of the pollution in the Bohai Sea and the East China Sea could be attributed to the robust livestock cultivation and aquatic farming practices, considering the prevalence of TCs, SAs, and QNs as major contaminants. Conversely, the South China Sea and the Yellow Sea exhibit a different scenario with SAs, QNs, and MLs as the primary contaminants, suggesting a more complex interplay of factors involving human activities, livestock husbandry, and aquaculture, all contributing to environmental hazards.

25.3 Antibiotics in Sediments

Twenty–five antibiotics detected in sediment samples from China in this study were selected for comparison, including four TCs, nine SAs, five QNs, four MLs, and three CAPs (Table 25.2). The highest antibiotic detected in the sediment samples of Bohai Bay in 2016 was OTC with a concentration of 1.6–48.4 ng/g [18]. The antibiotic levels detected in Dalian's section of the Yellow Sea were found to be comparably low, with SMM being the most prevalent, reaching a peak concentration of 7 ng/g [19]. In the waters of the East China Sea, particularly within Hangzhou Bay, the primary contaminant identified was SDZ, whose concentration levels varied from non-detectable (ND) to 21.7 ng/g [7]. The main antibiotic in the Yangtze Estuary was OFL with a maximum concentration of 458.2 ng/g [20]. In the Beibu Gulf region of the South China Sea, antibiotic levels tended to be modest, with Tetracycline (TC) registering as the most prevalent, reaching a peak concentration of 7.13 ng/g. [12]. The main contaminants in the Pearl River Estuary were MLs and QNs, the highest detected antibiotic was NOR with a concentration ranging of from LOD to 20.5 ng/g [13].

The most prevalent antibiotic category detected in the coastal sediments in the Bohai Bay was TCs, while in the Yellow Sea, the East China Sea and the South China Sea which were SAs, QNs and SAs, QNs and MLs, respectively. In general, the measured antibiotic levels in the East China Sea's coastal sediments showed a higher prevalence compared to sediment samples sourced from diverse geographic areas. The types of contaminants in coastal sediments are roughly the same as those in water. The main contaminants of coastal water and sediments in the Bohai Sea, the East China Sea and the South China Sea which were TCs, QNs and SAs, QNs and MLs, respectively. As the collection, extraction and analysis of samples were not the same, there is no obvious quantitative relationship between coastal waters and sediments, suggesting that further studies are needed.

25.4 Discussion and Future Perspectives

Despite presenting an extensive examination of antibiotics present in China's coastal waters and sediment, further concerns about antibiotics remain needed. Our suggestions for addressing these concerns are as follows.

(1) Medical Products Administration could provide financial subsidies to achieve further reform of the medical and health care system. Encouraging to use antibiotics with appropriate dosage to reduce the emergence of bacterial resistance.
(2) The government could increase investment in research and development of alternatives to antibiotics. For instance, replacing antibiotics with bacteriophage, Chinese herbal medicine and probiotics in pig production [21–23].

Further comprehensive monitoring of the antibiotics in the coastal area should be carried out. It was noteworthy that when multiple antibiotics were present the joint toxicity was enhanced simultaneously [24]. A nationwide laboratory, being operated as the European Antimicrobial Resistance Surveillance System (EARSS) [25], should be established to monitor the antibiotics and antibiotic resistance genes in the coastal regions and major cities.

25.5 Conclusions

Among China's four principal marine regions, the highest antibiotic concentration was detected within the Yellow Sea, followed by a relatively elevated level in the East China Sea, then the South China Sea, with the Bohai Sea registering the lowest levels.

Table 25.2 Concentrations (ng/g) of antibiotics from China's coastal sediments

Compound		The Bohai Sea	The Yellow Sea	The East China Sea		The South China Sea	
		Bohai Bay (2016, n[a] = 104)[18]	Dalian (2013, n = 20)[19]	Hangzhou Bay (2020, n = 5)[7]	Yangtze Estuary (2014, n = 7)[20]	Beibu Gulf (2018, n = 26)[12]	Pearl River Estuary (2013, n = 14)[13]
TCs	DC		1.10–1.54		< LOQ[b]-18.6		
	TC	ND[c]	1.66–1.74		< LOQ[d]-6.8	< LOQ-7.13	
	OTC	1.6–48.4	ND		0.3–14.0		
	CTC	9.1	ND		< LOQ-12		
SAs	SDZ		ND-1.68	ND-21.7	ND-0.5	< LOD-0.09	ND
	SMD		ND-1.76		< LOQ-4.8		< LOD-3.24
	SMX		ND		ND-1.1	< LOD-0.02	ND
	SPD			ND-4.9	ND-9.1	< LOD-0.14	
	SDM		ND			< LOD-0.09	
	TMP					< LOD-0.02	
	SMM		ND-7			< LOD	
	SMR		ND-3.67		ND-0.4	< LOD-0.35	
	STZ		ND-1.89		ND- < LOQ	< LOD	
QNs	CIP	2.4–9.8			< LOQ-42.9	< LOD-0.78	
	ENR	0.9–5.7		ND-8.1	< LOQ-4.8	< LOD-1.11	< LOD-1.43
	ENX					< LOD-2.76	
	NOR	1.2–6.4		ND-4.6	< LOQ-69.3	< LOD-4.06	< LOD-20.5
	OFL	0.2–6.0		ND-5.4	< LOQ-458.2	< LOD-0.40	< LOD-13.7
MLs	ATM			ND-1.2		< LOD-0.11	
	CTM			ND-9.9		< LOD-0.07	

(continued)

Table 25.2 (continued)

Compound		The Bohai Sea	The Yellow Sea	The East China Sea		The South China Sea	
		Bohai Bay (2016, n[a] = 104)[18]	Dalian (2013, n = 20)[19]	Hangzhou Bay (2020, n = 5)[7]	Yangtze Estuary (2014, n = 7)[20]	Beibu Gulf (2018, n = 26)[12]	Pearl River Estuary (2013, n = 14)[13]
	ETM-H2O					< LOD-0.57	< LOD-14
	RTM	ND		0.04–3.9	< LOQ-3.6	< LOD-0.10	< LOD-13.5
CAPs	CAP		ND-2.31		< LOQ	< LOD-0.04	
	FF		ND-1.21		< LOQ	< LOD-0.10	
	TAP				< LOQ		

[a] Numbers of samples
[b] Limit of detection
[c] Not detected
[d] Limit of quantitation

SAs, QNs and MLs are commonly detected in coastal waters. Antibiotics' prevalence and diversity predominantly stem from the regional economic development status and the therapeutic approaches employed for human, veterinary, and aquatic health issues. Possible origins of the pollution in the Bohai Sea and the East China Sea could be attributed to the robust livestock cultivation and aquatic farming practices, considering the prevalence of TCs, SAs, and QNs as major contaminants. Conversely, the South China Sea and the Yellow Sea exhibit a different scenario with SAs, QNs, and MLs as the primary contaminants, suggesting a more complex interplay of factors involving human activities, livestock husbandry, and aquaculture, all contributing to environmental hazards. The levels of antibiotics detected in the coastal deposits of the East China Sea displayed a significantly elevated status, surpassing those observed in distinct regions. The types of contaminants in coastal sediments are roughly the same as those in water. To ensure the coastal eco-environment safety, the supervision of the consumption and monitoring of antibiotics should be emphasized, and the usage and discharge of antibiotics should be restricted further.

References

1. Williams KJ (2009) The introduction of 'chemotherapy' using arsphenamine—the first magic bullet. J R Soc Med 102:343–348

2. Shao Y, Wang Y, Yuan Y, Xie Y (2021) A systematic review on antibiotics misuse in livestock and aquaculture and regulation implications in China. Sci Total Environ 798:149205
3. Ghosh S, Bornman C, Zafer MM (2021) Antimicrobial resistance threats in the emerging COVID-19 pandemic: where do we stand? J Infect Public Health 14(5):555–560
4. J, O.N. (2016) Tackling drug-resistant infections globally: Final report and recommendations. The Review on Antimicrobial Resistance.
5. Li H, Qiu LB, Chen BH, Wang H, Liu HY, Long YY, Hu LF, Fang CR (2022) Vertical distribution of antibiotics and antibiotic resistance genes in a representative municipal solid waste landfill China. Ecotoxicol Environ Saf 242:113919
6. Zhang QQ, Ying GG, Pan CG, Liu YS, Zhao JL (2015) Comprehensive evaluation of antibiotics emission and fate in the river basins of China: source analysis, multimedia modeling, and linkage to bacterial resistance. Environ Sci Technol 49(11):6772–6782
7. Li FF (2020) Distribution, ecological risks and priority control strategies of antibiotics in polluted coastal environment. China Univ Geosci (Beijing). [Doctoral thesis]
8. Li S, Shi WZ, Liu W, Li HM, Zhang W, Hu JR, Ke YC, Sun WL, Ni JR (2018) A duodecennial national synthesis of antibiotics in China's major rivers and seas (2005–2016). Sci Total Environ 615:906–917
9. Zhang YP (2018) Pollution characteristics of antibiotics and antibiotic resistance genes in the coastal area of Bohai Bay. Tianjin University [Master thesis]
10. Lu S, Lin CY, Lei K, Wang BD, Xin M, Gu X, Cao YX, Liu XT, Ouyang W, He MC (2020) Occurrence, spatiotemporal variation, and ecological risk of antibiotics in the water of the semi-enclosed urbanized Jiaozhou Bay in eastern China. Water Research 184:116187
11. Yan CX, Yang Y, Zhou JL, Liu M, Nie MH, Shi H, Gu LJ (2013) Antibiotics in the surface water of the Yangtze Estuary: occurrence, distribution and risk assessment. Environ Pollut 175:22–29
12. Wu Q (2021) Occurrence, bioaccumulation and trophic transfer of antibiotics in the Beibu gulf, South China sea. Guangxi University [Master thesis]
13. Liang XM, Chen BW, Nie XP, Shi Z, Huang XP, Li XD (2013) The distribution and partitioning of common antibiotics in water and sediments of the Pearl River Estuary South China. Chemosphere 92(11):1410–1416
14. Zhong ST, Gan BY, Tang YT, Wen F, Yang J, Zou JH (2018) Analysis of antibiotic pollution in rivers: case of Xijiang river basin of Guangdong. Strait Pharm J 30(08):61–64
15. Liu GY, Ge K, Wang ZH, Gao Y, Li A, Song L (2017) Research progress on antibioticpollution in coastal marine waters. J Bohai Univ (Natural Science Edition) 38(04):331–336
16. Murray NJ, Ma ZJ, Fuller RA (2015) Tidal flats of the Yellow Sea: a review of ecosystem status and anthropogenic threats. Austral Ecol 40(4):472–481
17. Zhang RJ, Yu KF, Li A, Wang YH, Pan CG, Huang XY (2019) Antibiotics in coral reef fishes from the South China Sea: occurrence, distribution, bioaccumulation, and dietary exposure risk to human. Sci Total Environ 135288
18. Liu XH, Zhang HB, Li LZ, Fu CC, Tu C, Huang YJ, Wu LH, Tang JH, Luo YM, Christie P (2016) Levels, distributions and sources of veterinary antibiotics in the sediments of the Bohai Sea in China and surrounding estuaries. Mar Pollut Bull 109(1):597–602
19. Na GS, Fang XD, Cai YQ, Ge LK, Zong HM, Yuan XT, Yao ZW, Zhang ZF (2013) Occurrence, distribution, and bioaccumulation of antibiotics in coastal environment of Dalian China. Mar Pollut Bull 69(1–2):233–237
20. Shi H, Yang Y, Liu M, Yan CX, Yue HY, Zhou JL (2014) Occurrence and distribution of antibiotics in the surface sediments of the Yangtze Estuary and nearby coastal areas. Mar Pollut Bull 83(1):317–323
21. You JM, Peng XT, Chen J, Zou TD, Wang ZR (2021) Application and research progress of bacteriophage in piglet diets. Feed Ind 42(16):1–5
22. Wang XY, Dong XX, Zhang Y, Zhang HL, Zhang BY (2022) Research progress on the application of Chinese herbal medicine to replace antibiotics in pig production. Heilongjiang Anim Sci Vet Med 42–45+51

23. Li GJ (2022) The effect of adding probiotics instead of antibiotics to feed on productive performance of pigs. Fujian J Anim Husb Vet Med 44(02):21–22
24. Li Z, Li M, Zhang ZY, Li P, Zang YG, Liu X (2020) Antibiotics in aquatic environments of China: a review and meta-analysis. Ecotoxicol Environ Saf 199:110668
25. Metz-Gercek S, Mittermayer H (2008) The European surveillance activities EARSS and ESAC in the context of ABS International. Wien Klin Wochenschr 120:264–267

Chapter 26
Spatio-Temporal Characteristics and Cause Analysis of Nutrient Salt Levels and Eutrophication in the Oujiang River Estuary

Jian Qian, Dongrong Zhang, Hengtao Xu, Lihong Chen, Fangqin Zheng, Yiran Jia, and Bo Lin

Abstract Estuaries are ecologically vulnerable water bodies and highly affected by pollutants from anthropogenic activities. Excessive nutrient input can cause water eutrophication and deterioration of vital marine ecosystems. However, few studies have compared nutrient salt concentrations and water eutrophication in the Oujiang River Estuary between seasons and their spatial distributions. This study used the single-factor standard index and eutrophication index (EI) methods to analyze the spatial and temporal patterns of nutrient salt levels and eutrophication occurrences in the estuary. Water quality monitoring data was obtained for each season of 2017. The dissolved inorganic nitrogen (DIN) and dissolved inorganic phosphorus (DIP) in the surface and bottom water exceeded the Chinese national standard to varying degrees, but that of DIN was significantly higher. The degree of eutrophication was relatively high. The seasonal DIN and DIP concentrations and EI were as follows: autumn > winter > spring > summer. Spatially, the DIN, DIP, and EI distribution exhibited high values in the estuary and coastal areas, whereas lower values were observed in the open sea and distant shores. The variation in the spatio-temporal characteristics is mainly influenced by ocean currents and terrigenous pollution. Our results provide scientific data support for strengthening environmental protection and sustainable utilization of important water bodies in the Oujiang River and Dongtou Islands.

Keywords Estuary · Nutrient · Spatio-temporal analysis · Eutrophication · Oujiang River Estuary

J. Qian (✉) · D. Zhang · H. Xu · L. Chen · F. Zheng · Y. Jia · B. Lin
Second Institute of Oceanography, Ministry of Natural Resources, Hangzhou, Zhejiang, China
e-mail: qianjian626@sina.com

J. Qian · D. Zhang · H. Xu · L. Chen
Key Laboratory of Nearshore Engineering Environment and Ecological Security of Zhejiang Province, Hangzhou, Zhejiang, China

K. Y. Foo et al. (eds.), *Water Nexus Approach for Sustainable Development*, Environmental Science and Engineering, https://doi.org/10.1007/978-3-031-89238-7_26

26.1 Introduction

Estuaries serve as the primary hubs for the interplay of various factors, including runoff, marine influences, atmospheric conditions, biological processes, geological dynamics, and human interventions. They act as the principal channels through which terrestrial materials are conveyed to the marine environment. Owing to their distinct geographical and physical characteristics, estuaries often face the most significant natural and human-induced impacts among coastal zones, thus playing a pivotal role in the development of marine eco-civilization in China [1].

Pollutant input may result in a range of ecological and environmental issues, including diminished water quality in estuaries, marine eutrophication, and degradation of the marine ecosystem. Various types of pollutants generated by human activities (represented by nitrogen and phosphorus nutrient salts) are eventually discharged into near-shore waters. Excessive nutrient salts in seawater trigger eutrophication of water bodies, potentially resulting in the emergence of red tides and other ecological disasters once the threshold of the seawater's self-purification capacity is surpassed. Extensive eutrophication resulting from nitrogen enrichment has been observed in the coastal waters of China, with more than 20% of these waters now categorized as "affected by eutrophication"[2].

The Oujiang River commences in Lishui City, Zhejiang Province, and terminates its journey by emptying into the East China Sea via Wenzhou City. With a total length of 388 km and a watershed area of 17,900 km^2, the Oujiang River stands as the second most prominent river in Zhejiang Province, occupying a significant position among China's estuaries, trailing only behind the Yangtze, Yellow, Pearl, and Qiantang estuaries. The Oujiang River Estuary is a subtropical monsoon climate area, with four distinct seasons and obvious seasonal changes in hydrological features [3]. In recent years, scientific studies have been carried out on the aquatic environment near the Oujiang River Estuary [4–7], but studies on nutrient salts and eutrophic conditions in combination with year-round and spatial distributions are still relatively rare.

This investigation delved into the seasonal changes and spatial distribution patterns of nutrient salts and eutrophic conditions in the seawater surrounding the Oujiang River Estuary. We used survey data on the current status of the water environment during spring, summer, autumn, and winter in 2017 to provide a research data support for the ecological environmental protection of the Oujiang River Estuary and the sustainable use of the marine environment.

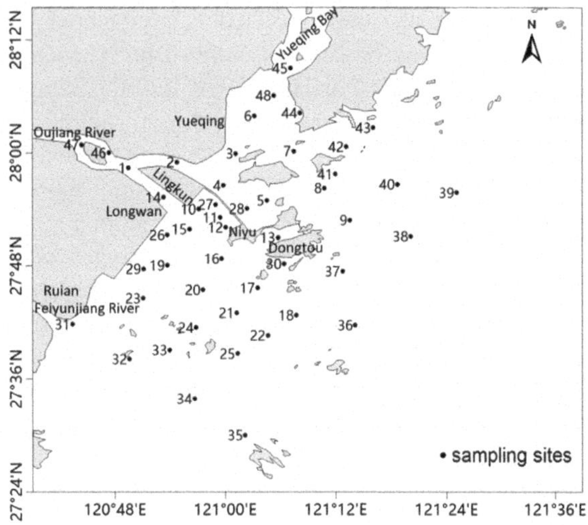

Fig. 26.1 Location of sampling sites (n = 48)

26.2 Methodology

26.2.1 Study Area

The study area encompassed the sea area of the Oujiang River Estuary and Dongtou Islands (Fig. 26.1), from Yueqing Bay in the north to the mouth of Feiyun River in the south (longitude 120° 44′–121° 26′ E to latitude 27° 29′–28° 10′ N) and centered on Niyu Island with an outer edge of ~ 30 km in all directions. Outside the Oujiang River Estuary are the Dongtou Islands, which are located in the southeastern sea area of Zhejiang Province where the mouths of the Oujiang and Yueqing Bay converge. The area contains many islands and rich marine biological resources. Following the Zhoushan fishing ground, the Dongtou fishing ground ranks as the second-largest in Zhejiang Province. In recent years, the realization of the Wenzhou Peninsula Project, beach reclamation projects, cross-sea bridge projects, farming activities, sewage discharge, and harbor shipping have caused changes in the ecological environment of the sea.

26.2.2 Sampling

In 2017, four rounds of sampling were conducted in the Oujiang River Estuary. Samples were collected in February 2017 (winter), May 2017 (spring), August 2017 (summer), and October 2017 (autumn) from 48 sampling sites (Fig. 26.1).

The process of sample collection, pretreatment, and analysis adhered strictly to the standardized procedures outlined in "The specification for marine monitoring" (GB17378.3-2007 and GB17378.4-2007). Sampling was conducted as follows: Surface water samples were gathered from stations with depths less than 10 m; for stations between 10 and 25 m in depth, both surface and bottom water samples were taken; and for stations with depths of 25 m or more, samples were collected from the surface, the 10-meter layer, and the bottom.

Sampling factors included chemical oxygen demand (COD), dissolved inorganic nitrogen (DIN), and dissolved inorganic phosphorus (DIP). COD was assayed through the application of the alkaline potassium permanganate method. DIP content was determined using phosphomolybdenum blue spectrophotometry. DIN was calculated as the total of ammonium nitrogen (NH_4^+-N), nitrate nitrogen (NO_3^--N), and nitrite nitrogen (NO_2^--N). NH_4^+-N content was determined using the hypobromite oxidation method. NO_3^--N content was determined using the zinc–cadmium reduction method. NO_2^-N content was determined using the diazo azo method.

26.2.3 Evaluation Methods

Evaluation of nutrient salts.

The single-factor standard index method is widely used to evaluate the nutrient salts in seawater. The equation for this method is as follows:

$$Pi = C_i / C_{io} \tag{26.1}$$

where Pi is the standardized index of the ith factor; that is the single-factor standardized index. If $Pi > 1$, this indicates that the factor exceeds the corresponding water quality evaluation criteria. C_i is the measured concentration of the ith factor. C_{io} is the evaluation standard value of the ith factor, the evaluation standard adopts the Chinese national standard "Sea water quality standard of China (GB 3097-1997)." The standard DIP values for the first, second, third, and fourth categories were 0.015, 0.030, 0.030, and 0.045 mg/L, respectively. The standard DIN values for the first, second, third, and fourth categories were 0.20, 0.30, 0.40, and 0.50, respectively.

Calculation of eutrophication index (EI).

Over the past decades, many scholars have devoted themselves to water eutrophication research and proposed various evaluation methods. In China's near-shore waters, current eutrophication evaluation models and methods are mainly based on the initial generation of the nutrient-based evaluation system, specifically, utilizing mathematical expressions that determine the eutrophication index (EI) through the consideration of nitrogen, phosphorus, COD, and chlorophyll concentrations [8]. Internationally, the second generation of the National Estuarine Eutrophication Assessment

System (NEEA/ASSETS) in the United States [9] and OSPAR-COMPP(OSPAR-COMPP) in the European Union[10] have become the most important eutrophication evaluation systems, which have become the mainstream evaluation method used in nearshore waters. However, China is still in the transition stage from the first generation evaluation system to the second generation evaluation system [11].

To make the results of this study comparable with those of other sea areas in China, we used the eutrophication index method (E-method) commonly used in China to analyze and evaluate the eutrophication status of the Oujiang River Estuary.

Since 2010, the Bulletin of Marine Environmental Quality of China issued by the State Oceanic Administration of China has used the following equation:

$$E = \text{DIN} \times \text{DIP} \times \text{COD}/4500 \times 10^6 \tag{26.2}$$

where E is the eutrophication index, DIN is dissolved inorganic nitrogen (mg/L), DIP is dissolved inorganic phosphorus (mg/L), and COD is chemical oxygen demand (mg/L). Eutrophication was defined as $E \geq 1$, light eutrophication as $1 \geq E \geq 3$, medium eutrophication as $3 < E \geq 9$, and heavy eutrophication as $E > 9$ [12].

26.3 Results and Discussion

26.3.1 Seasonal Variation of Nutrient Salts

We analyzed the nutrient salt survey data in the Oujiang River Estuary throughout the year and applied the single-factor standard index method to the exceedances of DIN and DIP (Tables 26.1 and 26.2, respectively).

In spring, DIN values varied between 0.221 and 2.310 mg/L, with an mean value of 0.753 mg/L (Table 26.1). Stations exceeding the standard for seawater quality of classes I, II, III, and IV accounted for 100%, 98.7%, 87.2%, and 71.8% of the overall stations, respectively. In summer, DIN values varied between 0.159 and 2.227 mg/L, averaging at 0.605 mg/L. The proportion of stations exceeding Class I, II, III,

Table 26.1 Seasonal descriptive statistics of DIN

Item		Spring	Summer	Autumn	Winter
Min (mg/L)		0.221	0.159	0.372	0.265
Max (mg/L)		2.310	2.227	2.130	1.991
Mean (mg/L)		0.753	0.605	1.032	0.893
Exceedance rate (%)	First class (I)	100	97.2	100	100
	Second class (II)	98.7	81.9	100	96.1
	Third class (III)	87.2	59.7	98.7	94.7
	Fourth class (IV)	71.8	41.7	98.7	90.8

Table 26.2 Seasonal descriptive statistics of DIP

Item		Spring	Summer	Autumn	Winter
Min (mg/L)		0.005	0.004	0.035	0.016
Max (mg/L)		0.111	0.091	0.089	0.064
Mean (mg/L)		0.030	0.027	0.049	0.035
Exceedance rate (%)	First class (I)	76.9	68.1	100	100
	Second class (II)	30.8	31.9	100	59.2
	Third class (III)	30.8	31.9	100	59.2
	Fourth class (IV)	16.7	13.9	44.2	9.2

and IV standards was 97.2%, 81.9%, 61.1%, and 41.7%, respectively, among all monitored stations. In autumn, the DIN values varied between 0.372 and 2.130 mg/L, with a mean value of 1.032 mg/L. Stations exceeding the standard for classes I–IV accounted for 100%, 100%, 98.7%, and 98.7% of the total number of stations, respectively. In winter, the measured DIN values varied between 0.265 and 1.991 mg/L, with an mean value of 0.893 mg/L, and stations exceeding the standard of Class I to IV accounted for 100%, 96.1%, 94.7%, and 90.8% of all stations, respectively. From the seasonal average, DIN concentration in the Oujiang River Estuary was in the order autumn > winter > spring > summer, with the highest in autumn and lowest in summer.

In spring, DIP values varied between 0.005 mg/L and 0.111 mg/L, with 76.9%, 30.8%, 30.8%, and 16.7% of the stations exceeding the Class I, Class II, Class III, and Class IV standards, respectively (Table 26.2). In summer, DIP values varied between 0.004 and 0.091 mg/L, with 69.4%, 31.9%, 31.9%, and 13.9% of the stations exceeding the standard for Class I to Class IV, respectively. In autumn, DIP values varied between 0.035 and 0.089 mg/L, with 100%, 100%, 100%, and 45.5% of the total stations exceeding the standard for Class I to Class IV, respectively. In winter, the measured DIP values varied between 0.016 and 0.064 mg/L, with 100%, 63.2%, 63.2%, and 9.2% of the stations exceeding Class I to Class IV standards, respectively. Seasonally, DIP concentration was in the order autumn > winter > spring > summer, with the highest value in autumn, the second highest in winter, and relatively low DIP concentrations in spring and summer.

The exceedance rate of DIN was significantly higher than that of DIP. Most water bodies in the autumn and winter study areas exceeded the fourth-class standard for seawater quality. Thus, DIN is the most important factor impacting eutrophication in the Oujiang River Estuary.

26.3.2 Spatial Distribution of Nutrient Salts

The results of the single-factor evaluation showed that DIN and DIP degrees generally differed between seasons. Thus, we further analyzed the spatial distribution characteristics of these two nutrient salt factors. Figures 26.2, 26.3, 26.4 and 26.5 demonstrate the seasonal distributions of the surface and bottom DIN and DIP concentrations in the seawater near the Oujiang River Estuary.

As seen in Figs. 26.2 and 26.3, DIN planar distribution in the waters surrounding the Oujiang River Estuary is characterized by a high distribution in the estuary and a low distribution in the outer sea; planar distribution patterns do not differ greatly between the surface and bottom layers. The concentration of DIN in the majority of the sea areas studied surpassed the DIN Class III water quality standard concentration. The region with the high DIN concentration was found in the sea area between the Oujiang River Estuary and the Dongtou Islands. In the autumn and winter 2017, the surface and bottom layer water bodies had a DIN concentration inferior to IV water quality.

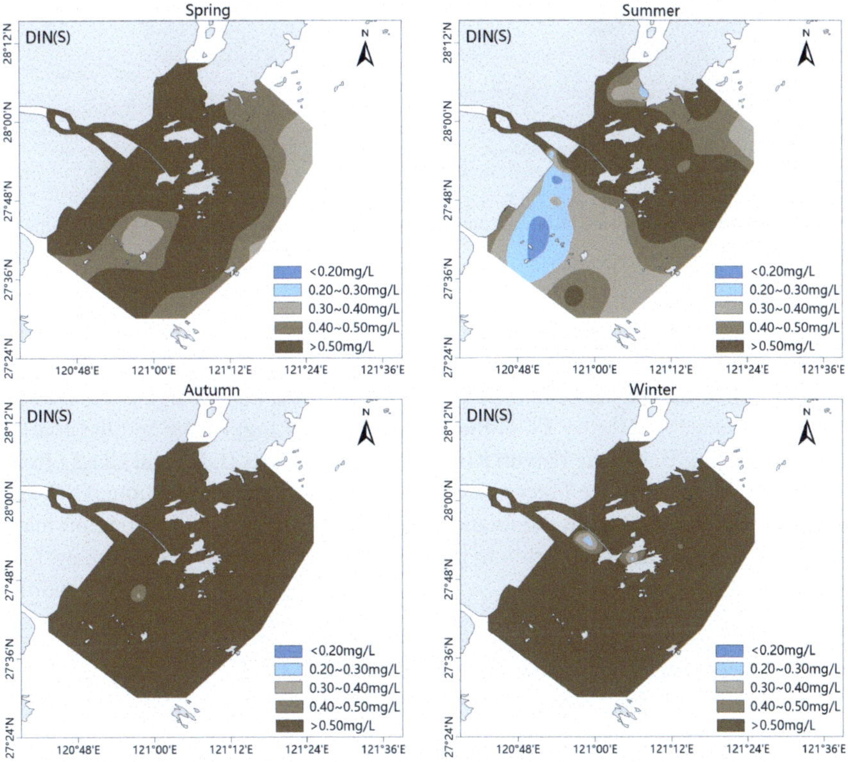

Fig. 26.2 Plane distribution of DIN concentrations in the surface layer of the sea near the Oujiang River Estuary

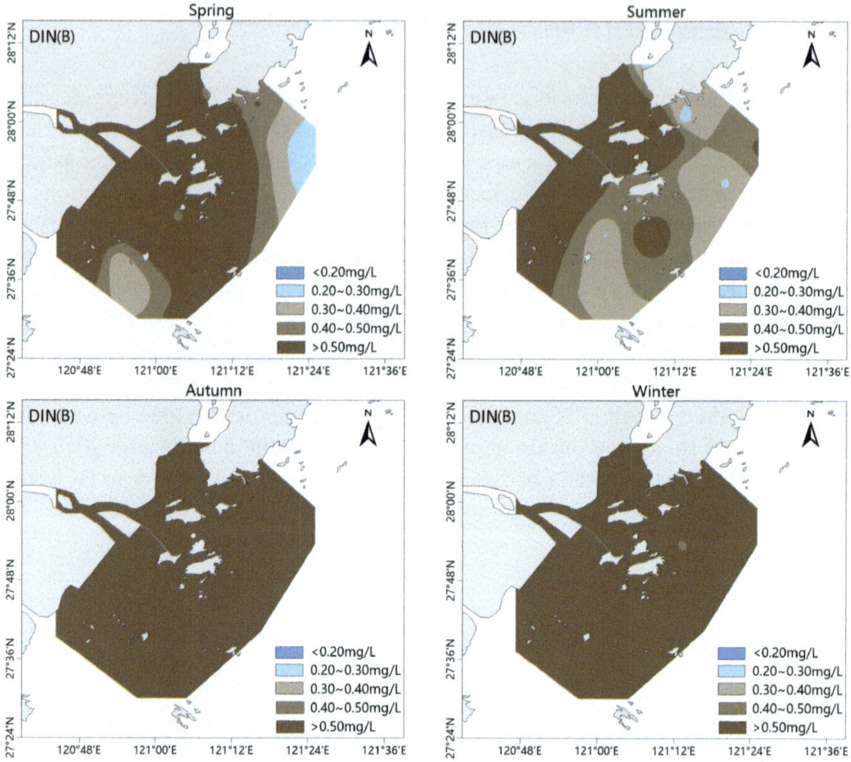

Fig. 26.3 Plane distribution of DIN concentrations in the bottom layer of the sea near Oujiang River Estuary

DIP plane distribution in the Oujiang River Estuary was'step-like,' which gradually decreased from the estuary to the outer sea (Figs. 26.4 and 26.5). In spring and summer, DIP in the sea between the Oujiang River Estuary and Lingkun Island was inferior to Class IV, category IV in the sea around the Lingni Dike and the coastal sea from Yueqing Bay to the Feiyun River Estuary, and Class II, III, and Class I from Dongtou main island to the outer sea was DIP (in that order). In autumn, DIP was inferior to Category IV from the mouth of the Oujiang River to the Dongtou main island, and Class IV from Dongtou to the outer sea.

26.3.3 *Seasonal and Spatial Variation in the EI*

Eutrophication refers to a water quality degradation phenomenon that arises due to an overabundance of nutrients, primarily nitrogen and phosphorus. As seen in our analysis in Sect. 3.1, the amounts of DIN and DIP in the Oujiang River Estuary

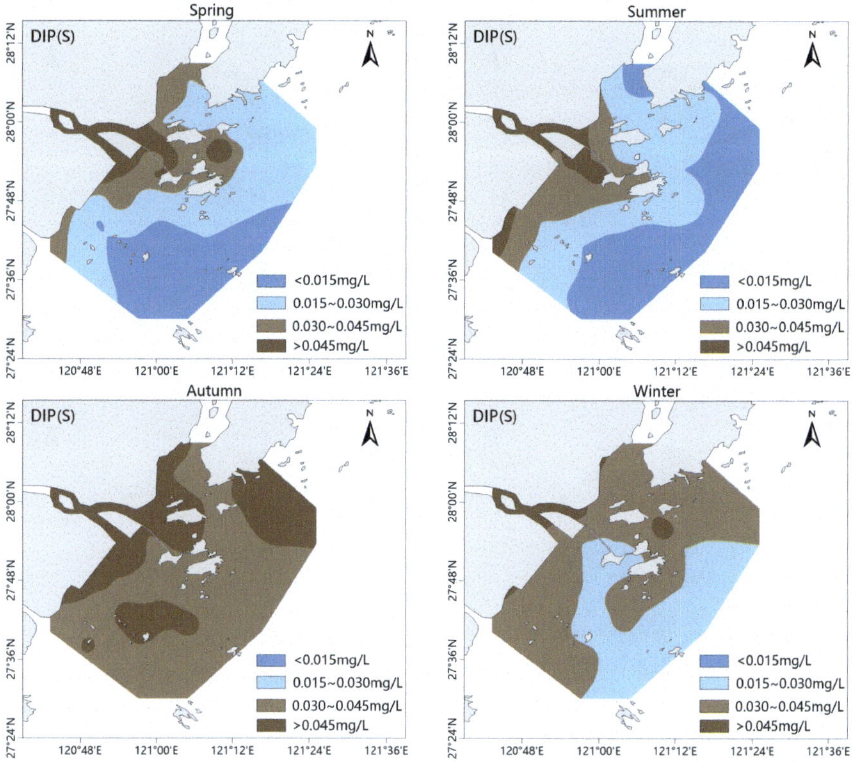

Fig. 26.4 Plane distribution of DIP concentrations in the surface layer of the sea near Oujiang River Estuary

waters are generally excessive, which is the main cause of eutrophication. The EI distribution in the Oujiang River Estuary is shown in Fig. 26.6.

In spring, the Oujiang River and Yueqing Bay estuary waters showed heavy eutrophication; the Dongtou Island sea area and the outer sea had medium and mild eutrophication, respectively. In summer, the Oujiang River Estuary has heavy eutrophication; the Yueqing Bay mouth to Lingkun Island waters had medium eutrophication, Dongtou Island to the mouth of the Feiyun River inshore waters were lightly eutrophic, and the outer waters were not eutrophic. In autumn, the Oujiang River Estuary and Dongtou Island in the northeastern part of the sea were heavily eutrophic, and the rest of the sea was moderately eutrophic. In winter, heavy, medium, and mild eutrophication occurred from the Oujiang River Estuary, Yueqing Bay mouth, and Feiyun River estuary near the sea coast to the outer sea. Seasonal differences in eutrophication were evident, with the highest eutrophication in autumn, followed by winter and spring, and a relatively low degree of eutrophication in summer. Eutrophication distribution showed a pattern of gradual decrease from the estuary to the outer sea and from the nearshore to the far shore.

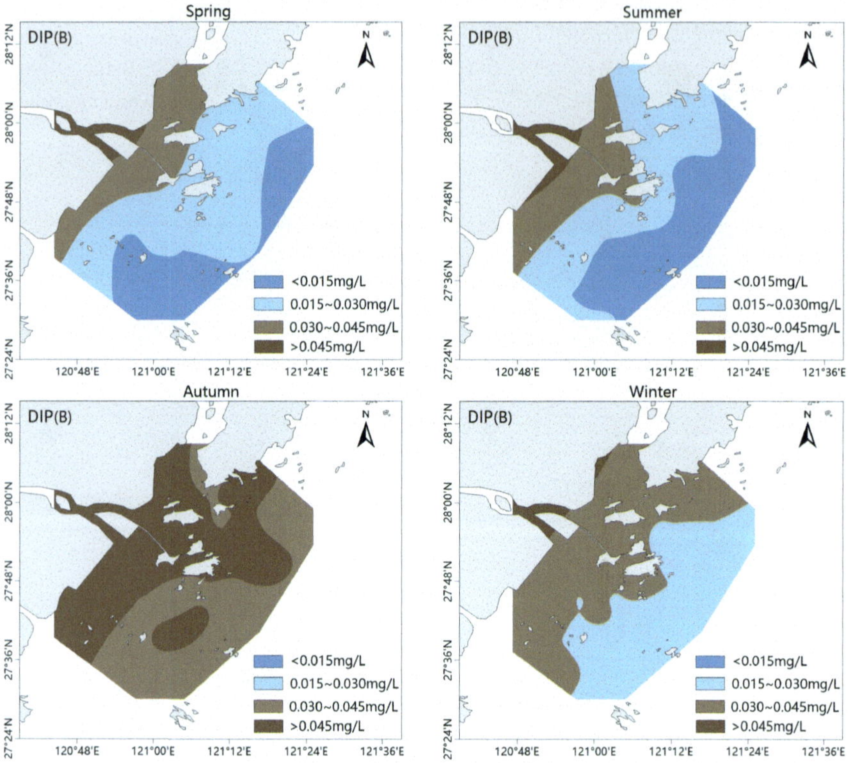

Fig. 26.5 Plane distribution of DIP concentrations in the bottom layer of the sea near Oujiang River Estuary

26.3.4 Cause Analysis of the Variation in Spatio-Temporal Characteristics

The study area is located in the south of Zhejiang Province, with Yueqing Bay to the north, and the Oujiang and Feiyun rivers to its west, which are affected by the northbound Taiwan warm current during summer and the southbound Zhejiang coastal current during winter. The Zhejiang coastal current has low salinity, high sand content, little transparency, and abundant nutrient salts, whereas the Taiwan warm current has high salinity, high transparency, and low nutrient salts [13, 14].

From the seasonal perspective, from May (spring) 2017, the Taiwan warm current moved northward along the coasts of Zhejiang and Fujian under the influence of monsoon, its intensity gradually increased, and it approached the coast of the mainland, whereas the intensity of the Zhejiang coastal current gradually decreased. At this time, the salinity of the sea increased and the nutrient salt decreased significantly. In the summer, the Taiwan warm current had the highest intensity, highest salinity, and lowest nutrient content. In September (autumn), the Taiwan warm current retreated

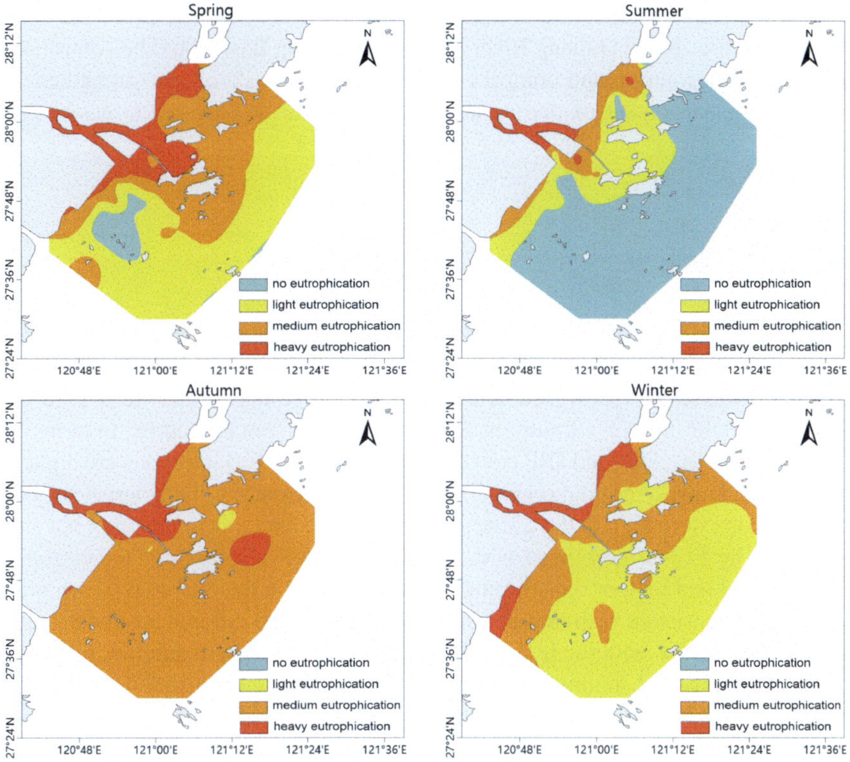

Fig. 26.6 Seasonal distribution of EI of the sea near the Oujiang River Estuary

and gradually formed the Jiangzhe coastal current. By autumn, it had reached its maximum value, and the salinity of the sea decreased and the nutrient content increased significantly. In the winter, the coastal current intensity was the highest, resulting in high nutrient salt concentrations in the sea. Simultaneously, the runoff of rivers along the study area into the sea also has an impact on the seasonal variation of nutrient concentration in this area, but because both the Oujiang and Feiyun rivers are mountain streams, the scope and extent of their impact are considerably limited [15].

From the spatial perspective, the sea is affected by the pollution of Oujiang River, Feiyun River, and Yueqing Bay, which makes the nutrient salt and eutrophication situation present the characteristics of high estuary and near shore, and low offshore and far shore. In particular, Oujiang River is the second largest river in Zhejiang Province, and the flux of its major pollutants into the sea accounts for a large proportion of land-based pollution input in the Wenzhou sea area [4]. Yueqing Bay in the north has been undergoing moderate and severe eutrophication due to the combined impact of domestic sewage, agricultural fertilizers, and aquaculture products [16].

Therefore, the total amount of pollutants produced by daily life, industrial, and agricultural activities in the Oujiang River basin and Yueqing Bay should be controlled, the pollution prevention and control efforts in the region should be strengthened, and the scale and layout of marine aquaculture should be adjusted to ensure the sustainable development of the ecological environment.

26.4 Conclusions

In this study, the spatial and temporal patterns of nutrient salts and eutrophication status in the Oujiang River Estuary were analyzed using the common Chinese methods of the single-factor standard index and EI. Similar results were obtained for DIN and DIP in terms of nutrient salt spatial and temporal patterns. In terms of seasonal variations, DIN and DIP concentrations in the studied sea area were highest in autumn and lowest in summer; in terms of spatial distribution characteristics, both concentrations were high in the estuary and near shore, and low in the outer sea and far shore. The EI was correlated with the DIN and DIP values, thus, the eutrophication spatial and temporal distribution characteristics in the Oujiangkou sea area were largely consistent with that of nutrient salts. Seasonal differences in eutrophication were marked, with the highest degree of eutrophication in autumn, followed by winter, spring, and summer. The varying spatio-temporal distribution pattern of nutrient levels in the study area is influenced by the seasonal trends of the Zhejiang coastal and Taiwan warm currents, as well as the pollution inputs from the Oujiang River, Feiyun River, and Yueqing Bay. These results provide a reference for ecological environmental protection and sustainable utilization of the marine environment in the Oujiang and Dongtou Islands.

Acknowledgements This study was supported by the Scientific Research Fund of the Second Institute of Oceanography, Ministry of Natural Resources, China (Grant numbers JG1610 and JG2216).

References

1. Lin Y, Zhang H, Ye G, Jiang J, Jiang Q (2023) Marine eco-civilization performance evaluation in Oujiang Estuary, Zhejiang Province, China. Environ Dev Sustain 25:176–197. https://doi.org/10.1007/s10668-021-02047-y
2. Wang Y, Liu D, Xiao W, Zhou P, Tian C, Zhang C, Du J, Guo H, Wang B (2021) Coastal eutrophication in China: trend, sources, and ecological effects. Harmful Algae 107:102058
3. Chai PL (2020) Distribution Characteristics of Heavy Metals of the Oujiang Estuary Sediments and their Response to Human Activities. Nanjing University, Nanjing
4. Ma ZK, Liu YL, Qiu JK (2021) Quality assessment of aquatic environment and fluxes of major pollutants to the Oujiang River Estuary. Ocean Dev Manag 38:92–96. https://doi.org/10.20016/j.cnki.hykfygl.2021.05.016

5. Wang S, Huang W, Xu S, Wan ZH (2023) Water pollution characteristics and water quality evaluation of Oujiang Estuary (2014–2018). Polish J Environ Stud 32:1–8. https://doi.org/10.15244/pjoes/169395
6. Li YC, Liu JG (2019) A numerical study on salinity stratification at the Oujiang River Estuary, China. Acta Oceanol Sin 38:40–50. https://doi.org/10.1007/s13131-019-1497-0
7. Zheng W, Li R, Qin W, Chen B, Wang M, Guan W, Zhang X, Yang Q, Zhao M, Ma Z (2022) Tidal water exchanges can shape the phytoplankton community structure and reduce the risk of harmful cyanobacterial blooms in a semi-closed lake. J Ocean Limnol 40:1868–1880. https://doi.org/10.1007/s00343-022-2032-2
8. Lin XJ, Gao S, Zhang TY, Liu GM (2018) Researching progress and application status of eutrophication evaluation method of seawater. Adv Earth Sci 33:373–384
9. Bricker SB, Clement CG, Pirhalla DE, Orlando SP, Farrow DRG (1999) National estuarine eutrophication assessment: effects of nutrient enrichment in the nation's estuaries. Silver Spring: NOAA-NOS Special Projects Office and National Centers for Coastal Ocean Science. https://repository.library.noaa.gov/view/noaa/1693
10. OSPAR Commission (2005) Common procedure for the identification of the eutrophication status of the OSPAR maritime area. OSPAR Publication, pp 36
11. Wang YH, Qin HW, Wang QL, Lv HR, Ma HY, Liang SK, Guo H (2023) Evaluation of spatial and temporal distribution of water quality and eutrophication in Bohai Bay during 2006–2020. Marine Environ Sci 42:693–703. https://doi.org/10.13634/j.cnki.mes.2023.05.006
12. Ministry of Ecology and Environment of the People's Republic of China (MEE) (2023) Bulletin of Marine Ecology and Environment Status of China in 2022 (in Chinese)
13. Wen TT (2010) Preliminary Study of Nutrients in Yellow Sea and the East China Sea and the Cold Eddy Northeast of Taiwan, Ocean University of China
14. Ye P, Xuan J, Huang D (2022) Evolution and dynamics of a summertime penetrating front off the Zhejiang-Fujian coast, China. Science China (Earth Sciences) 65:556–569
15. Jiang GC, Wang YH, Dong HL (1986) The variations of nutrients in the adjacent area of Oujiang Estuary. Donghai Marine Sci 69–76
16. Chen L, Xu ZL, Chen S, Cai SW, Lv BQ, Bao XP (2010) Causal analysis and distribution of eutrophication index in the Yueging Bay in spring-summer of 2007. J. Shanghai Ocean Univ. 19:91–97

Chapter 27
Using Photosynthetic Bacteria to Remove NH_4^+ Pollutants from Aquaculture Wastewater

Shuai Liang, Weiyu Du, Ying Wang, Lianshun Wang, Guojun Yang, Yanan Lu, and Yuting Cong

Abstract Water pollution is a significant challenge facing the world today, highlighting the critical importance of researching efficient wastewater treatment technologies. Photosynthetic bacteria have emerged as a hot topic in wastewater treatment due to their advantages such as ease of cultivation, strong adaptability, and non-toxicity without secondary pollution. In this study, we screened and identified a strain of photosynthetic bacteria that exhibited excellent performance in denitrification experiments. This finding lays a solid foundation for the large-scale application of photosynthetic bacteria in wastewater treatment.

Keywords Photosynthetic bacteria · Aquaculture · Waste water treatment · Biological nitrogen removal · Biomass

27.1 Introduction

Since the 1970s, China's aquaculture industry has experienced rapid growth, consistently ranking as the world's top producer for 31 years with around 60% of global output. While this has boosted high-quality aquatic product availability and contributed to economic growth, it has also led to environmental challenges. High-density cultivation methods adopted to meet demand have resulted in widespread pollution of aquaculture water bodies. Limited expertise among practitioners has led to environmental disregard and unregulated discharge of nitrogen-containing

S. Liang · W. Du · Y. Wang · L. Wang (✉) · G. Yang (✉) · Y. Lu · Y. Cong
National Demonstration Center for Experimental Aquaculture Education (Dalian Ocean University), College of Fisheries and Life Science, Ministry of Education, Dalian, China
e-mail: wanglianshun@dlou.edu.cn

G. Yang
e-mail: ygj@dlou.edu.cn

Dalian Key Laboratory of Breeding, Reproduction and Aquaculture of Crustaceans, Dalian, China

Key Laboratory of Environment Controlled Aquaculture, Ministry of Education, Dalian, China

K. Y. Foo et al. (eds.), *Water Nexus Approach for Sustainable Development*, Environmental Science and Engineering, https://doi.org/10.1007/978-3-031-89238-7_27

wastewater, exacerbating environmental deterioration. Urgent action is needed to address China's environmental pollution crisis. Common pollutants in aquaculture wastewater are nitrogen compounds, including NH_4^+, NO_2, and NO_3. Among them, ammonia nitrogen is the most important water quality indicator because it can cause various problems for aquatic organisms such as respiratory difficulties and decreased disease resistance when it enters their bodies with feed in the water. Severe cases can lead to toxicity and even death of aquatic organisms. Therefore, efficiently treating aquaculture wastewater is a very serious ecological and environmental issue. In terms of wastewater treatment, microbial treatment methods have gained widespread attention worldwide due to their high efficiency, low cost, and simple operation [1].

Purple Sulfur Bacteria (PSB), as a commonly used wastewater treatment agent, has a widespread distribution. It can typically be found growing in soil, ponds, wetlands, aquaculture sludge, and even livestock manure [2], PSB possesses characteristics such as no secondary pollution, no excess, and no residue [3].

In addition, PSB contains abundant proteins, which can be used as feed additives to provide additional nutrients for the growth of aquatic organisms. The key function of PSB is to effectively reduce harmful substances such as ammonia nitrogen and nitrite nitrogen in water bodies, thus possessing the ability to regulate the aquatic environment. Currently, PSB has a wide range of applications in environmental remediation, energy recovery and reuse, agriculture, and other fields [4].

In light of this, the authors selected a strain of PSB from the turtle ponds in the natural reserve area of Chaoyang, Liaoning, and conducted wastewater degradation experiments to investigate its ability to degrade ammonia nitrogen in organic wastewater. This study aims to provide theoretical support for the future large-scale industrial production of PSB and its application in environmental remediation.

27.2 Materials and Methods

27.2.1 Main Instruments

Double-Person Single-Sided Purification Workbench, UV–Visible Spectrophotometer, Steam Sterilizer, Light Incubator, pH Meter. Incandescent Lamp (60 W), Glass Bottles with Stoppers (30 mL).

Enrichment Culture Medium: NaCl:1 g, $MgSO_4 \cdot 7H_2O$:0.1 g, NH_4Cl:0.5 g, KH_2PO_4:0.875 g, CH_3COONa: 1.5 g, Yeast Extract: 0.5 g, Distilled Water: 500 mL, pH: 7.0. After sterilization, add 1 g of Sodium Bicarbonate ($NaHCO_3$).

Isolation Culture Medium: The isolation medium is the same as the enrichment medium, followed by the addition of 1.7% agar.

Aquaculture Wastewater: The wastewater samples were collected from the aquaculture discharge of a shrimp farming factory in Dalian, Liaoning Province, and from natural sea cucumber ponds. The concentrations were 65.8 mg/L and 36.8 mg/L, respectively.

27.2.2 Experimental Procedure

Enrichment of photosynthetic bacteria.

Sampling and Enrichment: Samples of mud were collected from the turtle ponds in Zhong hua Turtle Nature Reserve, Chaoyang City, Liaoning Province. The samples were sealed in bags and transported back to the laboratory at Dalian Ocean University. Subsequently, 5 g of the sample was placed into a 30 ml glass bottle with a stopper. The bottle was then filled with sterilized enrichment culture medium and placed in a light incubator set at 35 °C with a light intensity of 2000 lx for 5 days. Upon the liquid in the bottle turning red, indicating the presence of PSB, 10 ml of the enriched liquid was extracted for further enrichment culture.

Isolation of photosynthetic bacteria.

Isolation and Purification: Using the double-layer agar plate method, inoculate 0.2 ml of the enriched culture onto solid isolation culture medium. Then, cover it with another layer of agar culture medium. Incubate at 35 °C with continuous illumination until red colonies appear. Designate the isolated strain as SL-1. Pick single colonies and streak them onto plates again, repeating the process until the strain is fully purified.

Experiment on the Degradation of Ammonia Nitrogen in Aquaculture Wastewater by Strain SL-1.

The experiment was conducted using the sodium hypobromite method, with measurements taken at a wavelength of 543 nm using a 1 cm glass cuvette. A standard curve was established using distilled water as a reference. In the experimental groups, 60 ml of wastewater was taken for both types of wastewater, to which 1% inoculum was added. Samples were taken every 24 h with two replicate samples at each time point, continuously for 5 days. The concentration of ammonia nitrogen in the inoculated wastewater was determined based on the standard curve, with the untreated wastewater serving as the control. Degradation rates were calculated accordingly. $R = (Ct - C0)/C0 \times 100\%$(R is the degradation rate; Ct is the control concentration; C0 is the processing concentration) To ensure the smooth progress of the experiment, this study summarized the experimental methods according to the steps outlined in Fig. 27.1. Additionally, the wastewater treatment process was conducted to ensure consistent environmental conditions, as depicted in Fig. 27.2.

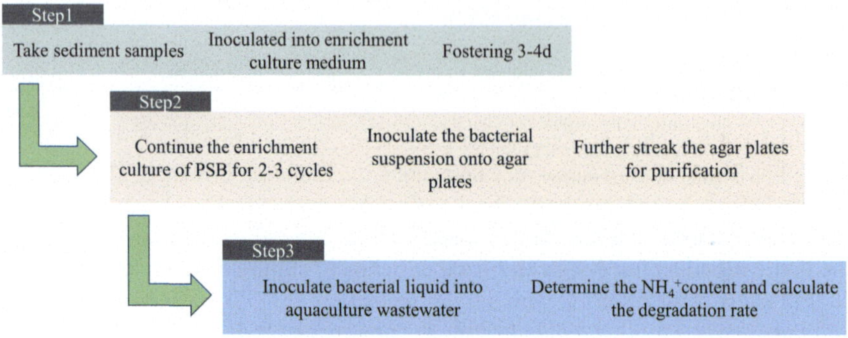

Fig. 27.1 Overall experimental flowchart

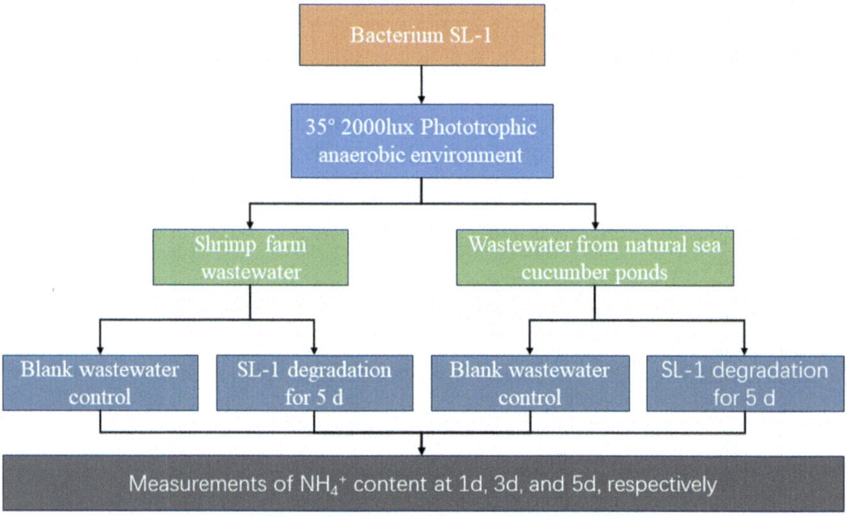

Fig. 27.2 Flow chart of wastewater degradation experiment

27.3 Results and Analysis

27.3.1 Morphological Identification of SL-1

After 5 days of light cultivation, the enrichment culture medium notably turned deep red, as shown in Fig. 27.3. Upon streaking onto agar plates, the colonies appeared as red opaque, flat, and streak-like, as depicted in Fig. 27.4. The morphological characteristics, size, and appearance of the isolated strain were consistent with Gram staining being negative and closely resembled the description of *Rhodopseudomonas* palustris in the Bergey's Manual of Systematic Bacteriology [5]. Therefore, the isolated strain can be tentatively identified as *Rhodopseudomonas* palustris.

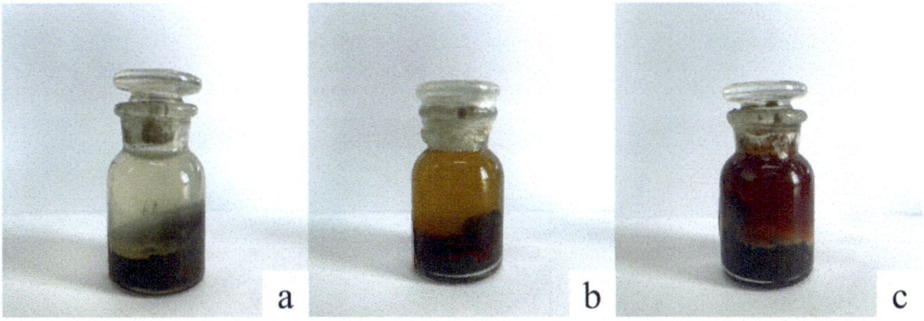

Fig. 27.3 Presents the changes in the enrichment culture medium on the first day (**a**), the third day (**b**), and the fifth day (**c**)

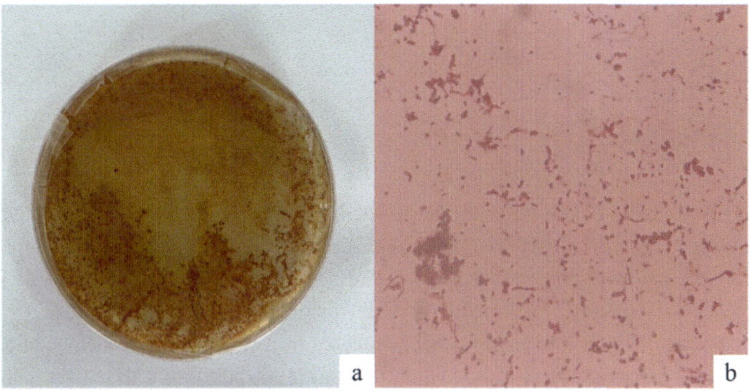

Fig. 27.4 Shows the results of plate coating and Gram staining of photosynthetic bacteria, where **a** is plate coating and **b** is Gram staining

27.3.2 Degradation Characteristics of Aquaculture Wastewater by Strain SL-1

From Fig. 27.5, it can be observed that the degradation ability of strain SL-1 towards both types of aquaculture wastewater gradually increases over the experimental period and stabilizes by the fifth day. In the environments of both highly polluted organic wastewater, SL-1 exhibited only 6% and 4% degradation capabilities on the first day. However, by the fifth day, this ability had increased to 68% and 70%, respectively. If the treatment time is further extended, the degradation capacity of SL-1 towards ammonia nitrogen will become even stronger. This indicates that strain SL-1 possesses excellent denitrification capabilities in treating aquaculture wastewater.

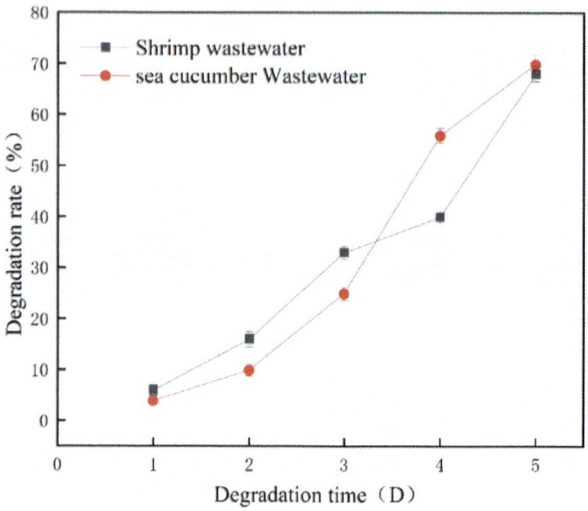

Fig. 27.5 Wastewater treatment effect diagram

27.4 Conclusions

In this study, a strain of photosynthetic bacteria, SL-1, was isolated from turtle ponds in Chaoyang City, Liaoning Province. The bacteria were cultured using a medium prepared according to the research of a previous scholar [6], and the cultivation results were satisfactory. Furthermore, SL-1 was placed in an NH_4^+ culture medium, and the content of NH4 + was measured every 24 h over a period of 5 days, as shown in Table 27.1. Subsequently, the degradation rate was calculated using the formula, and the treatment effect was found to be very significant.

According to [3], the unique physiological characteristics of photosynthetic bacteria allow them to hydrolyze organic nitrogen in wastewater into amino acids under anaerobic conditions, followed by deamination to produce ammonia nitrogen. The generated ammonia nitrogen, along with the original ammonia nitrogen in the wastewater, is assimilated into the bacterial cells to support growth and reproduction. Additionally, photosynthetic bacteria with denitrification activity utilize nitrate or nitrite in the wastewater as the final electron acceptor, employing external simple organic compounds as hydrogen donors for denitrification. Denitrification

Table 27.1 Ammonia Nitrogen degradation rate of strain SL-1 in different wastewaters

Strain No.	Shrimp farm wastewater NH_4^+-N		Sea cucumber wastewater NH_4^+-N	
	Degradation rate (%)	Concentration (mg/L)	Degradation rate/%	Concentration (mg/L)
CK		65.8 ± 0.00		36.8 ± 0.00
SL-1 (1d)	6 ± 1.00	61.79 ± 0.92	4 ± 1.02	35.34 ± 0.74
SL-1 (3d)	33 ± 0.49	46.64 ± 0.71	10 ± 0.68	29.46 ± 0.54
SL-1 (5d)	68 ± 0.88	37.11 ± 0.41	56 ± 0.45	21.61 ± 0.62

can occur through assimilatory and dissimilatory pathways, where nitrate and nitrite are converted into ammonia nitrogen for bacterial assimilation or reduced to gaseous nitrogen and released from the water body, respectively. The dissimilatory pathway is the main route for denitrification [7, 8].

In summary, this study collected sediment from Chaoyang, Liaoning, and isolated and purified a strain named SL-1 with effective ammonia nitrogen degradation capabilities. SL-1 exhibited excellent performance in degrading two types of aquaculture wastewater, demonstrating its strong potential. The strain also showed good applicability in different types of wastewater, providing a scientific basis for future large-scale experiments in aquaculture, environmental protection, and wastewater treatment fields.

Acknowledgements This work was supported by the Liaoning Provincial Science and Technology Mission Project (2022JH5/10400099), the Research and Demonstration of Selenium-enriched Sweet Potato Production Technology (070522122), the Doctoral Scientific Research Foundation of Liaoning Province (2022-BS-278), General Program Funded by Liaoning Province Education Administration (LJKZ0706; LJKMZ20221104), Dalian Youth Science and Technology Star Project Support Program (2021RQ092), the Open Project of the Key Laboratory of Environment Controlled Aquaculture, Ministry of Education (2021-MOEKLECA-KF-02).

References

1. Cao K, Zhi R, Zhang G (2020) Photosynthetic bacteria wastewater treatment with the production of value-added products: a review. Biores Technol 299:122648
2. Puyol D, Barry E, HüLSEN T et al (2017) A mechanistic model for anaerobic phototrophs in domestic wastewater applications: photo-anaerobic model (PAnM). Water Res 116:241–253
3. Chen J, Wei J, Ma C et al (2020) Photosynthetic bacteria-based technology is a potential alternative to meet sustainable wastewater treatment requirement? Environ Int 137:105417
4. Lu H, Zhang G, Zheng Z et al (2019) Bio-conversion of photosynthetic bacteria from non-toxic wastewater to realize wastewater treatment and bioresource recovery: a review. Bioresour Technol 278:383–399
5. Vos P, Garrity G, Jones D et al (2011) Bergey's manual of systematic bacteriology: volume 3: The Firmicutes. Springer Science & Business Media
6. Xue-Jiao H, Chong Y, Ya-Xue L et al (2014) Research progress of photosynthetic bacteria in water pollution governance. China Biotech 34(11):119–124
7. Niederman RA (2013) Membrane development in purple photosynthetic bacteria in response to alterations in light intensity and oxygen tension. Photosynth Res 116:333–348
8. Nagadomi H, Hiromitsu T, Takeno K et al (1999) Treatment of aquarium water by denitrifying photosynthetic bacteria using immobilized polyvinyl alcohol beads. J Biosci Bioeng 87(2):189–193

Chapter 28
Effect of Ship Contamination Degree on Ship Resistance Based on CFD

Ruhong Li and Zhongcheng Wang

Abstract The occurrence of hull fouling is an unavoidable aspect of vessel navigation. The accumulation of fouling on a vessel's surface leads to increased resistance to navigation, decreased propulsion efficiency, heightened energy consumption, and compromised vessel performance. To investigate the impact of varying levels of hull surface roughness and fouling attachment on navigation resistance, this study employs computational fluid (CFD) dynamics methods and conducts numerical simulations using FLUENT software. By referencing methods for estimating resistance under rough surface conditions and categorizing the degree of hull fouling, the severity of fouling attachment is determined based on the type of biological attachment and the density of fouling. Four distinct hull surface conditions are delineated, and corresponding models and conditions are established to analyze the effects of differing degrees of surface roughness caused by hull fouling on total vessel resistance and resistance characteristics. The findings indicate that poorer hull surface conditions result in greater increases in navigation resistance. Furthermore, under consistent surface conditions, the increase in navigation resistance correlates with higher speeds. Methods such as reducing speed and surface fouling removal are viable approaches for reducing navigation resistance across all surface conditions.

Keywords Hull dirt bottom · Ship resistance · Fouling organism · CFD

28.1 Introduction

In the process of vessel navigation, the surface of the hull becomes rougher due to the attachment or corrosion of fouling organisms over time and distance traveled. This increased roughness of the hull plates leads to an inevitable increase in hull resistance, particularly affecting frictional resistance and resulting in decreased speed. Failure to remove fouling in a timely manner can further lead to paint degradation and structural corrosion of the vessel's outer plates due to biofouling, ultimately causing

R. Li (✉) · Z. Wang
Merchant Marine College, Shanghai Maritime University, Shanghai, China
e-mail: 202330110121@stu.shmtu.edu.cn

K. Y. Foo et al. (eds.), *Water Nexus Approach for Sustainable Development*,
Environmental Science and Engineering, https://doi.org/10.1007/978-3-031-89238-7_28

damage to the vessel's appearance and structure. Moreover, the natural detachment of marine organisms attached to the vessel's surface, which contains heavy metal components, can further pollute the marine ecosystem. Several scholars have studied the phenomenon of hull fouling and its impact on resistance. Chen et al. [1] discussed how seawater corrosion and marine biofouling increase hull surface roughness [1]. Hochkirch [2] conducted numerical calculations on model scales and actual vessels [2]. Demirel et al. [3] used numerical simulation methods to analyze the effects of biofouling on vessel resistance and its components [3]. Yao [4] employed a two-dimensional flow field to simulate the effects of external environmental factors on hull resistance [4]. Wang [5] analyzed the research of CFD in each speed correction of real ship test [5]. Li [11] analyzed the resistance characteristics of the special ship type by the mechanism [11]. Song analyzed the resistance performance under different scales, conditions, hull surface conditions and the scale effect of ships, and studied the influence of rough surface on the flow field [6, 13].

The phenomenon of fouling is closely related to the marine regions in which a ship navigates and the growth of marine organisms such as barnacles and shellfish in these areas, as well as to the temperature, salinity, and depth of the seawater. The rate of growth and changes in actual fouling are influenced by a variety of factors, making research into fouling both complex and challenging. Numerical simulation theories have been developed and refined over a long period and are now quite mature; the application of computational fluid dynamics (CFD) in the shipbuilding industry is widely recognized. To deeply study the impact of different levels of fouling on ship resistance, this paper will establish a corresponding hull model and set up a case in the FLUENT software for numerical simulation, validating the feasibility of applying CFD to simulate real ship fouling and to predict changes in ship resistance caused by fouling.

28.2 Theoretical Analysis

28.2.1 Resistance of the Ship During Navigation

When ships navigate on the water surface, their hulls move through two different fluid mediums: water and air. Consequently, the total resistance experienced by the hull is divided into water resistance and air resistance.

Water resistance pertains to the force exerted by water on the part of the hull below the waterline, primarily consisting of still water resistance and wave-making resistance. Within still water resistance, bare hull resistance and appendage resistance are two significant components. Bare hull resistance refers to the resistance faced by the hull navigating in still water, while appendage resistance arises from protrusions on the hull, such as bilge keels, causing additional resistance. The components of bare hull resistance include: first, the frictional resistance Rf, which is the resultant force in the direction of movement caused by viscous shear stress due to the viscosity of

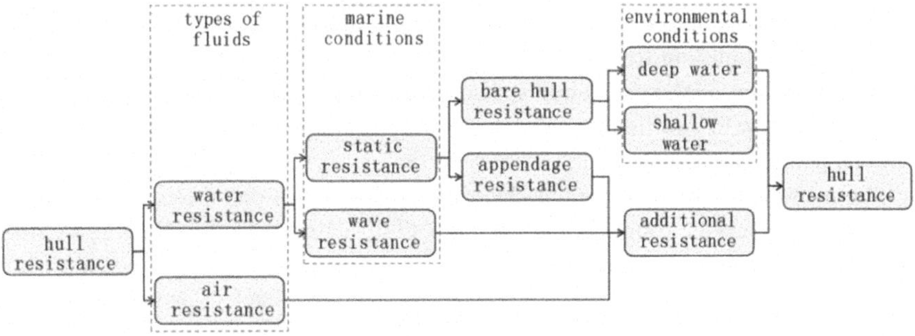

Fig. 28.1 Classification of resistance of ships

seawater creating a boundary layer around the hull as it moves; second, the viscous pressure resistance Rp, resulting from changes in pressure distribution due to the hull's surface curvature and the effects of water viscosity. Furthermore, as the hull navigates on the free surface, it generates waves, leading to increased pressure at the bow with wave peaks and decreased pressure at the stern with wave troughs, creating a differential in hydrodynamic pressure from bow to stern. The additional resistance caused by wind, waves, and the ship's motion, known as wave-making resistance Rw, is complex and usually not calculated directly but analyzed through the variation curve of the stalling ratio with wind scale.

Air resistance pertains to the reaction force exerted by the air on the part of the hull above the water, primarily related to wind force, wind direction, and superstructures. When the angle of the wind on the beam is not considered, the air resistance a ship encounters while navigating primarily depends on the magnitude of the relative wind speed. Since the fouled bottom exists in the water, it does not influence the air resistance [7]. The classification of resistance to the hull is shown in Fig. 28.1.

28.2.2 The Influence of Dirty Bottom on the Roughness of Hull

The formation of surface roughness on the underwater part of a vessel involves several comprehensive factors. Firstly, the unevenness of the ship's hull plates due to curvature and irregular seams, as well as the uneven surface coating caused by poor physicochemical properties of the paint and blistering of old coatings, and scratches formed by rust corrosion from seawater and sea breeze contribute to physical damage. Secondly, during long-term navigation or anchorage in anchorages, marine organisms such as algae, barnacles, oysters, and molluscs adhere to the ship's surface below the waterline, resulting in marine biofouling, which makes the underwater hull dirty and rough. Both factors affect surface roughness.

Surface roughness of the ship's hull is generally defined according to the rules of the British Shipbuilding Research Association. The definition stipulates that within a length of 50 mm on the ship's hull surface, the difference between the highest peak and the lowest valley of the rough surface is measured and calculated. Fifty to one hundred points are taken on each ship, with ten measurements taken at each point, and the average value is defined as the average roughness parameter, expressed in μ/50 mm. The surface roughness of the ship's hull is influenced by factors such as the age of the ship and surface processing. Pre-treated steel roughness during ship construction ranges from 40 to 75 μm. For newly constructed ships that have been fully painted, the surface roughness is approximately 100 μm. For older ships with longer service lives, the range of hull roughness varies from 500 μm to 1000 μm. For ships aged 6 to 8 years, the roughness is approximately 500 μm, and for ships aged 12 to 14 years, the roughness is approximately 1000 μm [1].

28.2.3 Frictional Resistance Compensation Coefficient

In the actual calculation, the total friction drag coefficient takes the smooth plate friction drag coefficient Cf plus a roughness subsidy coefficient ΔCf independent of the Reynolds number, therefore, the hull friction resistance including the influence of surface roughness is calculated as:

$$R_f = \left(C_f + \Delta C_f\right) \cdot 1/2\rho v^2 s \tag{28.1}$$

The 14th International Ship Model Test Pool Conference suggested that the friction resistance coefficient Cf adopted the 1957 ITTC formula, where the corresponding roughness subsidy coefficient is calculated by the following formula:

$$\Delta C_f = \left[105(k_s/L)^{1/3} - 0.64\right] \times 10^{-3} \tag{28.2}$$

This formula is suitable for ships with a length of less than 400 m. Where L is the total length of the ship; ks is the roughness expression height, that is, the average surface protrusion height is measured within 50 mm of the ship surface, and the attachment of different Marine life on the hull increases the height of the hull surface.

For the calculation of the real ship resistance, considering the roughness, the MARIN pool uses the real ship resistance calculation value plus the results obtained by the subsidy factor Ca as the real ship resistance forecast value considering the roughness [6].

$$C_a = 0.044\left[(k_s/L)^{1/3} - 10Re^{-1/3}\right] + 0.125 \times 10^{-3} \tag{28.3}$$

28.3 Materials and Methods

28.3.1 Model Building

To explore the varying effects of fouling levels on vessel resistance, this study establishes three models and multiple sets of fixed conditions for comparison based on mild, moderate, and severe fouling scenarios.

Model 1 represents the real ship model, referencing the 2022 Shanghai Construction Completion Main Civil Ship Main Dimensional Elements Table [8]. The selected model is a bulk carrier with a main engine type of 6G70ME-C9.5Tierlll, and its main scale information is shown in Table 28.1. Utilizing existing prototype ship data, a ship model is constructed using MAXSURF software to generate the ship's surface, smooth the ship's lines, complete the hull form, and refine the numerical model. Finally, the ship model is imported into FLUENT, and the final solid model of the bulk carrier is shown in the figure. Model 2 represents the heavy fouling hull model, with a semi-circular protrusion added to the real ship model with a diameter of 40 mm and a height of 20 mm, covering the bow and bottom sections of the hull. The method of the model bottom coating part refers to Wang [9], Accurately simulate the dirty bottom profile. Due to the complexity of importing the real ship modeling into the FLUENT software, with too many discrete entities, setting up the grid becomes extremely difficult, making it disadvantageous for calculation. Moreover, the appearance change of the real ship model is not significant. Therefore, simplification of the model and regularization of the fouling coverage area are carried out to establish Model 3. The protrusions are integrated into a cylinder with a diameter of 800 mm and a height of 30 mm, covering an area of 3.3% of the ship's surface below the design waterline, with an average protrusion height of 1 mm. The purpose of using this method is to transform the surface appearance changes caused by fouling attachment into a model that can be numerically simulated for calculation. The construction-completed version of the model is shown in Fig. 28.2.

In the simulation, the ship has no superstructure and no wind in the navigation process, and the basic resistance of the real ship is only the static water resistance without wind.

Table 28.1 Main ship type parameters of bulk carriers

Length between perpendiculars Lpp/m	Molded breadth B/m	Molded depth D/m	Design draft/m	Structural draft/m	Range/n mile	Main engine power/kw	Service speed/kn
286. 00	45. 00	24. 90	16. 10	18.30	25 000	SMCR15748	14.5

hull model 1 hull model 2 surface features hull model 3

Fig. 28.2 Simulation model

28.3.2 Working Condition Design

Condition one simulates the surface of a new ship by altering the roughness parameter of the hull wall below the design waterline in the boundary conditions for the actual ship model (Model 1), setting the roughness height to 75 μm. Condition two simulates light fouling by changing the height of the hull wall's roughness below the design waterline in the boundary conditions for the actual ship model (Model 1), selecting a range of variation between 250 to 1000 μm. Condition three simulates moderate fouling, which is characterized by a relatively uniform thickness of the fouling layer. In the boundary conditions, it changes the roughness parameter of the hull wall below the design waterline for the actual ship model (Model 1), with the roughness height set at 1000 μm. Condition four simulates heavy fouling by altering the roughness parameter of the hull wall below the design waterline in the boundary conditions for the heavily fouled ship model (Model 3), setting the roughness height to 75 μm. All working conditions were performed at the full load condition, setting the depth at which the hull is submerged in the fluid domain to 16.1 m.

28.3.3 Meshing and Calculation Parameter Settings

This study utilizes an actual ship model for simulation. In Model 3, the grid near the protruding cylindrical part of the ship's surface is refined to capture relevant physical features, while the grid away from the ship's surface is coarser to reduce the number of grids and computational load. The grid division is illustrated in the Fig. 28.3. Model 1 has 2.09 million elements, while Model 3 has 17.8 million elements.

As the research focuses on the underwater part of the ship's surface and does not involve a free liquid surface, steady-state calculations are used without the need for the VOF model. The SST k-omega turbulence model is chosen. Since the underwater part of the ship's hull is symmetric, the flow around the ship has a certain degree of symmetry. To reduce the total number of grids and computation time, only half of the

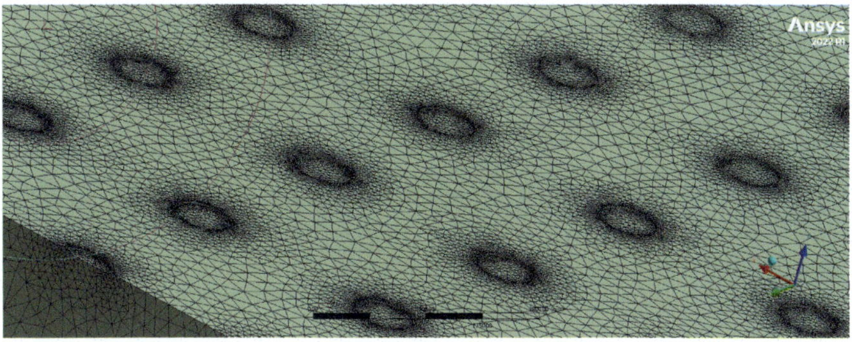

Fig. 28.3 Meshing situation of the ship module calculation domain

ship model is used for calculation, with the symmetric plane set as symmetry. The boundary conditions are set as follows: the inlet boundary is set as a velocity inlet (water-inlet), the outlet boundary is set as an outlet, and the ship's surface boundary is set as a non-slip wall condition. The boundary condition for roughness is based on the standard roughness model, with the default value of the roughness constant Cs set to 0.5 [12]. Speed selection of 8, 10, 12, 14.5, 14.8 kn. The calculation method for the flow field in Model Four uses the Coupled algorithm to enhance convergence, solving for the flow field and the forces acting on the ship's surface.

28.3.4 Analysis of Computational Results

The same numerical simulation method is used to calculate all kinds of working conditions to calculate the change of each resistance of the four groups of models at each speed, the total resistance, viscous force and pressure force are recorded, and the total resistance value of the four groups of models is summarized and analyzed (Fig. 28.4).

The simulation results show that the surface roughness of the ship surface has the most significant influence on the friction resistance. For the hull resistance, the roughness of the ship surface will affect the existence of the boundary layer, and then change the friction resistance, so the increased resistance of the surface roughness occupies a considerable percentage. Based on the simulation results, at various speeds, when the hull surface roughness of a lightly fouled vessel is 250 μm, compared to a new vessel at the same speed, the resistance increases by 9.96% to 11.2%, as expected [10]. At the design service speed (14.5 kn), the simulated resistance for moderate fouling conditions is 2225.8 kN, representing an increase of 25.94% compared to the new vessel. For the heavy fouling model, the simulated resistance is 2226 kN, resulting in a 25.97% increase compared to the resistance of a new vessel at the same speed. At maximum speed (14.8 kN), the simulated resistance for moderate fouling conditions is 2309 kN, indicating a 25.86% increase compared to the new

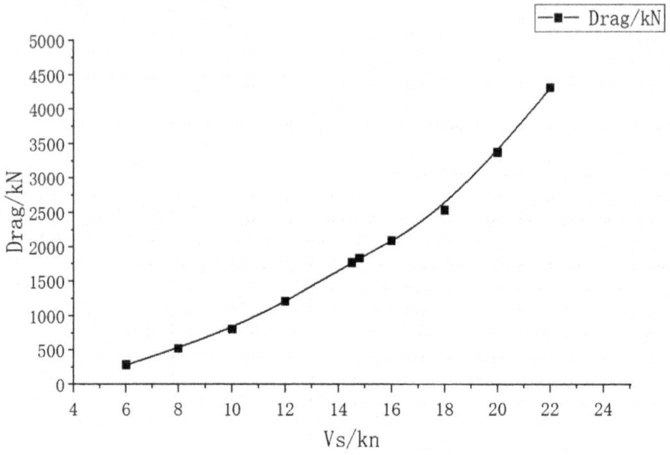

Fig. 28.4 New shipbuilding resistance to change curve with speed

vessel at that speed. Similarly, for the heavy fouling model, the simulated resistance is 2311 kN, resulting in a 25.97% increase compared to the resistance of a new vessel. At the same speed, higher roughness and more severe fouling lead to higher resistance values. In the roughness range of 75 μm to 1000 μm, there is a decreasing trend in resistance with increasing roughness. Since hull surface roughness increases with the age of the vessel, the annual increase in resistance due to surface roughness for newly-operational vessels is greater than that for older vessels.

In order to further analyze the influence of increasing navigation resistance because of the roughness and dirty bottom, the concept of increasing resistance value is introduced to describe the simulation results (Eq. 28.3) [4]. Where Rt-rough is the total resistance under the influence of rough surface and dirty bottom, and Rt-initial is the total resistance on the surface of a new ship at the same speed (roughness of 75 μm). The ends of ΔRr are plotted as in Fig. 28.5.

$$\Delta Rr = Rt - rough—Rt - initial \tag{28.4}$$

The line graph illustrates the increasing trend of ΔRr with roughness and fouling severity under five different sailing speeds, indicating that larger roughness and more severe fouling lead to higher ΔRr values across all sailing speeds. During the stages of mild fouling and moderate fouling attachment on the hull surface, the trend of ΔRr value growth is relatively gentle. Resistance growth is not significant at low sailing speeds, but becomes increasingly evident with higher speeds. In the roughness range of 75 μm to 1000 μm, there is a decreasing trend in the growth rate of ΔRr with roughness. The bar chart compares the variation of ΔRr values with sailing speed for each roughness or fouling level, revealing that under moderate fouling conditions, the ΔRr/Rt ratio ranges from 11 to 26%, and for heavy fouling models, it's approximately 26%. For all roughness and fouling levels, ΔRr increases parabolically with sailing

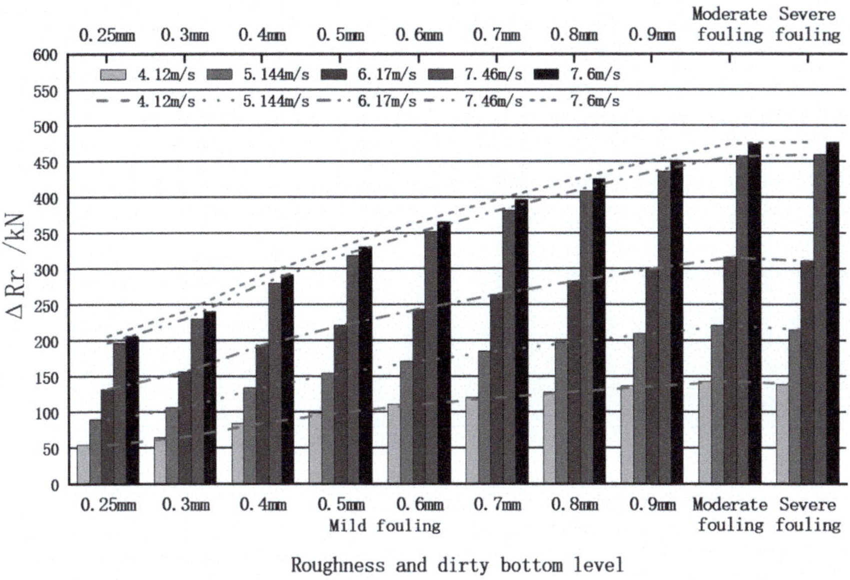

Fig. 28.5 Trend of the increasing resistance value ΔRr

speed, consistent with the resistance curve depicted in Fig. 28.4 (additional sets of high-speed conditions added), indicating that the resistance curve of a new vessel increases parabolically with sailing speed. This suggests that reducing speed to mitigate added resistance is applicable under any surface condition. Simulation results from moderate fouling to heavy fouling conditions indicate that the line graph and bar chart tend to flatten at 14.5 kn and 14.8 kn, with slight declines at lower speeds, suggesting the feasibility of modeling heavy fouling scenarios akin to condition three and altering wall roughness to predict ship fouling.

28.4 Conclusion

Through the comparison of multiple sets of data, the numerical simulation method adopted in this paper has been verified to be relatively reliable in calculating ship fouling. The simulation values of the moderate pollution condition model group are close to those of the self-designed heavy pollution model group, proposing a new simple modeling method that mimics ship fouling models.

By employing computational fluid dynamics to study the impact of hull surface roughness and fouling adhesion on navigational resistance, it is observed from the results of numerical simulations that the influence of fouling on ship resistance cannot be ignored. Under the same roughness conditions, the increased navigational resistance of the ship escalates with the increase in speed. The increment in total resistance tends to decrease as the speed increases. A component analysis of the total resistance revealed that changes in the hull surface roughness height primarily affect the magnitude of frictional resistance within the ship's total resistance, with the increase in frictional resistance accounting for a significant proportion of the total resistance increase. Observing changes in fouling growth and adopting timely and effective fouling removal methods can save ship energy consumption and maintain the marine ecological environment.

Acknowledgements Supported by National Key R & D program of China (No.2022YFB4300701).

References

1. Chen T, Chen GJ (2006) Seawater corrosion, Marine decontamination and ship operation economy. China Ship Repair (04):44–45+48
2. Hochkirch K, Mallol B (2013) On the importance of full-scale CFD simulations for ships. In: 1th International conference on computer and ITapplications in the maritime Industries, COMPIT
3. Demirel YK, Turan O, Incecik A (2017) Predicting the effect of biofouling on ship resistance using CFD. Appl Ocean Res 62:100–118
4. Yao XL, Yuan CQ, Zhao XK et al (2017) Effect of roughness and ship resistance of dirty adhesion based on CFD. Surf Tech 46(06):27–34. https://doi.org/10.16490/j.cnki.issn .1001–3660.2017.06.005
5. Wang JB, Wu Q, Yu HA (2020) Review of CFD technology in real ship speed forecast and trial correction. China Shipbuild 61(S2):100–112
6. Song KW, Guo CY, Sun C et al (2021) Calculation of real-scale ship resistance and scale effect study. J Huazhong Univ Sci Technol 49(06):74–80. https://doi.org/10.13245/j.hust.210614
7. Sheng ZB, Liu YZ (2017) Principle of Ship (Volume1). Shanghai JiaoTong University Press, Shanghai
8. Table of main scale elements of major civil vessels completed in Shanghai in 2022 (2023) Ships and Ocean Engineering 39(01):76–78
9. Wang MC, Song Q, Ding XX et al (2017) Analysis of the influence of dirty bottom position on vessel resistance based on CFD. China Water Transp 17(06):17–19
10. Song LG, Yang JG, Yu SB et al (2019) Analysis of reducing the dirty bottom and improving the energy efficiency of ships. China Water Transp (second half of the month) 19(04):4–5
11. Li YY, Liu P, Zou K (2023) Analysis of the resistance characteristics and mechanism of the ATB ship group. Ship Eng 45(S1):188–192. https://doi.org/10.13788/j.cnki.cbgc.2023.S1.038
12. Zhai JG, Xin RB, Wang ZY et al (2023) The water resistance calculation of 30000 t bulk carriers based on STAR-CCM +. China Shipbuild 45(22):34–39
13. Song KW, Wang LZ, Ge SW et al (2023) Study on the influence of roughness on real ship resistance and flow field. China Shipbuild 64(02):13–24

Chapter 29
Adsorption Studies of Sulfamethoxazole and Tetracycline Antibiotics in Seawater by Ceramicized Sludge Biochars

Mengmeng Lu and Xiaohan Ma

Abstract The abuse of antibiotics has led to a large amount of antibiotics residues in the environment, which seriously threatens the health of offshore ecosystems and human safety. Sludge biochar can effectively inhibit the proliferation and diffusion of antibiotics in the offshore environment through adsorption and degradation mechanisms. Among them, the resource development and utilization of ceramsite modified sludge biochar (CMSB) provides a new idea for the control of antibiotics pollution in the offshore environment. However, the adsorption effect and mechanism of CMSB on antibiotics in seawater are still unclear. Therefore, in this study, the adsorption characteristics of CMSB with different clay (kaolin, attapulgite) contents were investigated and compared under different initial antibiotics (sulfamethoxazole and tetracycline) concentrations. The results showed that the adsorption capacities of CMSB1, CMSB2 and CMSB3 for tetracycline solution were 1.06 mg g^{-1}, 1.36 mg g^{-1} and 1.39 mg g^{-1}, respectively, at a concentration of 5 mg L^{-1}. Compared with CSB (1.25 mg g^{-1}), CMSB3 increased the most, which was 11.2%. At the concentration of 10 mg L^{-1}, the adsorption capacity of CMSB1, CMSB2 and CMSB3 to tetracycline was significantly enhanced than the concentration of 5 mg L^{-1}, which was 2.27 mg g^{-1}, 2.49 mg g^{-1} and 2.71 mg g^{-1}. The adsorption capacity of CMSB to sulfamethoxazole showed a similar trend. This study provides experimental support and theoretical basis for environmental and economic remediation of antibiotic pollution in coastal seawater.

Keywords Antibiotics · Sludge · Biochars · Adsorption

M. Lu
Sanya Oceanographic Institution, Ocean University of China, Sanya, China

X. Ma (✉)
Institute of Coastal Environmental Pollution Control, Ministry of Education Key Laboratory of Marine Environment and Ecology, College of Environmental Science and Engineering, Sanya Oceanographic Institution, Frontiers Science Center for Deep Ocean Multi-spheres and Earth System, Ocean University of China, Qingdao, Qingdao, China
e-mail: maxiaohan@stu.ouc.edu.cn

K. Y. Foo et al. (eds.), *Water Nexus Approach for Sustainable Development*,
Environmental Science and Engineering, https://doi.org/10.1007/978-3-031-89238-7_29

29.1 Introduction

With the growth of population and the needs of material life, the use of antibiotics in medical and animal husbandry has been increasing yearly [1]. The vast majority of antibiotics taken by humans and animals cannot be fully absorbed and utilized by the body, but are directly excreted in the form of raw drugs or secondary metabolites through urine or feces, and enter the environment with domestic sewage or runoff [2]. The residual antibiotics in the environment are ultimately discharged into the ocean through rainwater erosion, surface runoff, and sewage discharge, causing serious antibiotics pollution in estuaries and nearshore areas. Antibiotics emitted into the nearshore environment not only have toxic effects on microorganisms and benthic flora and fauna in sediment, but can also enter human food and drinking water through the food chain, posing potential risks to human health. How to regulate the use of antibiotics, effectively control the environmental pollution of estuarine and coastal antibiotics, and reduce its environmental risk has become a major social and environmental problem to be solved urgently.

At present, the removal of antibiotics in water bodies is mainly carried out through physical control, chemical treatment, and biological treatment [3]. Among them, physical control technology has the characteristics of high removal efficiency, no intermediate products, and low cost, and has broad application prospects [4]. The technology of sludge pyrolysis/carbonization has gradually become an industry trend due to its ability to eliminate viruses, parasitic organisms, and harmful substances in sludge, significantly reduce the volume of sewage sludge, and produce biochar that can be used as a resource. Sludge biochar effectively inhibits the proliferation and diffusion of antibiotics in aquatic environments through mechanisms such as adsorption and degradation [5]. However, if sludge biochar powder is directly added to the estuarine environment, seawater flow may carry small particles of biochar, causing waste and making it easier to migrate within sediments and benthic organisms, resulting in secondary pollution and biological toxicity. Therefore, enhancing the stability and adsorption effect of sludge biochar through ceramic granulation and modification.

The objective of this study is to further analyze the adsorption properties of biochar in ceramicized sludge. The purpose of this study is to provide a scientific basis for the development of ceramicized modified sludge biochar (CMSB) technology for Marine antibiotic pollution control.

29.2 Materials and Methods

29.2.1 *Chemicals*

The chemicals Tetracycline and sulfamethoxazole were acquired from Sigma Aldrich. The sludge was collected from the Licun River sewage treatment plant in Qingdao, Shandong. $FeCl_3 \cdot 6H_2O$ was purchased from China National Pharmaceutical Group Chemical Reagent Co., Ltd. The concave convex rod stone was purchased from a mining product processing plant in Lingshou County, Hebei Province, and passed through a 120 mesh sieve for later use. Shell powder is purchased from a shell powder coating factory in Qingdao, Shandong. After air drying, it passes through a 120 mesh sieve.

29.2.2 *Preparation of Ceramic Granulated Modified Sludge Biochar*

The sludge is air dried and crushed through a 100 mesh sieve. The addition ratios of sludge, kaolin, and attapulgite in three modified ceramic granulation sludge materials are 2:1:1, 6:1:1, 18:1:1, respectively. Add 60 mL of 0.6 mol/L $FeCl_3 \cdot 6H_2O$ solution, stir quickly with ultrasound, knead into balls, and dry and weigh. Place the original ball in crucible and burn charcoal in a tube furnace (1000 °C, 5 °C min^{-1}, 0.5 h). After the pyrolysis is completed, the crucible is removed by lowering it to indoor temperature in a nitrogen environment. CMSBs prepared under different clay mineral ratios are labeled as CMSB1 (2:1:1), CMSB2 (6:1:1), and CMSB3 (18:1:1), respectively. In addition, the unmodified ceramicized biochar prepared from sludge raw materials using the same pyrolysis method as mentioned above is labeled as CSB.

29.2.3 *Adsorption Experiment*

The CMSB adsorption experiments of tetracycline and sulfamethoxazole were carried out at different initial concentrations. Using 0.22 μM polyethersulfone (PES) membrane filters natural seawater with a salinity of 31.6 ‰ and a pH value of 8.0 ± 0.02. Use filtered natural seawater as the background solution for the reserve solution of antibiotics (100 mg L^{-1}, and dilute the reserve solution of antibiotics to 5 mg L^{-1} and 10 mg L^{-1} with seawater. Using 0.1 μ Pre regulate seawater pH with M NaOH or HCl. Add CSB, CMSB, and 15 mL diluted antibiotics solution to a glass bottle. Use glass bottles containing only antibiotics as blank controls. All tests were conducted on three biological samples. Incubate all glass bottles in the dark at 150 rpm and 25 °C for 72 h. After adsorption, centrifuge the sample at 150 rpm for 30 min and collect the supernatant. The ATs concentration in supernatant was quantitatively analyzed by high performance liquid chromatography (1260, Agilent Technologies, USA) [6].

29.2.4 Data Analysis

The calculation of the antibiotic removal rate (R), adsorption capacity (Q_e), and adsorption coefficient (K_d) is based on the following formulas:

$$R = (C_0 - C_e)/C_0 \tag{29.1}$$

$$Q_e = (C_0 - C_e)V/m \tag{29.2}$$

$$K_d = Q_e/C_e \tag{29.3}$$

In the formula, C_0 is the initial concentration of antibiotics (mg L^{-1}); C_e is the concentration of antibiotics adsorbed (mg L^{-1}); V is the volume of the solution (mL); m is the mass of carbon containing substance (mg).

Use Origin 2023 software for data analysis and plotting. Single factor analysis of variance and Duncan test ($P < 0.05$) were conducted using SPSS 25.0 software.

29.3 Results and Discussion

29.3.1 Removal Rate of Antibiotics

The effect of different biochar on the removal rate of antibiotics are shown in Fig. 29.1. According to different types of biochar, CMSB3 has the best removal effect. For the concentration of tetracycline in seawater, CMSB3 has the highest removal rate (99.86% and 93.1%), followed by CMSB2 (98.72% and 86.04%), and CMSB1 has the lowest removal rate (76.83% and 78.22%). For the concentration of sulfamethoxazole in seawater, the removal efficiency of CMSB3 is the highest (94.76% and 90.66%), followed by CMSB2 (94.66% and 87.21%), and the lowest is CMSB1 (94.46% and 77.82%). Among them, the removal rate of CMSB1 is lower than that of CSB, possibly due to the high adhesion of clay substances in CMSB1, resulting in a lower specific surface area than CSB. According to different antibiotics concentrations, the removal rates of CSB and CMSB decrease with increasing initial antibiotics concentrations. For the concentration of tetracycline in seawater, except for CMSB1, the removal rate of biochar decreased from 91.81–99.86% to 81.7–93.1%. For the concentration of sulfamethoxazole in seawater, the removal rate of biochar is similar at 5 mg L^{-1}, ranging from 94.46% to 94.76%. At 10 mg L^{-1}, the

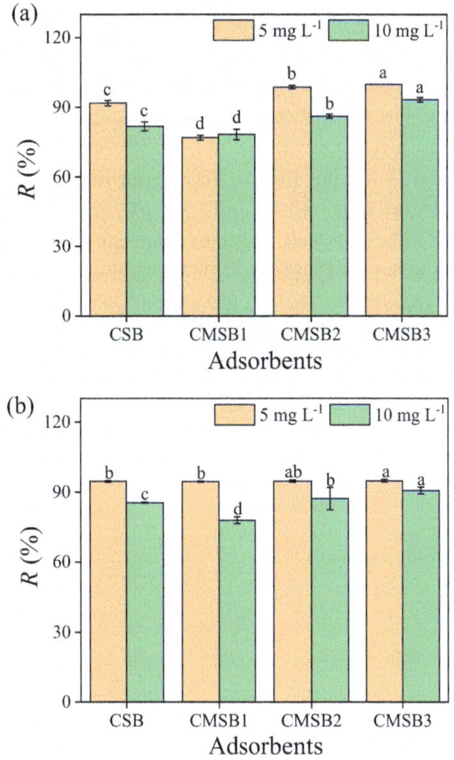

Fig. 29.1 The removal rates of antibiotics by CSB and CMSB. **a** tetracycline, **b** sulfamethoxazole

removal rate of biochar is 77.82% to 90.66%. The reason for this phenomenon may be that the adsorption sites on the biochar have reached saturation.

29.3.2 Adsorption Capacity of Antibiotics

The effect of CMSB on antibiotics adsorption capacity is shown in Fig. 29.2. When the tetracycline concentration is 5 mg L^{-1}, the adsorption capacity of CMSB3 for tetracycline in seawater reaches 1.39 mg g^{-1}, CMSB2 is 1.36 mg g^{-1}, and CMSB1 is the lowest at 1.06 mg g^{-1}. When the tetracycline concentration is 10 mg g^{-1}, the adsorption capacity of CMSB3 for tetracycline in seawater increases to 2.71 mg g^{-1}, CMSB2 is 2.49 mg g^{-1}, and CMSB1 is 2.27 mg g^{-1}. For the concentration of sulfamethoxazole in seawater, it shows the same trend as tetracycline. When the concentration of sulfamethoxazole is 5 mg L^{-1}, the adsorption difference of CMSB for sulfamethoxazole in seawater is not significant, which is 0.66 mg g^{-1}; When the concentration of sulfamethoxazole is 10 mg L^{-1}, the adsorption capacity of CMSB3 for sulfamethoxazole in seawater increases to 1.31 mg g^{-1}, CMSB2 is 1.26 mg g^{-1}, and CMSB1 is 1.12 mg g^{-1}. The maximum adsorption capacity of CMSB gradually decreases with the increase of clay content, which may be due to the adhesion of clay minerals on the surface of sludge biochar, blocking the pores on the surface. Although it increases the specific surface area of sludge biochar, the pore structure is smaller, and the adsorption effect of antibiotics decreases. Furthermore, Q_e will increase with the initial concentration of antibiotics. The starting amount of tetracycline went up from 5 mg per liter to 50 mg per liter, while CSB and CMSB increased by 90.53%, 115.04%, 82.80%, and 95.19%, respectively. The onset of sulfamethoxazole saw its concentration escalate from an initial 5 mg per liter to a heightened 50 mg per liter, while CSB and CMSB increased by 87.98%, 70.49%, 91.16%, and 98.49%, respectively. The upswing in quantity Q_e stems from the concentration gradient, with a stronger adsorption impetus observed between the high concentration antibiotics and the carbonaceous substrates. This enhanced interaction facilitates the permeation of antibiotics molecules into the microspaces within the carbonaceous matrices.

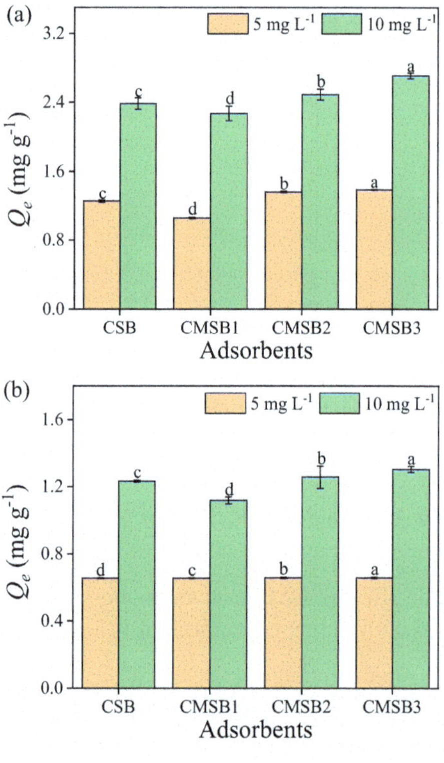

Fig. 29.2 The adsorption capacity of CSB and CMSB for antibiotics. **a** tetracycline, **b** sulfamethoxazole

29.4 Adsorption Coefficient of Antibiotics

The effect of CMSB on the adsorption coefficient of antibiotics is shown in Fig. 29.3. As the content of clay minerals increases, the K_d value gradually decreases. When the tetracycline concentration is 5 mg L^{-1}, the K_d value of CMSB3 is the highest (217.02 L g^{-1}), followed by CMSB2 (28.16 L g^{-1}), and CMSB1 is the lowest (0.98 L g^{-1}). When the concentration of sulfamethoxazole is 5 mg L^{-1}, the K_d value of CMSB3 is the highest (2.69 L g^{-1}), followed by CMSB2 (2.63 L g^{-1}), and CMSB1 is the lowest (2.52 L g^{-1}). This is in line with earlier findings suggesting that Qe and Kd exhibit a decline as clay content increases. In addition, the strong adsorption performance of CMSB is more significantly demonstrated under low concentrations of antibiotics. For a tetracycline concentration of 10 mg L^{-1}, the K_d value of CMSB3 was 4.05 L g^{-1}, a decrease of 98.13%, followed by a decrease of 93.52% in CMSB2 (1.82 L g^{-1}), and the lowest in CMSB1 (1.07 L g^{-1}), an increase of 9.52%. For a sulfamethoxazole concentration of 10 mg L^{-1}, the K_d value of CMSB3 was 1.46 L g^{-1}, a decrease of 45.77%, followed by CMSB2 (1.16 L g^{-1}), a decrease of 55.71%, and CMSB1 was the lowest (0.52 L g^{-1}), a decrease of 79.41%. This further indicates that CMSB3 has better adsorption capacity and affinity for antibiotics in seawater.

29.4.1 Potential Adsorption Mechanisms

The adsorption mechanism of CMSB for antibiotics and ARGs mainly includes pore filling, π–π interactions [7], negative charge assisted hydrogen bonding, and electrostatic interactions [8]. The CMSB material possesses a substantial proportion of oxygen-containing functional groups. These groups have the capacity to establish additional hydrogen bonds with sulfamethoxazole and tetracycline, thereby enhancing the material's capacity to adsorb these substances. CMSB exhibits a pronounced aromatization property, with its surface carboxyl, nitro, and ketone groups serving as electron acceptors. These groups can interact with aromatic compounds via π–π electron donor–acceptor bonds. Consequently, π–π interactions play a role in the adsorption of tetracycline and sulfamethoxazole onto CMSB. When CMSB and antibiotics surface charges attract each other, it exhibits "electrostatic adsorption" phenomenon. At pH 8, both sulfamethoxazole and tetracycline carry negative charges, while CMSB carries positive charges. Hence, antibiotics with negative charges will engage in electrostatic interactions with CMSB. The carboxyl and hydroxyl groups on CMSB can establish hydrogen bonds with the corresponding groups in antibiotics, thereby boosting CMSB's adsorption efficacy for these antibiotics. CMSB also possesses a highly porous structure with an extensive surface area, which offers an adequate number of sites for the adsorption of organic molecules.

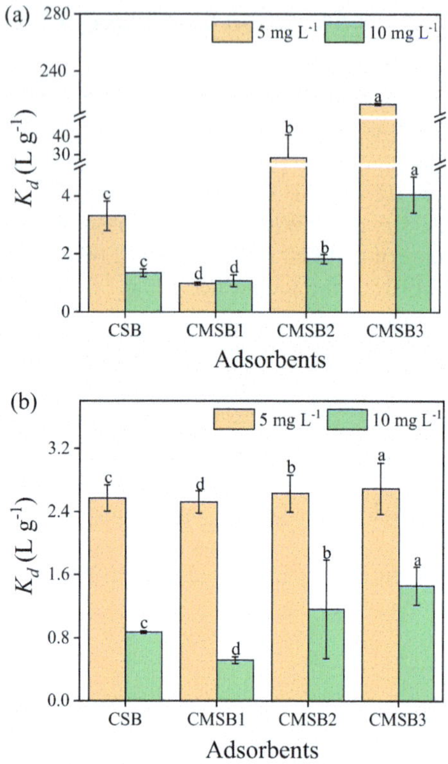

Fig. 29.3 The adsorption coefficients of CSB and CMSB for antibiotics. **a** tetracycline, **b** sulfamethoxazole

29.5 Conclusion

The *R*, *Qe*, and K_d values of CMSB3 adsorbing antibiotics are higher than those of CSB, CMSB1, and CMSB2. The adsorption mechanism of CMSB for antibiotics and ARGs mainly includes pore filling, π–π interactions, negative charge assisted hydrogen bonding, and electrostatic interactions. CMSB3 has broad potential as an excellent adsorbent for removing antibiotics from marine environments. This investigation offers empirical evidence and a conceptual framework for a technology aimed at both environmental and economic restoration from antibiotic contamination in coastal seawater.

Acknowledgements This research was supported by the Hainan Provincial Key Research and Development Project (ZDYF2022SHFZ018).

References

1. Roope LS, Smith RD, Pouwels KB, Buchanan J, Abel L, Eibich P, Butler CC, Tan PS, Walker AS, Robotham JV (2019) The challenge of antimicrobial resistance: what economics can contribute. Sci. 364(6435):eaau4679
2. Monahan C, Nag R, Morris D, Cummins E (2021) Antibiotic residues in the aquatic environment–current perspective and risk considerations. J Environ Sci Heal A 56(7):733–751
3. de Ilurdoz MS, Sadhwani JJ, Reboso JV (2022) Antibiotic removal processes from water & wastewater for the protection of the aquatic environment-a review. J Water Process Eng 45:102474
4. Ahmed MB, Zhou JL, Ngo HH, Guo W (2015) Adsorptive removal of antibiotics from water and wastewater: progress and challenges. Sci Total Environ 532:112–126
5. Wu C, Fu L, Li H, Liu X, Wan C (2022) Using biochar to strengthen the removal of antibiotic resistance genes: performance and mechanism. Sci Total Environ 816:151554
6. Zheng H, Wang Z, Zhao J, Herbert S, Xing B (2013) Sorption of antibiotic sulfamethoxazole varies with biochars produced at different temperatures. Environ Pollut 181:60–67
7. Cheng Y, Wang B, Shen J, Yan P, Kang J, Wang W, Bi L, Zhu X, Li Y, Wang S (2022) Preparation of novel N-doped biochar and its high adsorption capacity for atrazine based on π–π electron donor-acceptor interaction. J Hazard Mater 432:128757
8. Borthakur P, Aryafard M, Zara Z, David Ř, Minofar B, Das MR, Vithanage M (2021) Computational and experimental assessment of pH and specific ions on the solute solvent interactions of clay-biochar composites towards tetracycline adsorption: implications on wastewater treatment. J Environ Manage 283:111989

Chapter 30
Adsorption of Sulfamethoxazole and Tetracycline by Functionalized Biochars in Seawater

Xiaohui Sun and Xiaohan Ma

Abstract The pollution of antibiotic resistance genes (ARGs) in marine environment has become a major challenge threatening human health and ecological security. Researches on biochar have mainly focused on the adsorption effect of single biochar materials on antibiotics in offshore environment, while relatively few studies have been conducted on biochar modified with mixed materials. Therefore, different raw materials (sawdust biochar, shell powder and attapulgite) were selected according to the ratio of 1:1:1, 1:2:3 and 3:2:1 to prepare functionalized biochar (FBCs), which were named FBC, FBC2 and FBC3, respectively. Batch adsorption experiments were carried out on sulfamethoxazole (SMX) and tetracycline (TC) using FBCs at different initial concentrations of antibiotics (1 mg/L, 5 mg/L). The results indicated that FBC1, FBC2, and FBC3 adsorbed 0.73 mg/g, 0.63 mg/g, and 0.59 mg/g of SMX solution at a concentration of 1 mg/L, respectively, which is 113%, 85%, and 74% higher than that of FBC (0.34 mg/g). At a concentration of 5 mg/L, the enhanced adsorption capacity of FBC1, FBC2, FBC3 was not significant as at 1 mg/L, but their adsorption capacity of TC was significantly higher than that of SMX (7.05–33.05 mg/g) ($P < 0.05$). This research offers theoretical backing and empirical validation for the creation of effective antibiotic adsorbents within marine ecosystems.

Keywords Antibiotics · Biochar · Shell powder · Attapulgite · Adsorption

---

X. Sun
Sanya Oceanographic Institution, Ocean University of China, Sanya, China

X. Ma (✉)
Institute of Coastal Environmental Pollution Control, Ministry of Education Key Laboratory of Marine Environment and Ecology, College of Environmental Science and Engineering, Sanya Oceanographic Institution, Frontiers Science Center for Deep Ocean Multi-Spheres and Earth System, Ocean University of China, Qingdao, China
e-mail: maxiaohan@stu.ouc.edu.cn

K. Y. Foo et al. (eds.), *Water Nexus Approach for Sustainable Development*, Environmental Science and Engineering, https://doi.org/10.1007/978-3-031-89238-7_30

30.1 Introduction

Antibiotics are a group of secondary compounds generated by microorganisms (such as fungi and bacteria) as well as by plants and animals as part of their life processes, which have activities such as antipathogens and enable the disruption of the developmental functions of other cells. Since the 1940s, when they were first introduced into the field of medicine, antibiotics have become a core drug in modern healthcare with wide applications in health care, as well as in promoting the growth and preventing diseases in livestock and other food-producing animals [1]. According to statistics, China produces about 210,000 tonnes of antibiotics raw materials annually, which is much higher than developed countries such as Europe and the United States. However, antibiotics are not fully absorbed in humans and animals, and usually about 60–90% of the drug or its metabolitesare excreted through the excretory system in the prototype or transformed form. Antibiotics are commonly found in the water column, sediments, and organisms in the offshore aquaculture environment [2]. Sulfonamides and tetracyclines are the most prevalent classes. Concentrations of antibiotics such as SMX and TC were as high as 995.02 ng/L in marine farms in southeastern China. SMX can be readily transported and dispersed because of their substantial solubility and volatility. High concentrations of SMX (up to 914 ng/L) were identified in mariculture water samples from Vietnam. Concentration of TC varied considerably in different mariculture areas, and a study revealed that tetracyclines were predominantly used as antibiotics in aquaculture practices in Dalian, China. (2.11–9.23 ng/L). Contamination of antibiotics can affect the activity of microorganisms. For example, it can lead to an imbalance of microbial communities in water and soil, and lead to the formation of ARGs, which can also pose a threat to humans by being transmitted between bacteria via mobile genetic elements in the form of horizontal gene transfer. Therefore, controlling antibiotic pollution in the marine environment is important for aquatic organisms, marine ecosystems and human health.

Biochar shows high adsorption efficiency because of its intricate pore network, numerous reactive sites, and substantial surface area, and it can be used as a good adsorbent for antibiotics in the environment [3]. However, current studies mainly focus on the adsorption effect of single biochar materials on antibiotics in the offshore environment [4, 5], and relatively few studies on mixed-materials modified biochar. Meanwhile, the effect of the addition ratio of modified biochar raw materials on its adsorption effect is not clear. Therefore, it is crucial to explore the adsorption effect of modified biochar materials with different feedstock ratios on antibiotics in seawater.

The specific objective of this study is to investigate the adsorption effect of functionalized biochar with different raw material ratios on SMX and TC, and this investigation will offer a theoretical basis for developing marine antibiotic pollution treatment techniques employing modified biochar substances.

30.2 Materials and Methods

30.2.1 Chemicals

SMX and TC were purchased from Sigma-Aldrich Co. HPLC-grade organic solvents such as methanol, acetonitrile and acetone were purchased from Sigma-Aldrich, Australia.

30.2.2 Preparation of Functionalized Biochars

The woodchip biochar, attapulgite and shell powder were each mixed with 2% sodium alginate solution and magnetically stirred for 2 h. The mixed solution was cured using a syringe by dropwise addition to a beaker of $CaCl_2$ solution. Then washed and dried it for subsequent use. Three different functionalized biochars were prepared according to the above method with the addition ratios of wood chip biochar: shell powder: aubergine = 1:1:1, 1:2:3, and 3:2:1, respectively, and the above materials were named as FBC, FSP, FTP, FBC1, FBC2, and FBC3, respectively.

30.2.3 Adsorption Experiment

The six functionalized biochar materials (50 mg) were accurately weighed into 8 mL headspace vials, and 7 g of SMX or TC seawater standard solutions (1 mg/L and 5 mg/L) were added to construct two different concentrations of seawater-antibiotic adsorption systems. The vials were shaken continuously at 150 rpm for 72 h in a constant temperature incubator at 25 °C, and the supernatants were transferred to brown liquid phase vials (1–1.5 mL) through a 0.22 μm polytetrafluoroethylene (PTFE) membrane to determine the concentrations of SMX and TC by high performance liquid chromatography (1260, Agilent, USA).

30.2.4 Data Analysis

The elimination rates (R) of SMX and TC are computed using the following equation.

$$R = (C_0 - C_e)/C_0 \tag{30.1}$$

where C_0 is the initial concentration of SMX and TC and C_e is the concentration of SMX and TC after adsorption (mg/L).

The adsorption capacity (Q_e) and adsorption coefficient (K_d) are calculated according to the following equation.

$$Q_e = (C_0 - C_e) * V/m \quad (30.2)$$

$$K_d = Q_e/C_e \quad (30.3)$$

where C_0 is the initial concentration of SMX and TC (mg/L); C_e is the concentration of SMX and TC after adsorption (mg/L); V is the volume of solution (mL); and m is the mass of carbonaceous material (mg).

Data manipulation and visualization were conducted with Excel 2019. A one-way analysis of variance (ANOVA) was conducted via SPSS 25.0. LSD, Duncan, Dunett's, descriptive analysis and chi-square test were selected for ANOVA, and $P < 0.05$ in Duncan's test result indicated significant difference.

30.3 Results and Discussion

30.3.1 Removal Rate of SMX and TC

The effects of FBCs on the removal rate were shown in Fig. 30.1. The removal of SMX from seawater by FBCs ranged from 1.28% to 15.1%, and the adsorption rate of TC was significantly increased, with the highest removal rate of 95.7% for FBC2 at 1 mg/L concentration. Most of the materials showed a trend of decreasing removal rate with increasing initial antibiotic concentration. Between 1 mg/L and 5 mg/L, the removal of SMX by FBC1, FBC2 and FBC3 decreased by 27.6%, 7.8% and 20.2%, respectively. It may be that the adsorption capacity per unit mass of the adsorbent was saturated, and the adsorbent could not provide more active sites by further increasing the antibiotic concentration. However, the removal rate of FBC3 for 5 mg/L TC in seawater was 68.7%, which was significantly higher than that of 44% at 1 mg/L ($P < 0.05$). It indicates that in addition to active sites, the adsorption of antibiotics by functionalized biochar materials is also related to other mechanisms, such as π-π electron donating acceptor action (π-π EDA), electrostatic action, and hydrogen bonding, etc. [6].

30.3.2 Adsorption Capacity of SMX and TC

Comparing the adsorption effects of FBCs on two antibiotics SMX and TC, the adsorption capacity of TC was substantially greater than SMX ($P < 0.05$), indicating that FBCs have more excellent adsorption capacity and adsorption affinity for TC in seawater. The results indicated that the adsorption quantity rose as the concentration

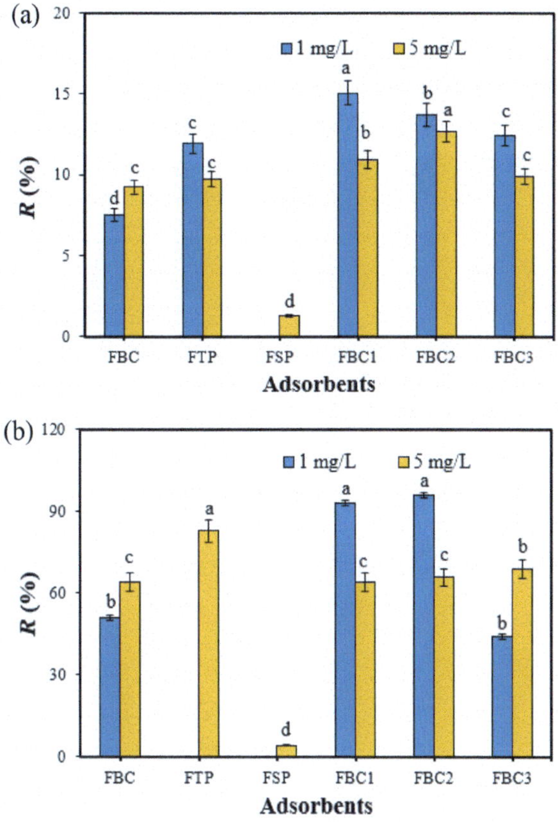

Fig. 30.1 Removal rate of SMX and TC by FBCs with different addition. **a** SMX, **b** TC

of antibiotics in the solution elevated (Fig. 30.2), under consistent biochar addition conditions. The adsorbed amount of FBCs ranged from 0.34 to 0.73 mg/g at the concentration of SMX was 1 mg/L, and the adsorbed amount significantly increased to 1.03–9.51 mg/g at 5 mg/L ($P < 0.05$), with 19-fold, 13.4-fold, 10.2-fold, 14.1-fold and 11.6-fold increased for FBC, FTP, FBC1, FBC2 and FBC3, respectively, which was due to the increased concentration of antibiotics in solution and easier contact between the adsorbent and the active sites on the biochar microspheres. On the basis of FBC, the modified functionalized biochar materials significantly enhanced the adsorption capacity of antibiotics, among which the adsorption effect of FBC1 and FBC2 was significantly increased. The adsorption amount of FBC on 1 mg/L SMX solution was 0.34 mg/g, and the adsorption amount of FBC1, FBC2, and FBC3 prepared according to the different raw material ratios were 0.73 mg/g, 0.63 mg /g and 0.59 mg/g, which increased by 113%, 85% and 74%, respectively. The enhancement of the adsorption capacity by FBC1, FBC2 and FBC3 was not as significant as that at 1 mg/L at 5 mg/L concentration.

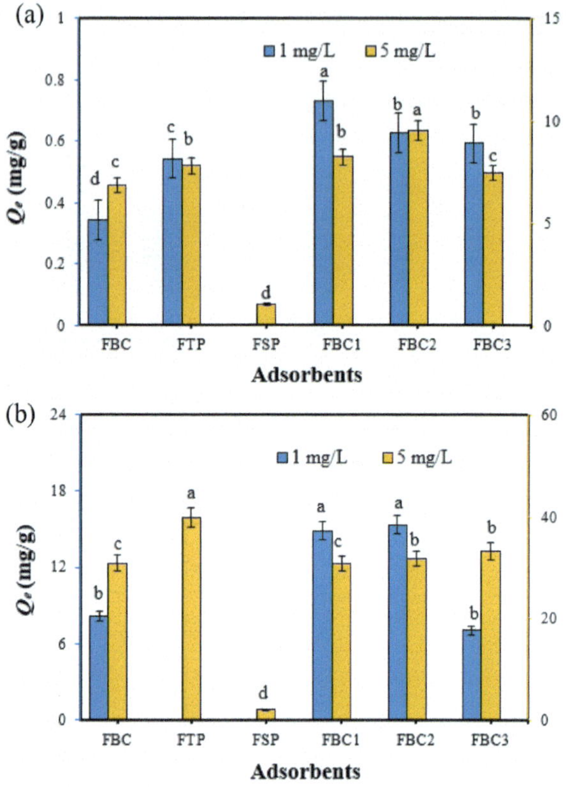

Fig. 30.2 Adsorption capacities of SMX and TC by FBCs with different addition. **a** SMX, **b** TC

30.3.3 Adsorption Coefficient of SMX and TC

Among the modified materials prepared from the three original materials, wood chip biochar (FBC), aubergine (FTP) and shell powder (FSP), the adsorption capacity of FTP was slightly stronger than that of FBC, and FSP showed almost no adsorption for SMX and TC. For the SMX solution with a concentration of 1 mg/L, the adsorption of SMX by FBC was 0.34 mg/g with a removal rate of 7.5% and an adsorption coefficient of 1.17 mL/g, and the adsorption by FTP was 0.54 mg/g with a removal rate of 11.9% and an adsorption coefficient of 1.87 mL/g, and the adsorption capacity of the two materials showed no significant variation ($P > 0.05$) (Fig. 30.3). The adsorption amount of FTP was 39.7 mg/g and the removal rate was 82.6% in TC solution at a concentration of 5 mg/L, which was significantly exceeded than FBC ($P < 0.05$), but there was almost no adsorption of FTP at a concentration of 1 mg/L. The difference in the adsorption potential of FTP on SMX and TC may be related to the nature of the antibiotics themselves, such as molecular weights, the functional groups they carry, and hydrogen bonding, etc. The adsorption capacity of FSP on 5 mg/L of SMX and TC were adsorbed at 1.03 mg/g and 1.93 mg/g, respectively, and almost no adsorption on 1 mg/L of antibiotics. The original removal rate of FBC

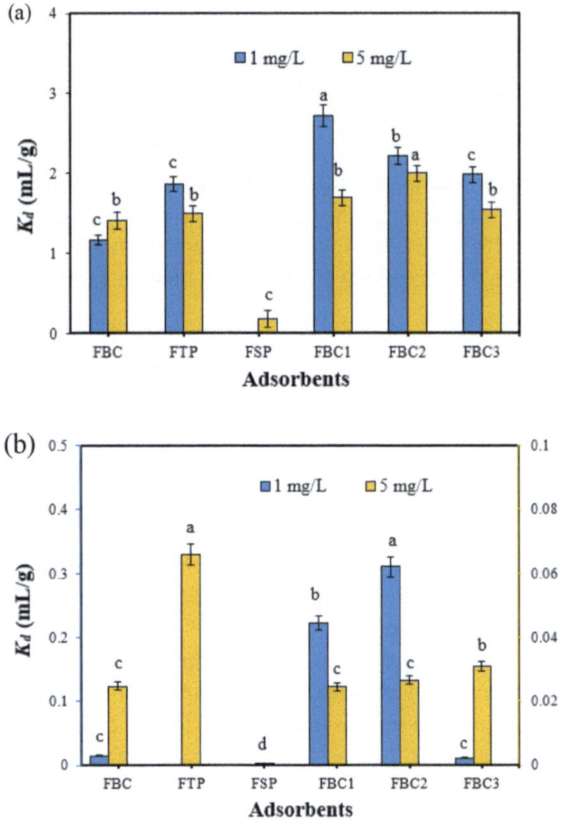

Fig. 30.3 Adsorption coefficients of SMX and TC by FBCs with different addition. **a** SMX, **b** TC

was 7.5%, and the removal rates of FBC1, FBC2 and FBC3 were 15.1%, 13.7% and 12.4%, respectively. The adsorption and removal rates of FBC1 and FBC2 increased significantly ($P < 0.05$) in the TC for a solution initialized at a concentration of 1 mg/L. The adsorption coefficients of FBC1 and FBC2 increased 13.4-fold and 19.5-fold with 81.9% and 82.3%, respectively, and 87.9% and 87.6%, respectively, with FBC2. The adsorption capacity of FBC3 for 1 mg/L TC was slightly smaller than the original material FBC, and the adsorption amount and adsorption coefficient for 5 mg/L TC were not significantly different from those of FBC ($P > 0.05$).

30.3.4 Potential Adsorption Mechanisms

The adsorption mechanisms of woodchip biochar on ionic antibiotics, such as SMX and TC, are mainly surface adsorption, microporous filling, electrostatic effect and ion exchange [7]. Aconite has silica hydroxyl groups on its surface and has a better adsorption performance for antibiotics due to its natural rod crystal structure and pore channels, which have a high adsorption capacity for negatively charged and cations shell powder is mainly composed of $CaCO_3$, which can adsorb antibiotics through physical effects such as microporous filling. FBC1 has the most average composition and FBC2 has the highest percentage of aconite addition, and the adsorption capacity of them is significantly higher compared to FBC, which is due to the fact that the addition of the mineral materials (aconite, shell powder), which leads to the enhancement of the ion-exchange capacity of the granular biochar material, and the surface-carrying functional groups facilitate the electrostatic and cation bridging effects [8], increasing the ability of biochar to adsorb antibiotics.

30.4 Conclusion

The adsorption capacity of the three functionalized biochar materials for antibiotics was significantly improved, and the adsorption of SMX by FBC1, FBC2 and FBC3 increased by 113%, 85% and 74% compared with that of FBC at the SMX concentration of 1 mg/L, and the elimination rate increased by 105.3%, 82.7% and 65.3%, respectively. The adsorption capacity of functionalized biochar materials for TC was markedly greater than SMX ($P < 0.05$), indicating that they have more excellent adsorption capacity and adsorption affinity for TC in seawater. The adsorption mechanisms of granularized biochar materials on antibiotics mainly include physical adsorption such as microporous filling, chemical adsorption such as π-π electron donor acceptor action (π-π EDA), electrostatic action, ion exchange and hydrogen bonding, with chemical adsorption dominating.

Acknowledgements This research was supported by characteristics of antibiotic and antibiotic resistance gene contamination in marine aquaculture environment and control technology research in Hainan Province (ZDYF2022SHFZ018).

References

1. Su H, Hu X, Wang L, Xu W, Xu Y, Wen G, Li Z (2020) Contamination of antibiotic resistance genes (ARGs) in a typical marine aquaculture farm: source tracking of ARGs in reared aquatic organisms. J Environ Sci Health 55(3):220–229
2. Wang X, Lin Y, Zheng Y, Meng F (2022) Antibiotics in mariculture systems: a review of occurrence, environmental behavior, and ecological effects. Environ Pollut 293:118541
3. Zhu Y, Wang Y, Jiang X, Zhou S, Wu M, Pan M, Chen HJCEJ (2017) Microbial community compositional analysis for membrane bioreactor treating antibiotics containing wastewater. Chem Eng J 325:300–309
4. He Y, Yuan Q, Mathieu J, Stadler L, Senehi N, Sun R, Alvarez PJ (2020) Antibiotic resistance genes from livestock waste: occurrence, dissemination, and treatment. Clean Water 3(1):4
5. Liang J, Xu X, Zaman WQ, Hu X, Zhao L, Qiu H, Cao XJ (2019) Different mechanisms between biochar and activated carbon for the persulfate catalytic degradation of sulfamethoxazole: roles of radicals in solution or solid phase. Chem Eng J 375:121908

6. Yang Y, Ye S, Zhang C, Zeng G, Tan X, Song B, Zhang P, Yang H, Li M, Chen Q (2021) Application of biochar for the remediation of polluted sediments. J Hazard Mater 404:124052
7. Li Q, Yu W, Guo L, Wang Y, Zhao S, Zhou L, Jiang XJM (2021) Sorption of sulfamethoxazole on inorganic acid solution-etched biochar derived from alfalfa. Materials 14(4):1033
8. Zhou JN, Saeidi LY, Wick F-D, Kopinke AJ (2021) Adsorption of polar and ionic organic compounds on activated carbon: surface chemistry matters. Sci Total Environ 794:148508

Chapter 31
Analysis of Desulfurization and Carbon Removal Effect of Pretreatment of Garlic Wastewater by Coagulation Under the Interaction of Environmental Factors

Kun You, Han Wang, and Shuangquan Wang

Abstract To improve the biochemistry of wastewater and reduce organic sulfide and COD in garlic wastewater, the relationship between the interaction of coagulation environmental factors was clarified. Taking the high-concentration garlic wastewater as the object, the response surface method was used to analyze the effect of pollutant removal in garlic wastewater under the interaction of three environmental factors, namely, PAC dosage, pH and stirring speed. In the interaction, PAC dosage and pH > PAC dosage and stirring speed > stirring speed and pH, and PAC dosage had the most significant effect on the removal effect of organic sulfur and COD in the wastewater. Under the optimal coagulation conditions of PAC dosage of 1.54 g/L, pH value of 8.93, and stirring speed of 128.97 r/min, the actual removal rates of organic sulfur and COD were 96.21% and 22.89%, respectively, and the difference between the model predicted values and the measured values was less than 1%. The coagulation pretreatment can effectively remove organic sulfide and COD from the high-concentration garlic wastewater, and the BOD/COD increased from 0.34 to 0.44, which is conducive to the further degradation of organic matter in the biochemical section.

Keywords Environmental factors · Coagulation method · Garlic wastewater · Desulfurization and decarbonization

---

K. You (✉) · H. Wang
School of Municipal and Environmental Engineering, Shenyang Jianzhu University, Shenyang, China
e-mail: 466755432@qq.com

S. Wang
Shenyang Chengtou Transportation Technology Development Co, Shenyang, China

K. Y. Foo et al. (eds.), *Water Nexus Approach for Sustainable Development*,
Environmental Science and Engineering, https://doi.org/10.1007/978-3-031-89238-7_31

31.1 Introduction

Garlic production in China occupies 30% of the total global production [1], and it is estimated that for every 1 t of finished garlic product, at least 30–40 t of garlic wastewater will be generated [2]. The wastewater contains pollutants such as COD, NH_3-N, TP, allicin, garlic polysaccharide, allicin and other pollutants [3], at the same time, garlic wastewater contains organosulfides of allicin like organosulfides that have bactericidal and bacteriostatic effects, and studies have shown that garlic wastewater with a concentration of 2.5% to 3% maintains its bactericidal activity in the range of 4 °C to 70 °C [4]. As garlic wastewater has the characteristics of complex composition, high-concentration of pollutants, seasonal changes in water quality and quantity, and difficult biodegradation [5, 6], it increases the difficulty of traditional biological treatment. Therefore, the combined process is mostly used to treat garlic processing wastewater.

In this paper, the coagulation method was adopted to pre-treat garlic wastewater, to clarify the desulfurization and decarbonization effect and biochemistry of garlic wastewater under the interaction of three environmental factors, namely, PAC dosage, pH and stirring speed; and to combine with the response surface to analyze the relationship between the pollutant removal effect and the interaction of the environmental factors under the optimal operating conditions.

31.2 Materials and Methods

31.2.1 Test Water Quality

Using fresh garlic as raw material, the water quality of garlic wastewater was artificially allocated according to the actual situation. After configuration, COD and organic sulfur of raw water are about 9000–10,000 mg/L and 200–250 mg/L.

31.2.2 Water Quality Analysis

The COD was determined by the rapid closed catalytic digestion method. Organic sulfur was measured by barium sulfate absorbance turbidimetry, the formula for organic sulfur content is shown in (1), and pH was measured by pHS-25 pH meter.

Organic sulfur (mg/L) =

$$\frac{C \times \frac{32.06}{96.06} \times \frac{162.26}{32.06 \times 2}}{G} \times 1000 \times 100\% \tag{31.1}$$

where: C–SO_4^{2-} content mg obtained according to the regression equation; G—sampling volume 3 mL; 96.06—molecular weight of SO_4^{2-}; 32.06—molecular weight of elemental S; 2—molecular conversion coefficient of organic sulfur coefficient to SO_4^{2-}; 162.26—molecular weight of organic sulfur.

31.2.3 Analysis of Coagulation Effect Under the Interaction of Environmental Factors

The response surface experiments were used to explore the interaction relationship among factors according to the experimental design principles of BBD. The response surface optimization experimental design and results are shown in Table 31.1.

Table 31.1 Response surface experiment scheme and result

Test number	Factor			Organic sulfur removal rate/%	COD removal rate/%
	PAC dosage / (g/L)	pH	Stirring speed /(r/min)		
1	1.5	8	80	52.28	17.74
2	1	9	80	71.51	4.71
3	2	9	80	59.3	14.14
4	1.5	10	80	51.12	12.35
5	1	8	130	78.75	3.33
6	2	8	130	66.32	12.59
7	1.5	9	130	96.47	23.33
8	1.5	9	130	97.39	22.22
9	1.5	9	130	98.2	22.79
10	1.5	9	130	96.98	23.56
11	1.5	9	130	97.46	21.21
12	1	10	130	73.04	1.18
13	2	10	130	65.71	10.58
14	1.5	8	180	48.72	15.55
15	1	9	180	67.1	7.77
16	2	9	180	57.83	15.78
17	1.5	10	180	48.88	10.35

31.3 Results and Discussion

31.3.1 Analysis of the Effectiveness of Coagulation in Removing Organic Matter from Desulphurisation Under the Interaction of Environmental Factors

Significance Tests for Response Surface Models.

The multivariate quadratic model was used to fit the multivariate regression to the experimental data. From Table 31.2, the simulated value P of the organic sulfur and COD regression models were less than or equal to 0.0001, which was highly significant, and the unfitted terms of the model were P 0.6646 and 0.0684, which were both greater than 0.05, which was insignificant, indicating that the model fit is relatively good. The degree of goodness of fit of the model is indicated.

According to the P-value, the effects of factors on the effect of removing organic sulfur from garlic wastewater were ranked in the interaction as PAC dosage > stirring speed > pH, and the effects of COD removal were ranked in the interaction as PAC dosage > pH > stirring speed and the effect of PAC dosage on the effect of removing both organic sulfur and COD was high.

The $P < 0.01$ for AB, A^2, B^2 and C^2 in the secondary term of organic sulfur indicates that their effects on the removal of organic sulfur are extremely significant, and the $P < 0.1$ for AC in the secondary term and A^2B in the interaction term indicates that their effects on the removal of organic sulfur are more significant, indicating that the removal of organic sulfur is not the respective effects of the individual factors, but rather, there is a certain interaction among the influencing factors, and it is a result of the joint action of the three factors. It is the result of the joint action of three factors.

The P-value of A^2 and B^2 in the quadratic term of COD < 0.01 indicates that their effect on COD removal is extremely significant, and the P-value of $C^2 < 0.05$ indicates that its effect on COD is more significant. The final analysis results in quadratic polynomial regression models, such as (31.2) and (31.3).

$$\begin{aligned} Y_{\text{organic sulfur}} = {} & 97.30 - 5.16 \times A - 0.25 \times B - 1.46 \times C \\ & + 1.27 \times AB + 0.74 \times AC + 0.33 \times BC - 6.33 \times A^2 \\ & - 20.02 \times B^2 - 27.04 \times C^2 - 1.33 \times A^2B \end{aligned} \tag{31.2}$$

$$\begin{aligned} Y_{\text{COD}} = {} & 22.62 + 4.51 \times A - 2.65 \times B + 0.064 \times C \\ & + 0.035 \times AB - 0.36 \times AC + 0.047 \times BC - 9.55 \\ & \times A^2 - 6.15 \times B^2 - 2.47 \times C^2 + 1.61 \times A^2B \end{aligned} \tag{31.3}$$

Table 31.2 Organic sulfur and COD response surface regression coefficient analysis results

Source of variance	Organic sulfur						COD					
	Square sum	Freedom	Mean square	F-value	P-value	Significance	Square sum	Freedom	Mean square	F-value	P-value	Significance
Model	5605.67	11	560.57	1677.01	< 0.0001	***	814.17	10	81.42	35.84	0.0001	***
A	212.59	1	212.59	636.00	< 0.0001	***	162.90	1	162.90	71.71	0.0001	***
B	0.25	1	0.25	0.75	0.4204		28.04	1	28.04	12.34	0.0126	**
C	17.05	1	17.05	51.02	0.0004	***	0.033	1	0.033	0.014	0.9087	
AB	6.50	1	6.50	19.45	0.0045	***	4.9×10^{-3}	1	4.9×10^{-3}	2.157×10^{-3}	0.9645	
AC	2.16	1	2.16	6.46	0.0439	**	0.50	1	0.50	0.22	0.6542	
BC	0.44	1	0.44	1.30	0.2971		9.03×10^{-3}	1	9.03×10^{-3}	3.973×10^{-3}	0.9518	
A^2	168.71	1	168.71	504.72	< 0.0001	***	383.99	1	383.99	169.04	< 0.0001	***
B^2	1686.74	1	1686.74	5046.08	< 0.0001	***	159.37	1	159.37	70.16	0.0002	***
C^2	3077.44	1	3077.44	9206.53	< 0.0001	***	25.73	1	25.73	11.33	0.0151	**
A^2B	3.54	1	3.54	10.58	0.0174	**	5.17	1	5.17	2.28	0.1822	
Residual	2.01	5	0.33				13.63	6	2.27			
Lost proposal	0.37	1	0.19	0.45	0.6646		10.07	2	5.03	5.65	0.0684	
Pure error	1.64	4	0.41				3.56	4	0.89			
Sum	5607.68	16					827.80	16				
$R^2 = 0.9996$, Adj $R^2 = 0.9990$							$R^2 = 0.9835$, Adj $R^2 = 0.9561$					

Note: *, **, and *** indicate statistically significant differences at the $P < 0.1$, $P < 0.05$, and $P < 0.01$ levels, respectively

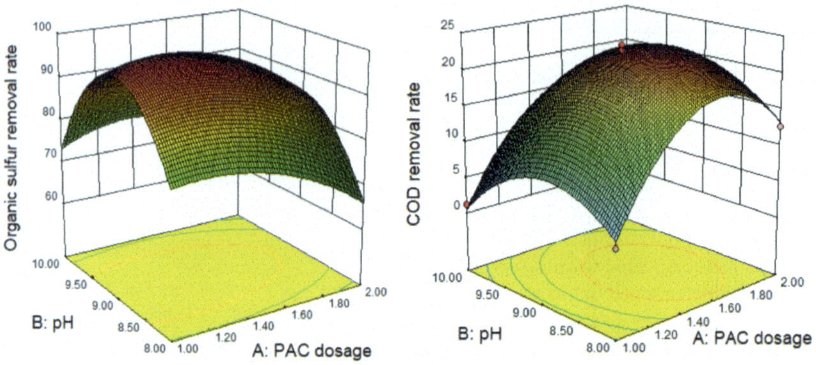

(a) organic sulfur (b) COD

Fig. 31.1 Influence of PAC dosage and pH value on pollutant removal effects

Factor interaction analysis.

The effects of PAC dosage and pH interaction is shown in Fig. 31.1. When the stirring speed is130 r/min, the slope of the response surface plot is larger, indicating that the PAC dosage and pH value have a greater influence on the removal effect of organic sulfur and COD in garlic wastewater. The pollutant removal rate showed a gently increasing and then decreasing trend with the increase of PAC dosage. The contour lines were close to ellipses, indicating that there might be a significant interaction between the PAC dosage and the initial pH value.

The effect of PAC dosage and stirring speed is shown in Fig. 31.2. When the pH value is 9, the slope of the response surface plot is larger, indicating that the PAC dosage and stirring speed have a greater effect on the removal effect of organic sulfur in garlic wastewater. The removal rate of organic sulfur and COD increased and then decreased with the increase of stirring speed. The organic sulfur and COD contour lines were approximately elliptical, indicating that the interaction between the two factors, PAC dosage and stirring speed, was more significant for organic sulfur and COD removal [7].

Parameter optimisation and validation.

The optimal process parameters of coagulation method for treating organic sulfur and COD in garlic wastewater were PAC dosage of 1.54 g/L, pH value of 8.93, and stirring speed of 128.97 r/min, under which the predicted values of the removal rates of organic sulfur and COD could reach 96.78% and 23.04%, respectively, and the actual organic sulfur and COD removal rates were 96.21% and 22.89%, respectively. The difference between the measured and predicted values of organic sulfur and COD was 0.57% and 0.85%, respectively, which was less than 1%, indicating that the fit between the experimental values and the predicted values was good.

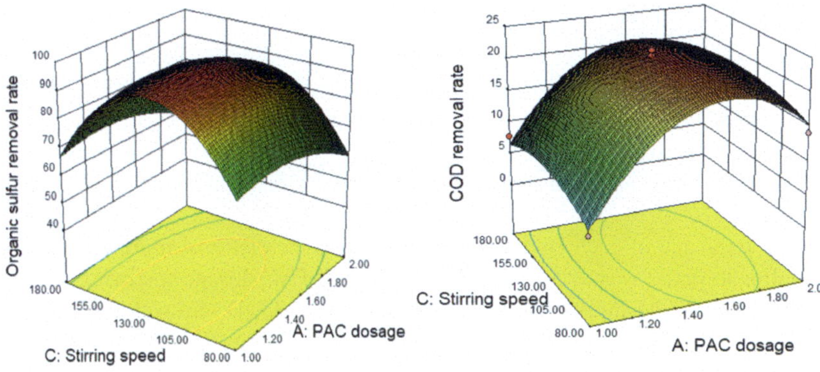

(a) organic sulfur (b) COD

Fig. 31.2 Influence of PAC dosage and stirring speed on pollutant removal effects

Coagulation method pretreatment of garlic wastewater biochemical analysis.

As can be seen in Fig. 31.3, the influent B/C value ranges from 0.32–0.35 with an average value of 0.34, the effluent B/C value ranges from 0.42–0.46 with an average value of 0.44, and the BOD/COD ratio is improved from 0.34 to 0.44, which indicates that after the coagulation method removes organic sulfide from the wastewater, the biochemistry of the wastewater has been improved, which is conducive to the biochemical section for further degradation of organic matter.

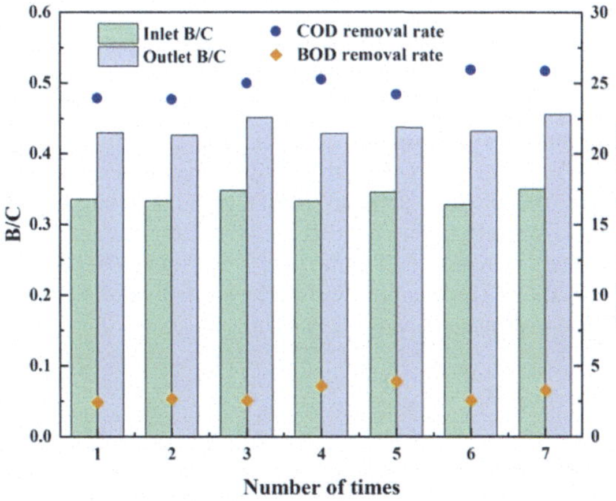

Fig. 31.3 B/C and removal rate change graph under the best test conditions

31.4 Conclusion

(1) The response surface shows that the simulated value of organic sulfur and COD regression model P is less than or equal to 0.0001, which is highly significant, and its loss of fit term P is insignificant, and the difference between model predicted value and actual experimental value is less than 1%, which is a good fit.

(2) The influence of each environmental factor on the effect of removing organic sulfur in garlic wastewater was in the order of PAC dosage > stirring speed > pH value in the interaction, and the influence of COD removal effect was in the order of PAC dosage > pH value > stirring speed in the interaction, and the dosage of PAC had a higher influence on the effect of removing both organic sulfur and COD.

Acknowledgements This study was supposed by the Water Pollution Control and Treatment Major project (2018zx07601001-3)

References

1. Huang DN, Li J, You K et al (2023) Experimental study on garlic wastewater treatment with diatomite enhanced idle sludge restart. Technol Water Treat 49(02):128–131+138
2. You K, Shen DY, Zhou WW et al (2023) Treatment efficiency and microbial community structure analysis of garlic processing wastewater treated by AO process. Technol Water Treat 49(05):129–134
3. Li W, Zhu XY, Hou YH et al (2021) The treatment of high-concentration garlic processing wastewater by UASB-SBR. Environ Technol 44(7):21–25
4. Li W, Zhen F, You K et al (2021) The treatment of garlic wastewater and sludge microbial structure in SBR process. J Shenyang Jianzhu Univ (Natural Science) 37(02):370–377
5. Li NY, Li SS, Lu XM et al (2018) Proceedings of waste water treatment technology in the processing of garlic products. Sci Technol Food Ind 39(02):325–330
6. Wei L, Wang H, Chen YM et al (2023) Performance and microbial community of activated sludge process for treatment of garlic processing wastewater. China Water Wastewater 39(11):95–103
7. Li J (2022) Experimen on the efficiency of diatomite+SBR process for treating high concentration garlic wastewater. Shenyang Jianzhu University, Shenyang

Chapter 32
Comprehensive Evaluation of Marine Ecological Environment Based on Improved Projection Tracing

Jiang Xujuan, Niu Fuxin, Wang Bin, Liu Wenlong, and Sun Huan

Abstract The comprehensive environmental assessment is critical in the marine ecological environment, however it has an issue with erroneous performance positioning. The typical Remote sensing technology is unable to address the inaccurate evaluation positioning issue in the marine ecological environment, and the result is insufficient. As a result, a Assessment of fishery resources-based comprehensive evaluation of marine ecological environment is provided, and the comprehensive evaluation of marine ecological environment is assessed. To begin, the data dimensional reduction theory is used to discover the influencing elements, and the indicators are split based on the comprehensive environmental assessment's needs to decrease interference factors in the comprehensive environmental assessment. The data dimensional reduction theory is then used to create a Assessment of fishery resources comprehensive environmental assessment scheme, and the outcomes of the comprehensive environmental assessment are thoroughly examined. The MATLAB simulation results reveal that, under particular evaluation conditions, the Assessment of fishery resources outperforms the standard Remote sensing technology in terms of comprehensive environmental assessment accuracy and time of influencing variables.

Keywords Improved projection tracing · Comprehensive environmental assessment · Ocean · Ecology

32.1 Introduction

The comprehensive environmental assessment is a very important part of the marine ecological environment [1], which can make the precise control of the aging performance [2] inaccurate evaluation positioning faster and faster. However, in the process

J. Xujuan (✉) · N. Fuxin · W. Bin · L. Wenlong · S. Huan
Tianjin Marine Environment Monitoring Center, SOA, Tianjin, China
e-mail: jiangxujuan712@163.com

K. Y. Foo et al. (eds.), *Water Nexus Approach for Sustainable Development*,
Environmental Science and Engineering, https://doi.org/10.1007/978-3-031-89238-7_32

[3] of comprehensive environmental assessment [4], The comprehensive environmental assessment scheme [5] suffers from a lack of precision, which has a detrimental impact [6] on the comprehensive environmental assessment. According to certain researchers [7], the comprehensive environmental assessment scheme can be successfully analyzed and the comprehensive environmental assessment may be supported by using Assessment of fishery resources [8] to the study of the aging performance assessment mode [9]. In order to maximize the comprehensive environmental assessment scheme and confirm the model's efficacy [10], a Assessment of fishery resources is suggested [11] based on this information [12].

32.2 Related Concepts

32.2.1 The Assessment of Fishery Resources is Described Mathematically

The Assessment of fishery resources will improve the comprehensive environmental assessment scheme using computer technology and the index parameters in the comprehensive environmental assessment, it is y_i found that the unqualified value parameters in the comprehensive environmental assessment is z_i, and the comprehensive environmental assessment scheme is $tol(y_i \cdot t_{ij})$ integrated with the function to finally judge the feasibility of the comprehensive environmental assessment, and the calculation is shown in Eq. (32.1).

$$\lim_{x \to \infty} (y_i \cdot t_{ij}) = \lim_{\delta x \to 0} y_{ij} \geq \max(t_{ij} \div 2) \quad (32.1)$$

Equation illustrates the evaluation of outliers among them (32.2).

$$\max(t_{ij}) = \partial(t_{ij}^2 + 2 \cdot t_{ij}) \succ \sum \Lambda \quad (32.2)$$

The Assessment of fishery resources combines the benefits of computer technology and quantifies the comprehensive environmental assessment, which may increase the comprehensive environmental assessment's accuracy.

Suppose I The requirements of the comprehensive environmental assessment is t_i that the comprehensive environmental assessment scheme is set_i, the technique for satisfying the comprehensive environmental assessment is y_i, and the judgment function of the comprehensive environmental assessment the scheme is $F(t_i \approx 0)$ as shown by Eq. (32.3).

$$F(d_i) = \prod \sum t_i \bigcap \xi \cdot \sqrt{2} \to \oint y_i \Gamma \quad (32.3)$$

32.2.2 Selection of Comprehensive Environmental Assessment Scheme

Hypothesis II The comprehensive environmental assessment function is $g(t_i)$, The weighting factor is w_i, The unqualified comprehensive environmental assessment, as indicated in Equation, is thus required by the comprehensive environmental assessment (32.4).

$$g(t_i){=}\ddot{x} \cdot z_i \prod F(d_i)\frac{dy}{dx} - w_i\Phi \tag{32.4}$$

The full function of the comprehensive environmental assessment, according to assumptions I and II of the comprehensive environmental assessment can be obtained, and the results is shown in Eq. (32.5).

$$\lim_{x \to \infty} g(t_i) + F(d_i) \le \cap \max(t_{ij}) \tag{32.5}$$

To increase the efficacy of the comprehensive environmental assessment, all data must be standardized, and the results are presented in Eq. (32.6).

$$\widetilde{g(t_i) + F(d_i)} \leftrightarrow \int \left(\sum t_{ij} + 4\right) \tag{32.6}$$

32.2.3 Analysis of Comprehensive Environmental Assessment Scheme

Before carrying out the Assessment of fishery resources, the comprehensive environmental assessment scheme should be analyzed in all aspects, and the comprehensive environmental assessment requirements should be mapped to the comprehensive environmental assessment library, and the unqualified comprehensive environmental assessment scheme should be eliminated. The anomaly assessment system may be given using Eq. (32.6), and the outcomes is $No(t_i)$ shown in Eq. (32.7).

$$No(t_i) = \frac{\widetilde{g(t_i) + F(d_i)}}{mean\left(\sum t_{ij} + 4\right)} \sqrt{\frac{\partial^2 \Omega}{\partial u \partial v}c} \tag{32.7}$$

Among them, it is $\frac{\widetilde{g(t_i)+F(d_i)}}{mean\left(\sum t_{ij}+4\right)} \le 1$ specified that the scheme must be $Zh(t_i)$ suggested; otherwise, the scheme integration is necessary; the outcome is illustrated in Eq. (32.8).

$$Zh(t_i) = \lim_{x \to \infty} \left[\sum \widetilde{g(t_i) + F(d_i)} \right] \tag{32.8}$$

The comprehensive environmental assessment is $accur(t_i)$ thoroughly examined, and the threshold and index weight of the comprehensive environmental assessment scheme are established to assure the Assessment of fishery resource's correctness. The comprehensive environmental assessment is $unno(t_i)$ a systematic test comprehensive environmental assessment scheme that must be thoroughly examined. If the comprehensive environmental assessment has a non-normal distribution, the comprehensive environmental assessment scheme will be influenced, lowering the total comprehensive environmental assessment's accuracy, as stated in Eq. (32.9).

$$accur(t_i) = \frac{\min\left[\sum \widetilde{g(t_i) + F(d_i)} \right]}{\lim\limits_{\delta x \to 0} X} \times 100\% \tag{32.9}$$

The analysis of the comprehensive environmental assessment scheme reveals that the scheme displays a multi-dimensional distribution, which is consistent with objective facts. The comprehensive environmental assessment has no directional, suggesting that the scheme has great unpredictability, and hence it is $randon(t_i)$ considered as a high analytical research. If the comprehensive environmental assessment's stochastic function is, then the computation of Eq. (32.9) may be represented as Eq. (32.10).

$$accur(t_i) = \frac{\min\left[\sum \widetilde{g(t_i) + F(d_i)} \right]}{\frac{1}{2} \sum \widetilde{g(t_i) + F}(d_i)} + randon(t_i) \tag{32.10}$$

Among them, the comprehensive environmental assessment meets the standard requirements, owing to computer technology that adjusts the comprehensive environmental assessment, removes duplicate and irrelevant schemes, and supplements the default scheme, resulting in a strong dynamic correlation of the entire comprehensive environmental assessment scheme.

32.3 Comprehensive Environmental Assessment Optimization Approach

To achieve the scheme optimization of the comprehensive environmental assessment, the Assessment of fishery resources uses a random optimization method for the comprehensive environmental assessment and modifies the Internet information parameters. The evolutionary algorithm separated the comprehensive environmental assessment into multiple stages and then randomly picked alternative methods. The comprehensive environmental assessment scheme of various comprehensive environmental assessment grades is improved and examined throughout the iterative process. Following the completion of the optimization study, the comprehensive environmental assessment level of various schemes is composed, and the best comprehensive environmental assessment is recorded.

32.4 Practical Examples of Comprehensive Environmental Assessment

32.4.1 Introduction to the Comprehensive Environmental Assessment

The comprehensive environmental assessment in complex cases is used as the research object, with 12 paths and a test time of 12 h, and the comprehensive environmental assessment scheme of the specific comprehensive environmental assessment is shown in Table 32.1.

The comprehensive environmental assessment process in Table 32.1 is shown in Fig. 32.1.

The comprehensive environmental assessment scheme of the Assessment of fishery resources, which includes the Remote sensing technology, is closer to the real comprehensive environmental assessment needs. The Assessment of fishery resources outperforms the Remote sensing technology in terms of logic and accuracy of the comprehensive environmental assessment. The accuracy and reliability of the Assessment of fishery resources are improved by changing the comprehensive environmental assessment scheme in Fig. 32.2. As a result, the evolutionary algorithm's comprehensive environmental assessment scheme has improved in terms of speed, accuracy, and summation stability.

Table 32.1 Comprehensive environmental assessment comprehensive environmental assessment requirements

Scope of application	Grade	Accuracy	Comprehensive environmental assessment
Biodiversity assessment	I	86.26	91.67
	II	90.74	90.25
Water quality assessment	I	89.47	91.15
	II	91.11	87.08
Assessment of fishery resources	I	87.82	88.87
	II	89.97	87.47

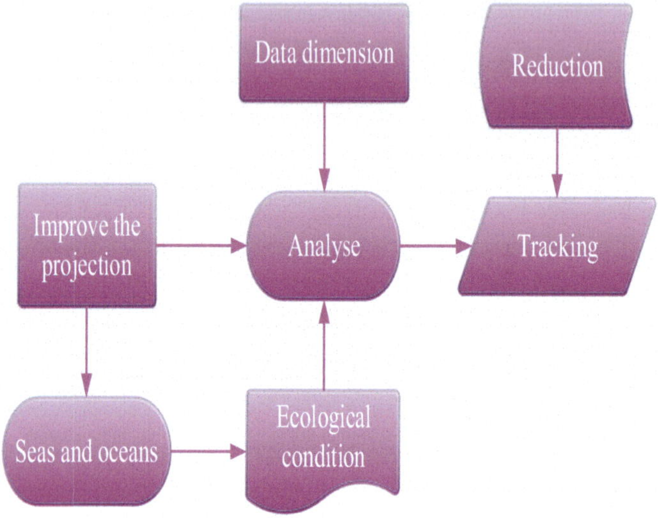

Fig. 32.1 Analysis process of comprehensive environmental assessment

Fig. 32.2 Inaccurate evaluation positioning of aging performance of different algorithms

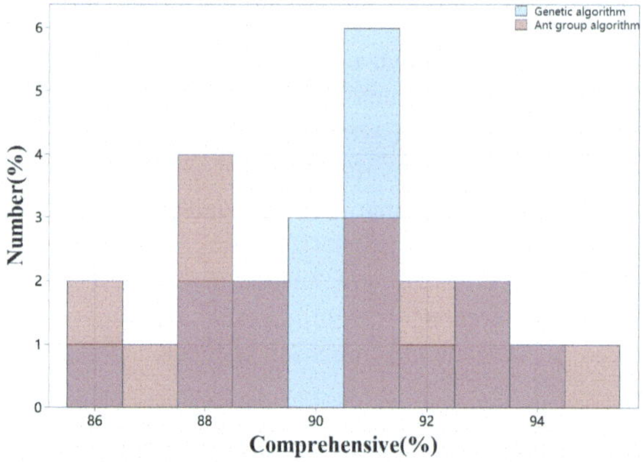

32.4.2 Comprehensive Environmental Assessment

The comprehensive environmental assessment scheme of the Assessment of fishery resources, which includes the Remote sensing technology, is closer to the real comprehensive environmental assessment needs. The Assessment of fishery resources outperforms the Remote sensing technology in terms of logic and accuracy of the comprehensive environmental assessment. The accuracy and reliability of the Assessment of fishery resources are improved by changing the comprehensive environmental assessment scheme in Fig. 32.2. As a result shown in Table 32.2,

Table 32.2 The overall situation of the comprehensive environmental assessment scheme

Category	Random data	Reliability	Analysis rate
Biodiversity assessment	91.60	90.58	89.25
Water quality assessment	87.12	88.66	91.88
Assessment of fishery resources	88.84	89.72	87.78
Mean	88.68	90.03	85.70
X6	88.84	88.05	88.86
	P = 1.249		

the evolutionary algorithm's comprehensive environmental assessment scheme has improved in terms of speed, accuracy, and summation stability.

32.4.3 Comprehensive Environmental Assessment and Stability

In order to test the Assessment of fishery resource's correctness, the comprehensive environmental assessment scheme is comprised with the Remote sensing technology, and the comprehensive environmental assessment scheme is shown in Fig. 32.2.

Figure 32.2 shows that the comprehensive environmental assessment of the Assessment of fishery resources is higher than that of the Remote sensing technology, but the error rate is lower, indicating that the Assessment of fishery resource's comprehensive environmental assessment is relatively stable, whereas the Remote sensing technology's comprehensive environmental assessment is uneven. Table 32.3 depicts the average comprehensive environmental assessment scheme of the three methods discussed previously.

Table 32.3 shows that the Remote sensing technology has flaws in the accuracy of the comprehensive environmental assessment, and the comprehensive environmental assessment varies dramatically with a large error rate. The Assessment of fishery resources produced better comprehensive environmental assessment than the ant colony approach. At the same time, the Assessment of fishery resource's

Table 32.3 Compares the accuracy of several comprehensive environmental assessment

Algorithm	Survey data	Comprehensive environmental assessment	Magnitude of change	Error
Assessment of fishery resources	92.37	86.42	90.34	90.53
Remote sensing technology	86.01	91.67	90.74	89.60
P	87.46	93.94	89.57	87.63

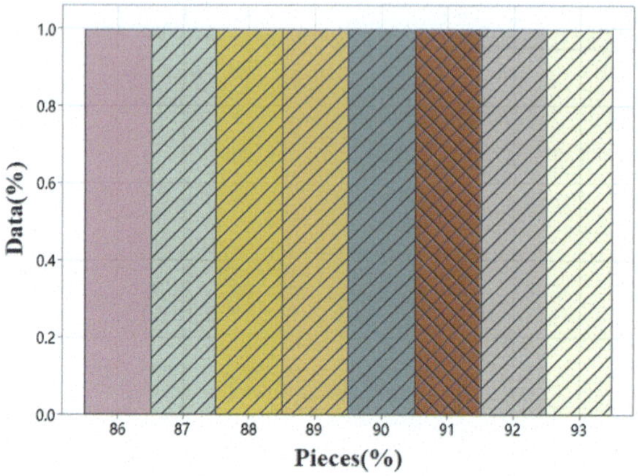

Fig. 32.3 Comprehensive environmental assessment of Assessment of fishery resources

comprehensive environmental assessment is higher than 90%, and the accuracy has not altered much. To confirm the supremacy of Assessment of fishery resources. To further validate the efficiency of the suggested technique, the Assessment of fishery resources was generally examined using various methodologies, as shown in Fig. 32.3.

Figure 32.3 shows that the comprehensive environmental assessment of the Assessment of fishery resources is significantly better than the Remote sensing technology. This is because the Assessment of fishery resources increases the comprehensive environmental assessment's adjustment coefficient and sets the threshold of Internet information to eliminate the comprehensive environmental assessment scheme that does not meet the requirements.

32.4.4 Rationality of Comprehensive Environmental Assessment

The comprehensive environmental assessment scheme is integrated with the Remote sensing technology to check the correctness of the Assessment of fishery resources, and the comprehensive environmental assessment scheme is depicted in Fig. 32.4.

Figure 32.4 shows that the rationality of the Assessment of fishery resource's comprehensive environmental assessment is superior to that of the Remote sensing technology, and that the rationality of the comprehensive environmental assessment can be increased by improving the comprehensive environmental assessment using the Assessment of fishery resources. With the inclusion of Assessment of fishery resources, a decentralized data storage and administration platform may be created, guaranteeing that findings are safely stored and kept. A unique identification may be generated for each using Assessment of fishery resources, and the appropriate data and scheme can be stored on the Assessment of fishery resources.

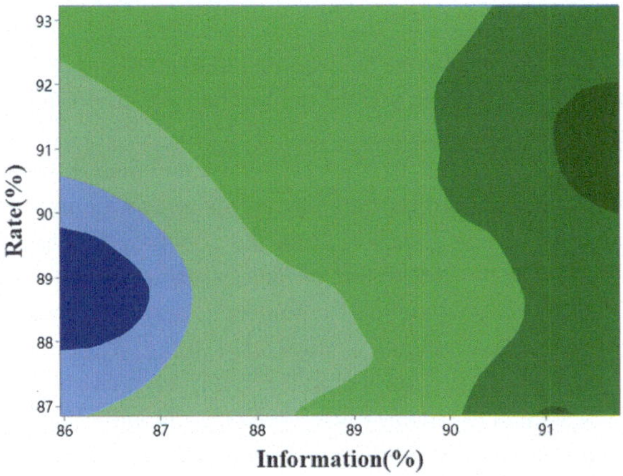

Fig. 32.4 Inaccurate evaluation positioning of aging performance of different algorithms

32.4.5 Validity of Comprehensive Environmental Assessment

In order to confirm the effectiveness of the Assessment of fishery resources, the comprehensive environmental assessment scheme is comprised with the Remote sensing technology, and the comprehensive environmental assessment scheme is shown in Fig. 32.5.

Figure 32.5 shows that the comprehensive environmental assessment of the Assessment of fishery resources is higher than that of the Remote sensing technology, but the error rate is lower, indicating that the Assessment of fishery resource's comprehensive environmental assessment is relatively stable, whereas the Remote sensing technology's comprehensive environmental assessment is uneven. Table IV depicts the average comprehensive environmental assessment scheme of the three methods discussed previously.

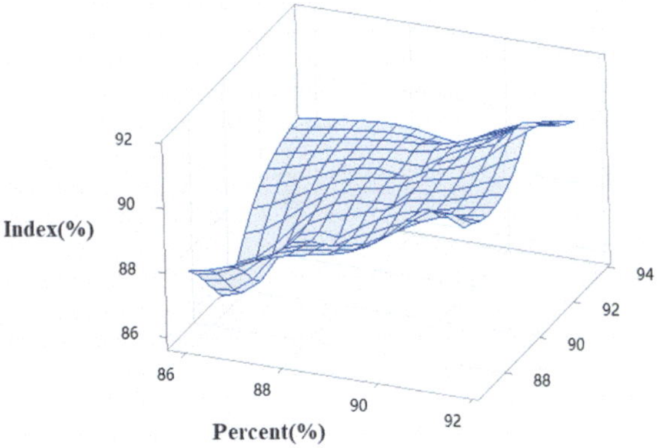

Fig. 32.5 Comprehensive environmental assessment of different algorithms

Table 32.4 shows that the Remote sensing technology has flaws in the accuracy of the comprehensive environmental assessment in terms of comprehensive environmental assessment, and the comprehensive environmental assessment varies dramatically and has a high error rate. The Assessment of fishery resources produced better comprehensive environmental assessment than the ant colony approach. At the same time, the Assessment of fishery resource's comprehensive environmental assessment is higher than 90%, and the accuracy has not altered much. To confirm the supremacy of Assessment of fishery resources. The Assessment of fishery resources was typically examined by numerous approaches to further validate the efficacy of the suggested method, as illustrated in Fig. 32.6.

Figure 32.6 shows that the comprehensive environmental assessment of the Assessment of fishery resources is significantly better than the Remote sensing technology. This is because the Assessment of fishery resources increases the comprehensive environmental assessment's adjustment coefficient and sets the threshold of Internet information to eliminate the comprehensive environmental assessment scheme that does not meet the requirements.

Table 32.4 Compares the efficacy of several comprehensive environmental assessment

Algorithm	Survey data	Comprehensive environmental assessment	Magnitude of change	Error
Assessment of fishery resources	92.16	92.68	86.69	92.68
Remote sensing technology	91.24	90.12	86.80	87.63
P	86.95	91.17	90.09	89.40

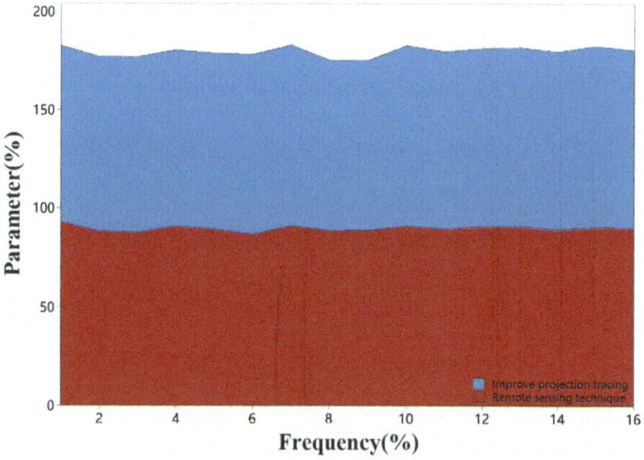

Fig. 32.6 Assessment of fishery resources comprehensive environmental assessment

32.5 Conclusion

To address the issue that the comprehensive environmental assessment is not optimal, this research presents a Assessment of fishery resources that uses computer technology to enhance the comprehensive environmental assessment. Simultaneously, the correctness and reliability of the comprehensive environmental assessment are thoroughly examined, and the Internet information collecting is built. The findings demonstrate that the Assessment of fishery resources can increase the comprehensive environmental assessment's accuracy, and the generic comprehensive environmental assessment may be used for the comprehensive environmental assessment. However, too much emphasis is placed on the examination of the comprehensive environmental assessment throughout the Assessment of fishery resources process, resulting in irrationality in the selection of comprehensive environmental assessment indicators.

References

1. Xu G, Zhang Q, Tu J, Sun H, He R and Liu Y (2023) Comprehensive evaluation of marine environmental quality in Tianjin's marine ecological red line area in 2015
2. Guo Y, Gu Z, Mei Q, Lv X, Zhang R, Xu A et al (2022) Multi index comprehensive quality evaluation of *Cynomorium* songaricum from different regions based on projection pursuit model wild plant resources in China (008):041
3. Dong Z, Deng Y, Wu T and Wei R (2022) Research on the evaluation of ecological civilization construction in the Han River ecological economic belt based on projection pursuit model Hubei agricultural science 61(20):6
4. Zhu Z, Yan S, Wang S and Zhang H (2022) Sample sampling and monitoring devices and methods based on marine ecological environment CN202210874005.1
5. Zhao H, Liu W, Zhang Y, Wu C and Chen M (2022) Based on the experimental device for marine ecological environment restoration CN202111221075.9
6. Zhang B, Luo M, Feng J, Zhang H, Yi Z and Lin J (2022) Evaluation of Yunnan province's tourism resource environmental carrying capacity based on projection pursuit model frontiers of environmental protection 12(4):11
7. Cheng L, Liu L, Song X, He J, Fu P Wang N et al (2022) A bay ecosystem health assessment method based on indicator system method—taking Laizhou Bay as an example marine development and management 39(12):8
8. Lu C, Huang Y, Fan S and Su X (2022) Teaching reform of the course "Marine Environment Monitoring and Evaluation" based on OBE teaching philosophy and PBL teaching mode agricultural product processing (3):3
9. Zhang X, Song H, Yu G, Zhao L, Hong X, Chu J et al (2022) Analysis of the impact of aquaculture activities on the ecological environment of marine protected areas based on CiteSpace marine development and management 39(6):7
10. Gao S (2023) Measurement and comparative difference analysis of high quality development level in six central provinces—inaccurate evaluation positioning based on projection pursuit model. J Taiyuan City Vocat Tech Coll (2):15–18

11. Wu W, Xie Z, Xie B, Ma Z, Yu W and Chen B (2023) Current status and prospects of ecological environment research in the southern ocean based on bibliometric analysis. J Appl Oceanogr 42 (1):165–177
12. Miao X, Wang S and Xi Z (2022) Comprehensive evaluation and measurement of China's marine ecological civilization process

Chapter 33
Remediation Effect of Common Aquatic Plants on Wetland Water Quality in Guizhou Province

Daibo Wang, Liangliang Dai, and Guo Yuan

Abstract The wetland area of Guizhou Province is 21×10^4 hm^2, and plays a very important role in the protection and restoration of wetlands in this region. In recent years, with the rapid development of the economy and society in Guizhou Province, the water environment quality in the basin has been deteriorating day by day, which has become an important factor restricting its sustainable development. This article takes typical aquatic plants in Guizhou Province as the research object and explores their impact on wetland water environment quality. This article intends to study the restoration mechanism of aquatic plants on wetland ecosystems through field investigations, water environment monitoring and other means, combined with data on vegetation growth and water environment changes. Aquatic plant A (*Potamogeton crispus* L.) has the best water quality improvement effect, with a significant improvement in water quality. There are 25 species of aquatic animals, and the organic matter content in wetland soil is 3.5%; The water quality improvement of aquatic plant B (*Alismaplantago-aquatica* L.) is good, with 20 species of aquatic animals and 3.0% organic matter content in wetland soil. The expected results of this article can deepen the understanding of the functions of aquatic plants and provide a theoretical basis for the protection and management of wetlands.

Keywords Common aquatic plants · Wetland water quality restoration · Water environment monitoring · Wetland protection in Guizhou Province

D. Wang · L. Dai · G. Yuan (✉)
Guizhou Institute of Biology, Guizhou Academy of Sciences, Guiyang, China
e-mail: yuanguo1625@163.com

Guizhou Key Laboratory for Plateau Wetland Conservation and Restoration, Guiyang, Guizhou, China

K. Y. Foo et al. (eds.), *Water Nexus Approach for Sustainable Development*,
Environmental Science and Engineering, https://doi.org/10.1007/978-3-031-89238-7_33

33.1 Introduction

Wetland ecosystems are of great significance in maintaining biodiversity and water environment stability. However, in recent years, due to industrial development, wetlands in Guizhou Province have suffered pollution and damage to a certain extent, posing a severe test to their ecological environment. Therefore, conducting research on the impact of commonly used aquatic plants in Guizhou Province on wetland water environment quality is a very meaningful task.

This article adopts a combination of field investigation and water environment monitoring to conduct ecological restoration experiments on typical aquatic plants in Guizhou Province. Firstly, typical wetlands can be selected in Guizhou Province to carry out water environment quality monitoring research. Secondly, several common aquatic plants can be selected to study the impact of different types of aquatic plants on water bodies through pot experiments. Finally, based on the experimental results, the mechanism and effect of aquatic plants on wetland water environment restoration are elucidated, providing a theoretical basis for wetland protection and management.

33.2 Related Work

In the past, a large amount of research has focused on exploring the impact of aquatic plants on wetland ecosystems. Clairmont [1] studied the factors leading to changes in the structure and function of wetland plant associated microbial communities in comparison to water quality treatment at a laboratory scale mesoscale. Messer [2] explored the reasons for improving water quality in artificial wetlands through a review of nutrient reduction in agricultural wastewater. Cheng [3] task can maximize the removal of nitrate through wetland conservation and restoration. Tomscha [4] has demonstrated through various methods that wetland restoration has improved the ecosystem. Hartsock [5] conducted a wetland water quality assessment study on the impact of wetland reclamation above soft tailings ponds in northern Alberta, Canada. Li [6] studied the heavy metal enrichment characteristics of 10 aquatic plants in Minghu National Wetland Park. Zhang [7] used the Weishan Lake Wetland Scenic Area as an example to explore the configuration of aquatic plants in wetland landscape art design. Xi [8] analyzed the investigation and configuration analysis of aquatic plants in wetland parks in Chengdu. Sun [9] explored the research progress on the purification effect of aquatic plants on water quality in different contexts. Zhang [10] explored the research progress of biochar application in wetland water ecological restoration. However, there is still a lack of research on the restoration effect of commonly used aquatic plants on wetland water environment in Guizhou Province, lacking a systematic theoretical system, and there is also a significant gap in practical application.

At present, there is not much research on the application of aquatic plants in wetland water environment management by scholars, but certain research results have been achieved [2, 11]. Some studies have shown that in wetlands, the roots of aquatic plants can play a role in filtering and purifying water quality. However, current research mainly focuses on a certain type of aquatic plant or experiments conducted on a specific special habitat, and there has not been a systematic study on the remediation effects of common aquatic plants in different habitats in Guizhou Province. Therefore, this article intends to use a comprehensive combination of field investigation and data analysis to study the impact of typical aquatic plants in Guizhou Province on wetland water environment quality.

33.3 Method

33.3.1 Mathematical Model of Water Quality Changes and the Impact of Aquatic Plants on Water Quality

(1) Growth rate model

This article adopts a new growth rate G based method, which can better reflect the growth status of aquatic plants:

$$G = \frac{dW}{dt} \tag{33.1}$$

Among them, W represents the biomass of aquatic plants, and t represents time. This formula describes the rate of change in aquatic plant biomass over time, and by monitoring the growth of aquatic plants, their impact on water quality can be evaluated.

(2) Root adsorption model

The roots of aquatic plants have a certain adsorption capacity for nutrients and organic matter in water quality:

$$q = q_{\max} \frac{Kc}{1 + Kc} \tag{33.2}$$

Among them, q represents the mass of pollutants adsorbed by the roots of aquatic plants per unit mass, $q_{\max}$ represents the maximum adsorption capacity of roots per unit mass, K is the adsorption constant, and c is the concentration of pollutants in water.

(3) Model for changes in water quality indicators

The water quality index change model is used to characterize the concentration distribution of different types of pollutants at different spatiotemporal scales. The variation of water environmental quality index is the key to evaluating its remediation effect. In general, this model can be attributed to the influence of various factors including diffusion, convection, and biochemical reactions.

The changes in water quality indicators can be described using diffusion models. In the process of aquatic plant remediation of water quality, the following diffusion equations can be used to describe the changes in water quality indicators over time and space:

$$\frac{\partial C}{\partial t} = D\mathrm{l_s} + S \tag{33.3}$$

Among them, C represents the concentration of water quality indicators, t represents time, D is the diffusion coefficient, $\mathrm{l_s}$ is the Laplace operator, and S is the influencing factor of aquatic plants on water quality.

(4) Water quality monitoring data analysis model

When processing water quality monitoring data, statistical models can be used for dynamic analysis. For example, a linear regression model $\mathrm{h_g}$ can be used to fit and predict the water quality index over time:

$$\mathrm{h_g} = \beta_0 + \beta_1 t + \varepsilon \tag{33.4}$$

Among them, β_0 and β_1 are regression coefficients, and ε is the error term. By fitting monitoring data, the trend of water quality indicators over time can be analyzed, and future water quality conditions can be predicted.

33.3.2 Exploration of the Mechanism of Aquatic Plant Restoration

Wetlands are an important ecosystem that plays a significant role in maintaining biodiversity and water environment stability. Aquatic plants are the main components of wetland ecosystems, playing a crucial role in water restoration and environmental purification. In response to this issue, this article intends to systematically study the impact mechanism of aquatic plants on wetland water environment from several aspects, including screening and introduction of aquatic plants, optimization of growth environment, research on root biological characteristics, and monitoring and analysis of water environment.

(1) Screening and Introduction of Aquatic Plants

In artificial wetlands, the selection and introduction of aquatic plants is a very important process. When selecting aquatic plants, comprehensive consideration should be given to their growth rate, root development, pollution and cold resistance, and other characteristics. The wetland ecological environment in Guizhou Province is very complex, therefore, suitable aquatic plants should be selected based on the characteristics of each type of wetland. Generally, aquatic plants with strong growth ability and water absorption ability are chosen, such as lotus root, water bamboo, calamus, etc. During the introduction process, attention should be paid to studying the ecological characteristics of the tree species to ensure that it can adapt to its wetland environment and achieve the best restoration effect.

(2) Experimental growth environment optimization experiment

The growth conditions of aquatic plants are directly related to their effectiveness in controlling water pollution. In practical applications, in order to ensure the necessary nutrients and water for its normal growth, it is necessary to optimize and regulate its growth environment. Its main methods include water level control, water quality regulation, and nutrient supply. Meanwhile, good sunlight is also a key factor affecting the normal growth of aquatic plants. Therefore, in the process of wetland management, it is necessary to reasonably allocate vegetation to ensure sufficient sunlight.

(3) Research on Biological Characteristics of Experimental Root System

The restoration of water environment by plant roots is of great significance. The roots of aquatic plants are larger than the surface and can effectively absorb nutrients and organic matter, thereby achieving the goal of purifying water quality. In addition, its root system can promote the growth and activity of microorganisms in the soil by releasing certain substances, thereby accelerating the decomposition and degradation of organic matter. At the same time, its root system plays a role in soil fixation, reducing soil erosion, and maintaining the stability of wetland ecosystems.

(4) Experimental water quality monitoring and analysis

Conducting water quality testing and analysis on wetlands is of great significance for restoring the ecological environment of wetlands. Regular testing of key parameters such as pH, dissolved oxygen, ammonia nitrogen, and nitrite in wetlands can timely detect abnormal conditions and provide a theoretical basis for the next step of governance. By analyzing monitoring data, it is possible to better evaluate the impact of aquatic plants on wetland environmental quality, thereby better understanding the mechanism of their impact on wetland environment. The research results of this article can provide a theoretical basis for the rational development and protection of wetland resources in China.

From a practical application perspective, the remediation mechanism of aquatic plants on the water environment is a complex system engineering involving interdisciplinary fields such as plant biology, hydrology, and soil science. Therefore, this article intends to conduct in-depth research on the growth characteristics of aquatic plants and the mechanism of water restoration, and adopt corresponding governance measures based on the actual situation to achieve effective protection and restoration of wetland ecosystems, thereby achieving the goal of continuously improving wetland environmental quality.

33.4 Results and Discussion

33.4.1 Impact of a Single Aquatic Plant Species on Water Quality

Experimental Design: A common aquatic plant A was selected and planted under different water quality, seasons, and water quality conditions.

Experimental evaluation indicators: The experiment monitors the water quality indicators such as dissolved oxygen, ammonium nitrogen, and nitrite in the aquaculture water area and control area.

Expected results of the experiment: The degradation effect of aquatic plants on pollutants in water can be observed under different water quality conditions.

Experiment numbers 1, 2, and 3 are in spring, Experiment numbers 4, 5, and 6 are in summer, Experiment numbers 7, 8, and 9 are in autumn, and Experiment numbers 10, 11, and 12 are in winter; Experiment numbers 1, 4, 7, and 10 indicate mild water pollution, Experiment numbers 2, 5, 8, and 11 indicate moderate water pollution, and Experiment numbers 3, 6, 9, and 12 indicate severe water pollution.

The comparison of dissolved oxygen, ammonia nitrogen, and nitrite under different experimental numbers is shown in Fig. 33.1. Experiment number 1: Dissolved oxygen 8.5 mg/L, ammonia nitrogen 0.2 mg/L, nitrite 0.05 mg/L; Experiment number 2: Dissolved oxygen 7.0 mg/L, ammonia nitrogen 0.5 mg/L, nitrite 0.10 mg/L; Experiment number 3: Dissolved oxygen 5.5 mg/L, ammonia nitrogen 1.0 mg/L, nitrite 0.20 mg/L.

33.4.2 Comprehensive Impact of Different Combinations of Aquatic Plant Species on Water Quality

Experimental Design: Different types of aquatic plants were selected for the experiment, such as duckweed and submerged plants.

Experimental evaluation indicators: The experiment monitors water quality changes in areas with different vegetation distributions and control areas, and compares the treatment results.

Expected results of the experiment: By comparing and analyzing the combined remediation effects of various aquatic plants on water bodies, the best composite ecosystem is selected.

Plant combination A is composed of floating leaves and submerged leaves, plant combination B is composed of emergent and floating leaves, plant combination C is composed of emergent and submerged leaves, plant combination D is composed of emergent, submerged and floating leaves, and plant combination E is the control area. Plant combination A: Dissolved oxygen 8.2 mg/L, ammonia nitrogen 0.3 mg/L, nitrite 0.06 mg/L, total phosphorus 0.05 mg/L; Plant combination B: Dissolved

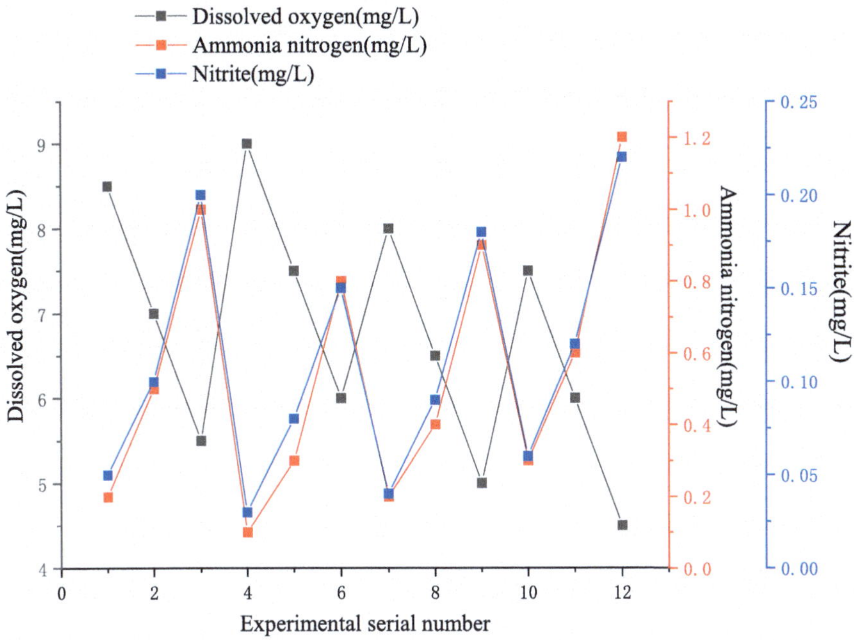

Fig. 33.1 Comparison of dissolved oxygen, ammonia nitrogen, and nitrite under different experimental numbers

oxygen 8.5 mg/L, ammonia nitrogen 0.2 mg/L, nitrite 0.04 mg/L, total phosphorus 0.04 mg/L; Plant combination C: dissolved oxygen 8.0 mg/L, ammonia nitrogen 0.4 mg/L, nitrite 0.07 mg/L, total phosphorus 0.06 mg/L. The experimental results of the comprehensive impact of different combinations of aquatic plant species on water quality are shown in Fig. 33.2.

33.4.3 Response of Aquatic Plants to Seasonal Water Quality Changes

Experimental design: The experiment monitors the water quality of each season by cultivating aquatic plants.

Experimental evaluation indicators: Compare the effects of aquatic plants in different seasons on water environment quality.

Expected outcome of the experiment: Understand the response patterns of aquatic plants to changes in water environment in different seasons, and provide a basis for the management of seasonal wetlands.

Experiment numbers 1, 2, 3, and 4 refer to spring, summer, autumn, and winter, respectively. The aquatic plant species is aquatic plant A. Experiment numbers 5, 6,

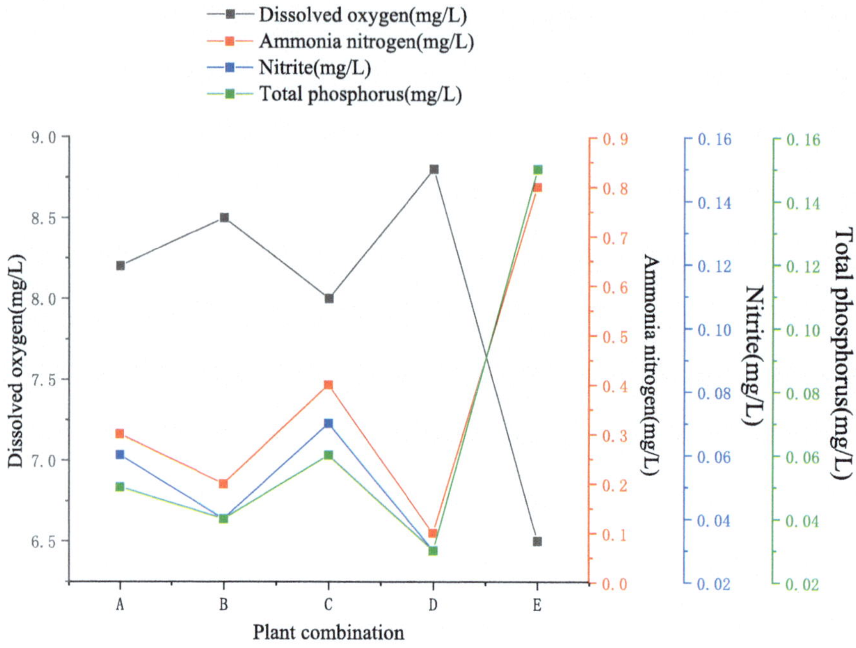

Fig. 33.2 Comprehensive impact experiment of different combinations of aquatic plant species on water quality

7, and 8 refer to spring, summer, autumn, and winter, respectively. The aquatic plant species is aquatic plant B.

Experiment number 1: Ammonia nitrogen 0.5 mg/L, nitrite 0.06 mg/L, dissolved oxygen 8.5 mg/L; Experiment number 2: Ammonia nitrogen 0.7 mg/L, nitrite 0.08 mg/L, dissolved oxygen 8.0 mg/L. The experimental results of the response of aquatic plants to seasonal water quality changes are shown in Fig. 33.3.

33.4.4 Joint Remediation of Water Quality by Aquatic Plants and Microorganisms

Experimental Design: In wetland areas, considering the interaction between aquatic plants and soil microorganisms, aquatic plants are planted in artificial wetlands.

Experimental evaluation indicators: The experiment monitors the microbial community and water environment index in the rhizosphere soil of aquatic plants.

Experimental research results: Experimental study on the impact of aquatic plant microbial interaction on the efficiency of wetland water environment remediation, exploring the mechanism of biological combined remediation. The experimental results of the joint remediation of water quality by aquatic plants and microorganisms

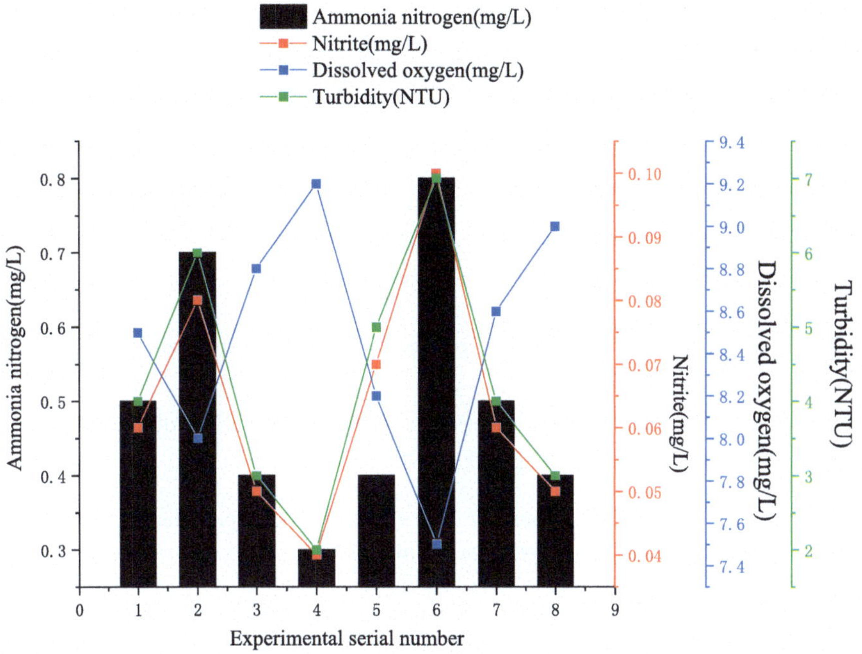

Fig. 33.3 Experimental results of response of aquatic plants to seasonal water quality changes

are shown in Table 33.1. Experimental group 1 has an ammonia nitrogen removal rate of 85%, a nitrite removal rate of 70%, and a total phosphorus removal rate of 60%.

Table 33.1 Experimental results of joint remediation of water quality by aquatic plants and microorganisms

Experimental group	Aquatic plant species	Microbial species	Ammonia nitrogen removal rate (%)	Nitrite removal rate (%)	Total phosphorus removal rate (%)
1	Aquatic plant A	Microbial X	85	70	60
2	Aquatic plant A	Microbial Y	75	65	55
3	Aquatic plant B	Microbial X	90	80	70
4	Aquatic plant B	Microbial Y	80	75	65
Control group	No aquatic plant	No microbial	20	15	10

Table 33.2 Exploration results of the impact of aquatic plants on wetland ecosystems

Aquatic plant species	Water quality improvement situation	Number of aquatic animal species	Organic matter content in wetland soil (%)	Wetland ecosystem health index
Aquatic plant A	Significant	25	3.5	Excellent
Aquatic plant B	Good	20	3.0	Good
Aquatic plant C	Common	15	2.5	Commonly
No plant control	Bad	10	2.0	Bad

33.4.5 Impact of Aquatic Plants on Wetland Ecosystems

Experimental Design: The experiment observes the overall effect of aquatic plants on wetland ecosystems from the perspectives of water quality, aquatic animals, and wetland soil.

Experimental evaluation indicators: The experiment comprehensively evaluates the health status of wetland ecosystems, including biodiversity and ecological chain structure.

Expected results of the experimental project: This article can comprehensively elucidate the synergistic effects of aquatic plants in wetland ecosystems, providing a more complete scientific basis for the protection and management of wetlands in China.

Aquatic plant A has the best water quality improvement effect, with a significant improvement in water quality. There are 25 species of aquatic animals, and the organic matter content in wetland soil is 3.5%; The water quality improvement of aquatic plant B is good, with 20 species of aquatic animals and 3.0% organic matter content in wetland soil; The water quality improvement of aquatic plant C is average, with 15 species of aquatic animals and 2.5% organic matter content in wetland soil. The exploration results of the impact of aquatic plants on wetland ecosystems are shown in Table 33.2.

33.5 Conclusions

This article adopts a combination of field investigation and water environment monitoring to conduct ecological restoration experiments on common aquatic plants in Guizhou Province. Based on the above research results, the mechanism and effect of aquatic plants on wetland water environment restoration can be elucidated, providing a theoretical basis for wetland protection and management. The implementation of this article can deepen the understanding of the functions of aquatic plants and provide scientific basis for the management of wetland water environment in Guizhou Province, which has significant theoretical and practical significance. The method used in this article also has shortcomings, such as a small number of individual wetland samples and insufficient research on the remediation effects of various aquatic plants. In the future, based on the previous work, this article can continue to increase the sampling volume and explore its ecological mechanism at a deeper level, thereby providing a theoretical basis for the protection and restoration of wetland ecosystems in China.

References

1. Clairmont LK, Slawson RM (2020) Contrasting water quality treatments result in structural and functional changes to wetland plant-associated microbial communities in lab-scale mesocosms. Microb Ecol 79(1):50–63
2. Messer TL, Trisha L, Nelson N et al (2021) Constructed wetlands for water quality improvement: a synthesis on nutrient reduction from agricultural effluents. Trans ASABE 64(2):625–639
3. Cheng FY, Van Meter KJ, Byrnes DK et al (2020) Maximizing US nitrate removal through wetland protection and restoration. Nature 588(7839):625–630
4. Tomscha SA, Bentley S, Platzer E et al (2021) Multiple methods confirm wetland restoration improves ecosystem services. Ecosyst People 17(1):25–40
5. Hartsock JA, Piercey J, House MK et al (2021) An evaluation of water quality at Sandhill wetland: implications for reclaiming wetlands above soft tailings deposits in northern Alberta, Canada. Wetlands Ecol Manage 29(1):111–127
6. Li J, Ding W, Weng G et al (2020) Heavy metal enrichment characteristics of 10 aquatic plants in Minghu National Wetland Park. J Water Ecol 41(1):86–91
7. Zhang X (2021) Aquatic plant configuration in wetland landscape art design—taking Weishan Lake Wetland Scenic Area as an example. Art Technol 034(009):217–218
8. Qi Y (2021) Survey and configuration analysis of aquatic plants in Chengdu Wetland Park. South Agric 015(005):65–67
9. Sun Z (2021) Research progress on the effect of aquatic plants on water purification in different situations. Appl Chem Ind 50(S01):287–291
10. Zhang C, Miao J, Gao J et al (2023) Research progress on the application of biochar to wetland water ecological restoration. Water Treat Technol 49(5):26–31
11. Li T, Jin Y, Huang Y (2022) Water quality improvement performance of two urban constructed water quality treatment wetland engineering landscaping in Hangzhou, China. Water Sci Technol 85(5):1454–1469

Chapter 34
Study on the Characteristics and Influencing Factors of Water Environment Quality in the Process of Algal Bloom in Lakes

Guilin Luo and Linfeng Tian

Abstract Taking the enclosed lake in Ningxia as the research object, the water pollution characteristics of the lake during the algal bloom in summer were continuously monitored, furthermore, the characteristics and influencing factors of water environment quality in the process of algal bloom were studied. The monitoring results showed that the main fluctuating pollutants in the process of algal bloom were COD_{Cr}, COD_{Mn} and chlorophyll, and the water quality index (WQI) ranged from 29.4 to 36.2, which was in a good grade. The correlation analysis showed that the monitoring concentration of chlorophyll a had positive correlation with WQI (R = 0.5781, $P < 0.05$), and the correlation of 11 monitoring indexes in 9 different regions was 15% ~ 44%, most monitoring indexes decrease with the increase of water temperature, while the concentration of chlorophyll changes with the change of COD_{Mn}, COD_{Cr} and DO, during the algal bloom process, 60.9% ~ 84.7% of the water pollution in the lake mainly came from two sources, among which TN, TP and F probably came from the recharge of the Yellow River, however, COD_{Mn}, COD_{Cr}, NH_3–N and chlorophyll a may come from the process of algal bloom, and 15.3% ~ 39.1% of the pollution source is uncertain.

Keywords Water environment · Pollution characteristics · Factors · Release

---

G. Luo
Ningxia Polytechnic Institute, Shizuishan, Ningxia, China

L. Tian (✉)
Ningxia Hui Autonomous Region Ecological and Environment Monitoring Center, Yinchuan, China
e-mail: tianlinfeng448@126.com

K. Y. Foo et al. (eds.), *Water Nexus Approach for Sustainable Development*,
Environmental Science and Engineering, https://doi.org/10.1007/978-3-031-89238-7_34

34.1 Introduction

Water is an important component of nature and the most active element in the environment. It is constantly moving and actively participates in a series of physical, chemical and biological processes in the natural environment, water Resources have certain time and space distribution in nature [1]. As an important part of terrestrial water resources, lakes are also the connecting points of the interaction of various elements of the terrestrial surface system, closely related to human aquaculture and life [2, 3]. Especially in northwest China, the precipitation is small, the evaporation is large, the salinity of the water body is high, the base flow of the lake is insufficient, the water body is easy to be polluted, once polluted, it is difficult to recover in a short period [4–6]. At the same time, with the population growth and rapid economic development of the lake basin and surrounding areas, resulting in the increase of pollutants entering the lake such as TN, TP and COD, the pollution of the lake water environment continues to worsen, the increase of pollutants into the lake causes the sharp decline of the water quality of the lake, at the same time, it leads to the algal blooms and the aggravation of eutrophication [7, 8]. In this paper, a lake in Ningxia is taken as the research object, and the method of water quality index (WQI) is adopted, combined with correlation analysis, cluster analysis, principal component analysis and other multivariate statistical methods, based on the 12 water quality data of the algal bloom in the lake, this paper attempts to probe into the water quality evolution law of the lake during this water quality change process, and identify the main driving factors of the water quality evolution, to reveal the spatial difference of water quality and its influencing factors, so as to provide scientific support and reference for the protection and management of lakes in western China.

34.2 Experimental Section

34.2.1 Sample Collection and Analysis

The study area is located in Shizuishan, Ningxia. Based on the lake's geographical shape and pollutant distribution, nine monitoring sites were set up for continuous monitoring for 20 days, the monitoring items mainly include 11 indexes such as water temperature, pH, conductivity, transparency, dissolved oxygen, COD_{Mn}, COD_{Cr}, NH_3–N, TP, F, TN, etc. The samples were monitored strictly in accordance with the environmental quality standard for surface water (GB 3838-2002).

34.2.2 Research Methods

1. Water quality index method: the water quality index method (WQI) used in this study is based on "Surface water environmental quality standard" (GB3838-2002), this method can effectively convert many physical and chemical parameters into a single value reflecting the water quality level, thus eliminating the difference between the parameters used in the evaluation alone, and can completely express the integrated water quality information of the water body. According to WQI, water quality was divided into five grades: excellent (0–20), good(21–40), moderate(41–60), poor(61–80) and extreme poor (81–100), the standard concentrations of the environmental parameters in the study are the standard limits of the basic items of the environmental quality standard for surface water in GB3838-2002, corresponding to the I, II, III, IV or V water bodies in GB3838-2002, respectively, the formula refers to Tian et al [9].

$$WQI = 1.38COD_{Mn} + 8.27NH_3 - N + 92.03TP + 7.08TN + 15.89F^- - 3.02$$

2. Geographic Information System Law: Geographic Information System (GIS) is a computer spatial or spatiotemporal information system that uses computer software to express, acquire, process, manage, analyze and apply the features of geospatial entities and phenomena, based on the core and theory of regional variables, the method of mathematical geology, which is based on spatial correlation and variogram, uses spatial interpolation to estimate the content of unknown area [10].
3. Social Science Statistics Package: the social science statistics package (SPSS) is primarily data-based data mapping that turns abstract numbers into figures that are easy to understand and analyze [11], it includes principal component analysis (PCA), cluster analysis and so on.

34.2.3 Quality Control

In order to ensure the accuracy and reliability of monitoring data, the whole process of experimental analysis was carried out according to the requirements of 《Manual of Quality Assurance for environmental water quality monitoring》.

34.3 Results and Discussion

34.3.1 Quality of Water Environment

The change of water environment quality is affected by various factors. Water temperature, pH and dissolved oxygen are the basic conditions and factors that lead to the change of water environment quality [12, 13]. The process of water environmental quality change lasted about 20 days. The monitoring results of 9 sites were evaluated according to the III standard of "Surface water environmental quality standard" (GB3838-2002), the water body is in the middle eutrophication level, the main pollutant indexes are COD_{Mn}, COD_{Cr}, NH_3–N, TP, TN and chlorophyll a. the standard deviation of COD_{Mn} monitoring data is 69% ~ 136%, the standard deviation of COD_{Cr} monitoring data was 301% ~ 746%, and that of chlorophyll a monitoring data was 1133% ~ 1858%. According to the monitoring results, the main pollution indexes causing the deterioration of water environment quality are COD_{Mn}, COD_{Cr}, TP, F and TN. The maximum exceeding standard rate of five pollutants reaches 95%, 100%, 100%, 15% and 95% respectively, and different areas have different levels of pollution (Fig. 34.1).

By calculating the water quality index (WQI), the process of water environmental quality change in the process of algal bloom was evaluated and analyzed. Through continuous monitoring for 20 days, the WQI concentration range was 29.4–36.2, and the mean value was 33, in good standing. Through comprehensive analysis, the monitoring concentration of chlorophyll a was correlated with WQI ($r = 0.5781$, $P < 0.05$) Fig. 34.2.

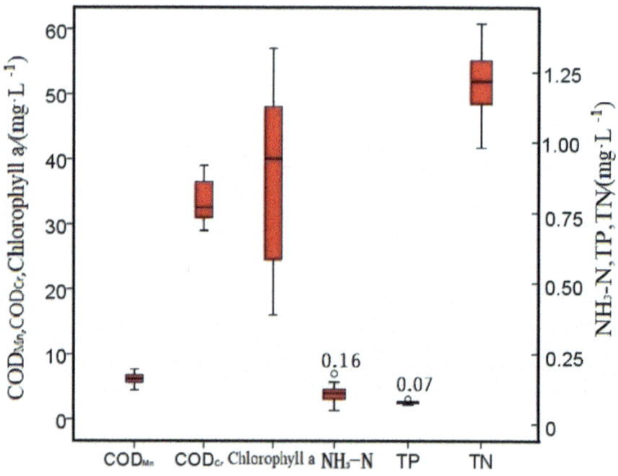

Fig. 34.1 Box diagram of integrated statistics of major pollutants

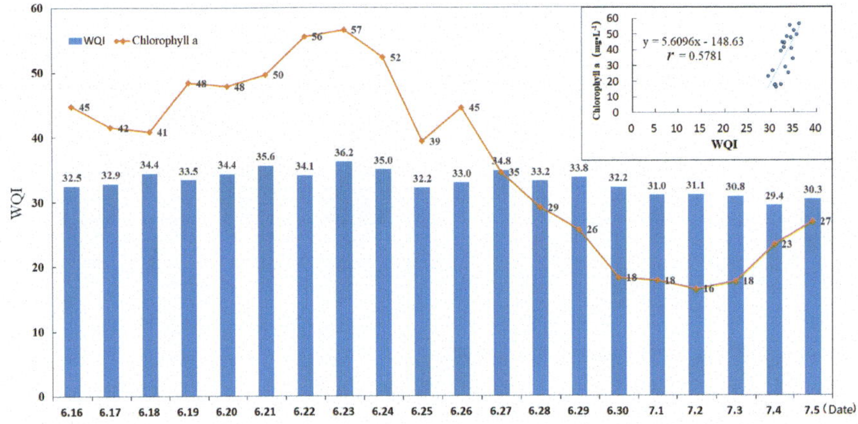

Fig. 34.2 Variation trend of DAILY WQI and chlorophyll a concentrations

34.3.2 Temporal and Spatial Variation of Pollutant Concentration

Geographic information system (GIS) can realize the spatio-temporal variation of pollutants in lakes, and provide scientific basis for the transfer of pollutants. Every 5 days as a cycle, divided into a, b, c, d 4 cycles, the spatial and temporal distribution of COD_{Mn}, COD_{Cr} and chlorophyll a (5-day mean concentration of pollutants), which have great fluctuation in the monitoring data of 9 sites, was analyzed (Fig. 34.3). The analysis results show that there is a high concentration monitoring area of the three pollutants in the whole pollution process, that is, Point 2, and the spatial distribution of the monitoring concentration of the three pollutants is basically the same, in some periods, the distribution of pollutants at other sites was similar, which indicated that the main pollutants in the lake might have some regionality and diffusivity, in the process of improving the water environment quality of the lake, the key polluted areas can be rectified accurately, and the water environment quality can be improved in a short time.

34.3.3 Correlation Analysis

The correlation among the monitoring factors at each monitoring site was analyzed by using the correlation of binary distance variables (Pearson correlation coefficient), this monitoring of 11 indicators in nine different areas of the water body showed varying degrees of correlation, the final formation of 66 pairs of data correlation factor group, there were 18 correlation factors at Sit-1(correlation rate 27%), 14 correlation factors at Sit-2(correlation rate 21%), and 9 correlation factors at Sit-3(correlation rate 14%), there were 23 correlation factors at Sit-4(correlation rate

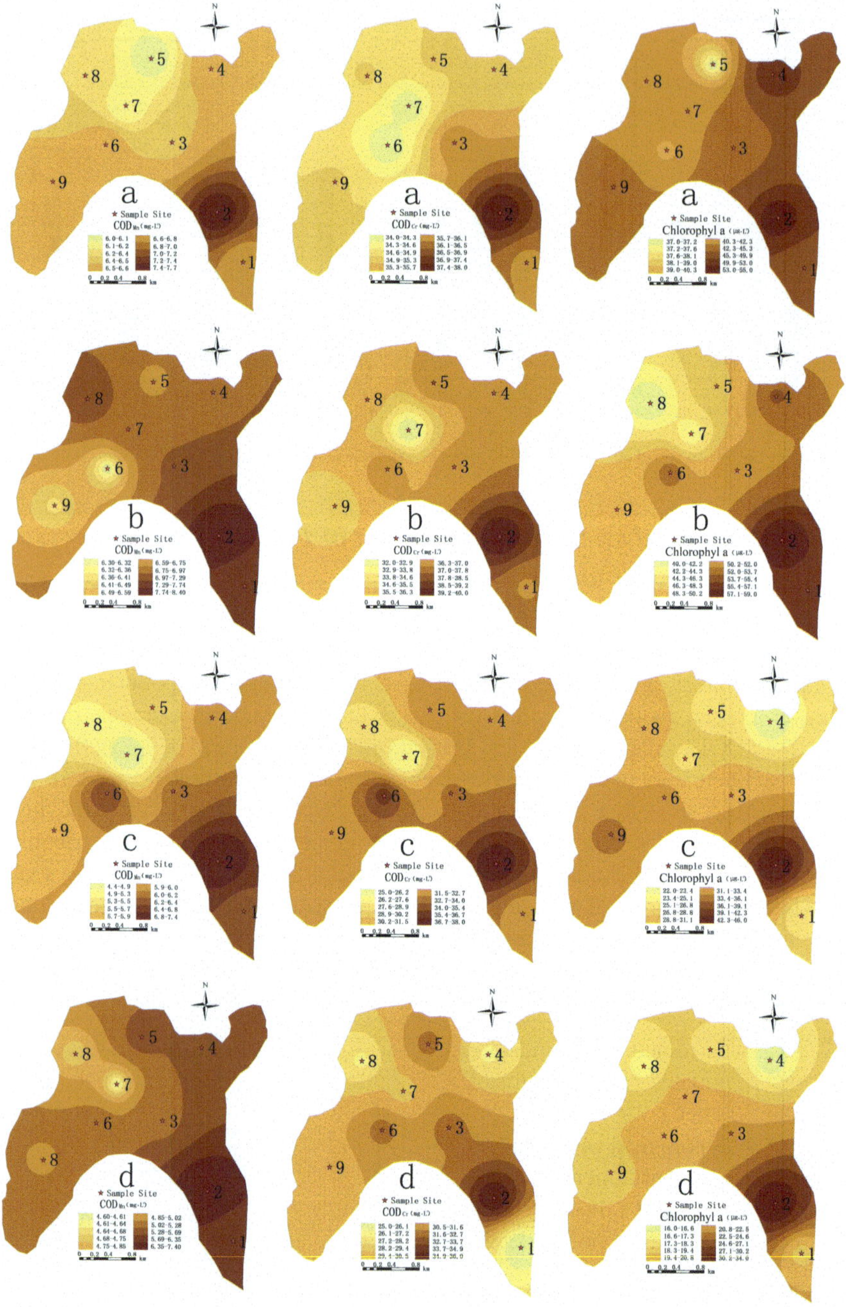

Fig. 34.3 Spatial-temporal distribution of major pollutants

35%), 10 correlation factors at Sit-5(correlation rate 15%), and 16 correlation factors at Sit-6(correlation rate 24%), there were 26 correlation factors at Sit-7(correlation rate 39%), 17 correlation factors at Sit-8(correlation rate 26%) and 23 correlation factors at Sit-9(correlation rate 35%). The negative correlation was found between water temperature and most of monitoring indexes. The reason may be that with the increase of water temperature in summer, all kinds of biological activities of water increased, and carbon, nitrogen and phosphorus were easy to transfer and transform There was a strong positive correlation between COD_{Mn} and COD_{Cr}, and chlorophyll had a strong positive correlation with COD_{Mn} and COD_{Cr}, analyzing the reason, algae itself is an organism, containing a lot of organic matter such as sugars and lipids, and the process of algae eutrophication is the growth of organic matter such as COD_{Mn} and COD_{Cr}, in the process of algal blooms, a lot of oxygen is released to the water by photosynthesis, which results in the increase of dissolved oxygen in the water, the reason is that nitrogen and phosphorus are algae nutrients, which exist in other forms in algae.

34.3.4 Cluster Analysis (HCA)

In this paper, the method of between-groups average linkage is used to analyze the standardization of variables, and the method of squared Euclidean distance is used to measure the distance, finally, the analysis tree was formed for different monitoring sites of different pollutants (Fig. 34.4). As can be seen from Fig. 34.4, the clustering of different pollutants is different in each monitoring site. Taking COD_{Mn} as the base pollutant, 9 monitoring sites were divided into 2 regions, Sit-2 clustered into one type, and the rest clustered into one type With COD_{Cr} as the base pollutant, 9 monitoring sites were divided into 3 regions, Site-7 was clustered into one, Site-2 was clustered into one, and the rest were clustered into one Based on TN, the 9 monitoring sites were divided into 3 regions: Site-4, Site-2 and Site-7, and NH_3–N, the 9 sites were divided into 4 regions: Site-2, Site-9, Site-3, and TP as the reference pollutant, the 9 monitoring sites were mainly divided into 2 regions, Site-2 clustered into one type, and the rest clustered into one type, site-7 is clustered into one class, and the other sites are clustered into one class.

34.3.5 Principal Component Analysis (PCA)

Before factor analysis, KMO test and Bartlett Sphere Test were performed. In this analysis, the test coefficient of KMO is less than 0.5 and statistical significance probability of Bartlett sphere test less than 0.5 were in accordance with the requirements of factor analysis. On the basis of data statistics, spatial distribution, correlation analysis and cluster analysis, the paper finally adopts principal component analysis, the main pollution sources and migration trend of COD_{Mn}, COD_{Cr}, NH_3–N, TP, F, TN and

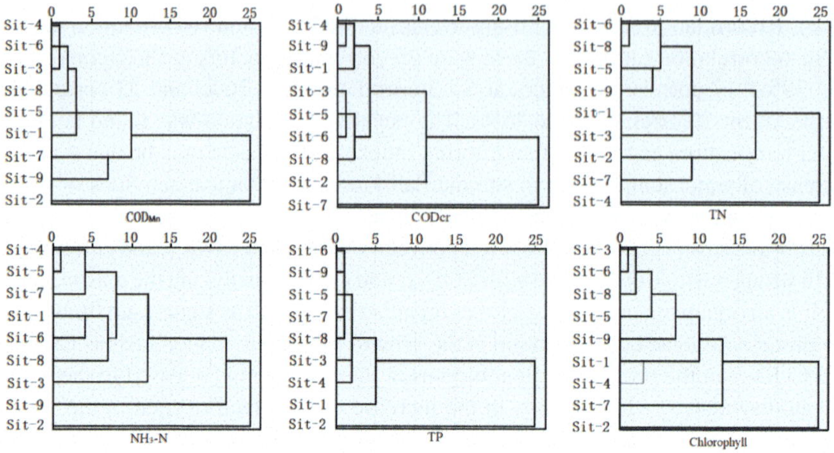

Fig. 34.4 Cluster analysis map of monitoring sites of different reference pollutants

chlorophyll a in 9 sites of the lake were analyzed and judged. Two principal components were extracted from Sit-1, which revealed 67% of the pollution source. Two principal components were extracted from Sit-2, which revealed 67% of the pollution source. Three principal components were extracted from Sit-3, which revealed 67% of the pollution source, the results showed that 74.7% of the total pollution sources were identified by Sit-4,84.7% by Sit-5,72.8% by Sit-5, and 84.7% by Sit-4, respectively Two principal components were extracted from Sit-6, which revealed 63.2% of the pollution source, and two principal components were extracted from Sit-7, which revealed 72.8% of the pollution source, the results showed that 60.9% of the pollution sources were identified, and two principal components were extracted by SIT-9, which revealed 68% of the pollution sources. PCA analysis showed that the main pollution sources were simple, 60.9% ~ 84.7% of the pollution mainly came from two aspects, TN, TP and F probably came from the Yellow River recharge, COD_{Mn}, COD_{Cr}, NH_3–N and chlorophyll-a may come from the process of algal blooms, and 15.3% ~ 39.1% of the pollution source is uncertain. Sit-3, Sit-4 and Sit-5 extract three principal components, this indicates that there may be other sources of contamination in the region that require further verification.

34.4 Conclusions

(1) In summer, the monitoring concentrations of pH, NH_3–N, TP, F and TN were relatively stable, while the water temperature, conductivity, transparency, DO, COD_{Cr}, COD_{Mn} and chlorophyll concentrations showed great fluctuation, the standard deviations of COD_{Cr}, COD_{Mn} and chlorophyll were 746%, 136% and 1858%, respectively. According to the monitoring results, the main pollution

indexes causing the deterioration of water environment quality are COD_{Cr}, COD_{Mn}, TP, F and TN. The maximum exceeding standard rate of five pollutants reaches 95%, 100%, 100%, 15% and 95% respectively, and different areas have different levels of pollution.

(2) The 11 indexes of 9 sites monitored by this research formed a total of 66 pairs of data correlation factor groups. From the correlation analysis results, the correlation rate of 9 sites reached 14% ~ 39%, water temperature and most monitoring indexes showed negative correlation, COD_{Mn} and COD_{Cr} showed strong positive correlation, chlorophyll showed strong positive correlation with COD_{Mn} and COD_{Cr}, dissolved oxygen and chlorophyll showed strong correlation, there was also a strong positive correlation between chlorophyll and total phosphorus and total nitrogen in some regions.

(3) The main source of pollutants in the algal blooms was relatively simple. 60.9% ~ 84.7% of the pollution mainly came from two sources. TN, TP and F probably came from the recharge of the Yellow River, COD_{Mn}, COD_{Cr} NH_3–N and chlorophyll a may come from the process of algal blooms, and 15.3% ~ 39.1% of the pollution source is uncertain.

(4) By collecting continuous and complete monitoring data in the whole process, the relationship between water temperature and water carbon, nitrogen and phosphorus was analyzed, and the process of water environmental quality change under the influence of various factors was expounded, therefore, the release process and degree of carbon, nitrogen and phosphorus, which are the main pollutants in summer, are revealed, and the theoretical basis is provided for the prevention and control of water pollution.

Acknowledgements This paper is one of the Ningxia Hui Autonomous Region results of the Natural Resources Foundation of China (WWF) project "Study on the catalytic oxidation of ozone by colloidal macromolecules in petrochemical secondary effluent" (2023AAC03370).

References

1. Hu RJ, Jiang FQ, Wang YJ et al (2007) On the importance of research on the lake in arid land of China. Arid Zone Res 24:137.https://doi.org/10.13866/j.azr.2007.02.001
2. Yang GS, Ma RH, Duan HT et al (2011) Number, area and spatial distribution of Chinese lakes. Chin Sci 41:394–401. https://doi.org/10.1007/s11430-010-4052-6
3. Wu ZW, Chun X, Liang WJ et al (2017) Review of lake evolution in arid semi-arid area. Wetland Sci Manage 13:55–60.https://doi.org/10.3969/j.issn.1673-3290.2017.01.12
4. Tian LF, Luo GL, Ma YX et al (2019) The causes of water pollution in the Shahu lake of Ningxia based on migration and transformation characteristics of regional water environment contaminants. Environ Monit China 35:85–92. https://doi.org/10.19316/j.issn.1002-6002.2019.06.12
5. Xiang B, Song J, Wang X et al (2015) Improving the accuracy of estimation of eutrophication state index using a remote sensing data-driven method: a case study of Chaohu lake, China. Water SA 41(5):753–761. https://doi.org/10.7524/j.issn.0254-6108.2018112615

6. Qian KM, Liu X, Duan M et al (2016) Distribution and its influencing factors of bloom-forming cyanobacteria in Poyang lake. China Environ Sci 36:261–267. https://docdown.cnki.net/docdown/fulltext/download
7. Zhang H, Zhao BT, Liu W (2016) SO_2 and CO_2 emission characteristics and mechanisms from algae biomass combustion. J Combust Sci Technol 22:262–268. https://doi.org/10.11715/rskxjs.R201504004
8. Liang XF, Zhao J, Cha JM et al (2016) Development and application of phosphoproteomic technologies in environmental toxicology. Acta Sci Circum 36:398–412. https://doi.org/10.13671/j.hjkxxb.2015.0655
9. Tian LF, Zhu X, Wang LM et al (2020) Long-term trends in water quality and influence of water recharge and climate on the water quality of brackish-water lakes: a case study of Shahu lake. J Environ Manage 276:111290
10. Tang GA, Yang X (2006) ArcGIS geographic information system spatial analysis experiment course. Science Press, Beijing in Chinese
11. Ni QS, Liu XD (2007) SPSS application test tutorial. Henan University Press, Henan in Chinese
12. Luo GL, Tian LF, Chen YX et al (2018) Statistics based study on the seasonal variation of main pollutants in Shahu lake, Ningxia. Environ Chem 37:2071–2080. https://doi.org/10.7524/j.ssn.0254-6108.2017102602
13. Tian LF, Hu JW, Qin FX et al (2011) Statistical analysis of distribution characteristics of heavy metals in surface water from Baihua lake. Res Environ Sci 24:16–24. https://doi.org/10.13198/j.res.2011.03.23.tianlf.005

Chapter 35
Water resource risk assessment for crop production and trade in Northwest China

Aohua Li and Jiaxin Tian

Abstract The increasing demand for agricultural products in arid and semi-arid regions of Northwest China highlights the critical challenge of water scarcity, exacerbated by shifting climatic conditions and expanding urbanization. This study aims to assess the impact of large-scale agricultural practices on water resource sustainability and explore effective strategies to enhance water efficiency in crop production. Employing the Penman-Monteith model alongside tools like CROPWAT 8.0 and ArcGIS, this research calculates the water footprint of various crops and analyzes the spatial and temporal dynamics of virtual water flows in the region. The results show that (1) in the Xinjiang Uyghur Autonomous Region and Inner Mongolia, with a water self-sufficiency rate of 100% and a low scarcity index of 0.3, substantially alleviate the water stress in Shaanxi, Gansu, and Ningxia, where self-sufficiency is about 96% and scarcity indices reach 0.8. (2) The Guanzhong area in Shaanxi, central Inner Mongolia, and southern Gansu report a high water scarcity index of 0.9, necessitating a 20% increase in crop imports and a 15% improvement in water resource management efficiency to maintain supply-demand balance. (3) Implementing water-saving technologies has reduced water usage by up to 25% in pilot areas, while the development of drought-resistant crop varieties has increased yield by 10% under the same water constraints, demonstrating vital strategies to alleviate water stress and promote sustainable agricultural development in Northwest China.

Keywords Water footprint · Virtual water · Blue-green water footprint · Water resources stress · Northwest China

A. Li (✉) · J. Tian
Xi'an University of Science and Technology, Xi'an, Shaanxi, China
e-mail: 1292344537@qq.com

K. Y. Foo et al. (eds.), *Water Nexus Approach for Sustainable Development*,
Environmental Science and Engineering, https://doi.org/10.1007/978-3-031-89238-7_35

35.1 Introduction

With the rapid development of China's economy, the southern and eastern regions have seen significant losses in arable land, and the centers of agricultural production have been progressively shifting northward [1–3]. This shift has increasingly emphasized the role of the northwest region in securing the nation's food supply. However, this area experiences scarce annual rainfall and high rates of evaporation, with water resources only accounting for 10.0% of the national total, while its water usage comprises 18.2% [4]. Due to regional differences in population and consumption capacity, the southern regions consume more agricultural products than the northern regions, leading to a pattern of "north-to-south" virtual water flow related to crop trade [5, 6]. This pattern could exacerbate water resource pressures in the northwest. Given the limited water availability, questions arise about the sustainability of large-scale agricultural production in the northwest and whether the virtual water flow model reallocating water resources could intensify the agricultural water pressures there [7]. These issues are critical to the food production and water security in the northwest region [8]. Thus, using virtual water and water footprint theories to comprehensively assess the risks of crop production and trade to agricultural water resource security in the northwest, and seeking fundamental solutions to alleviate water pressure, are vital for the sustainable development of agriculture and coordinated water resource management.

35.2 Materials and method

35.2.1 Description of the study area

In this article, the definition of the northwest region is based on the framework established by Qian Zhengying, with adjustments made to ensure the integrity of provincial areas. This includes considering parts of northeastern Inner Mongolia, southern Shaanxi, southern Gansu, and southern Qinghai. Additionally, referring to research by Sun Shikun and others, the study area is defined to include the six provinces (autonomous regions) of Xinjiang, Ningxia, Shaanxi, Gansu, Qinghai, and Inner Mongolia, encompassing a total of 63 cities (prefectures, leagues). The region spans an elevation range from – 162 m to 7249 m and covers an area of approximately 4.4255 million km^2, accounting for 46.1% of China's total land area. Located in the interior of the Eurasian continent, the northwest region lies between latitudes 31.36°N and 53.20°N, and longitudes 73.26°E and 126.04°E. The area features a diverse array of landscapes including mountains, plateaus, basins, deserts, oases, meadows, and grasslands. Rivers and lakes are relatively scarce, vegetation is sparse, but it is rich in resources such as oil and natural gas, making it China's largest reserve of energy resources. The specific spatial distribution is shown in Fig. 35.1.

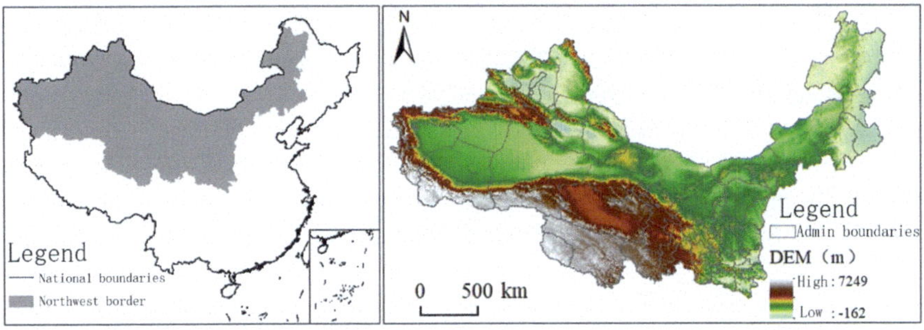

Fig. 35.1 Schematic diagram of the scope of the study area

35.2.2 Research Methods

The water footprint of a crop yield (WF, m^3/t) consists of the green water footprint (WF_{green}, m^3/t), blue water footprint (WF_{blue}, m^3/t), and grey water footprint (WF_{grey}, m^3/t). These metrics reflect the type and quantity of water used in crop production and the degree of freshwater pollution. The formulas are as follows:

$$WF = WF_{green} + WF_{blue} + WF_{grey} \tag{35.1}$$

Green Water refers to rainwater that can be utilized by crops, represented by the effective rainfall; Blue Water refers to the consumed surface or groundwater, represented by irrigation water; Grey Water refers to the water required to dilute pollutants to meet water quality standards, represented by the water demand for leaching dissolved nitrogen.

The calculation involves the following parameters:

$$WF_{green} = \frac{10 \times ET_{green}}{Y} \tag{35.2}$$

$$WF_{blue} = \frac{10 \times ET_{blue}}{Y} \tag{35.3}$$

$$WF_{grey} = \frac{(\alpha \times AR)/(C_{\max} - C_{nat})}{Y} \tag{35.4}$$

In the above calculations, ET_{green} and ET_{blue} are the amounts of crop transpiration from effective rainfall and irrigation water, respectively, unit in millimetres; The unit conversion factor of 10 transforms water depth (mm) into the water volume per unit area (m^3/hm^2); Y is the crop yield per unit area, t/hm^2; α is the leaching rate (percentage of pollutants entering the water body relative to the total chemical application), set at 10%; AR is the amount of fertilizer applied to the crops, kg/hm^2; C_{max} is the maximum allowable concentration of pollutants, set at 0.01 kg/

m^3 based on the Class III water standards of the "Surface Water Environmental Quality Standards" (GB3838-2002); C_{nat} is the natural background concentration of pollutants, assumed to be 0 in pristine conditions.

The water self-sufficiency rate (W_{SS}, %) is the ratio of the total water footprint of crop production (WF_s, m^3) to the consumption footprint (WF_{con}, m^3). This metric indicates the proportion of the consumption footprint that is met by local production, with a higher rate indicating greater self-sufficiency.

$$W_{ss} = \frac{WF_s}{WF_{con}} \times 100\% \tag{35.5}$$

The water resource pressure index (W_P) is the ratio of the total water footprint of crop production (WF_s, m^3) to the available water resources (W_A, m^3) in the region. The water resource pressure index (WP) is the ratio of the total water footprint of crop production (WFs, m^3) to the available water resources (WA, m^3) in the region. This index reflects the intensity of the impact of the crop production water footprint on available water resources. The levels of pressure are categorized as follows:

$$W_p = \frac{WF_s}{W_A} \tag{35.6}$$

According to the water resources Stress Index (WP), the pressure faced by water resources is divided into four levels: no pressure ($WP \leq 0.2$), moderate pressure ($0.2 < WP \leq 0.4$), severe pressure ($0.4 < WP \leq 1$) and extreme pressure ($WP > 1$).

35.2.3 Data Sources

The data and models used in this article include: ①Meteorological data for the northwest region for the years 2000, 2005, 2010, 2015, and 2019, sourced from the National Science & Technology Infrastructure—National Earth System Science Data Center (http://www.geodata.cn/data/). This data includes monthly rainfall, maximum monthly temperature, minimum monthly temperature, average monthly wind speed, monthly relative humidity, and daily sunlight duration. The spatial resolution is 1 km × 1 km. The data was processed using ArcGIS software to perform zonal statistics and extract meteorological parameters for each city (prefecture, league); ②The reference crop evapotranspiration and crop water requirements were calculated using the CROPWAT 8.0 software developed by the Food and Agriculture Organization (FAO) of the United Nations; ③Effective rainfall was estimated using the runoff curve number model from the United States Department of Agriculture's Soil Conservation Service; ④The growth periods and crop coefficients for various crops were referenced from the FAO's Irrigation and Drainage Paper No. 56; ⑤Data on crop yields, planted areas, fertilizer applications, and agricultural water use for each city (prefecture, league) were obtained from the provincial (autonomous region) statistical yearbooks and water resources bulletins for the years 2000–2019.

35.3 Results

35.3.1 Agricultural Water Resource Scarcity and Self-Sufficiency Rates

From 2000 to 2019, the agricultural water resource scarcity in the northwest region of China generally showed a decreasing trend, dropping from 36.2% in 2000 to 29.5% in 2019. Similarly, the water self-sufficiency rate in the region has consistently remained at 100% (as indicated in Tables 35.1 and 35.2). By 2019, the crop production water footprint was 1.42 times greater than the local consumption footprint, indicating that the northwest region not only meets the local demand but also has surplus production for export. The agricultural water resource scarcity and self-sufficiency rates vary among the provinces in the northwest region, directly correlating with the available water resources and differences in residential consumption structures. Qinghai Province exhibits the lowest agricultural water scarcity, but it also has a low self-sufficiency rate due to its relatively abundant water resources compared to other areas. The water available for agricultural production in Qinghai is not scarce, yet due to its climatic and geographical conditions, crop production is limited and cannot fully meet local demands, forcing the region to rely on external crop imports to satisfy local consumption needs. Xinjiang Uyghur Autonomous Region and Inner Mongolia Autonomous Region have lower levels of agricultural water scarcity with a self-sufficiency rate reaching 100%. Particularly in the Tacheng area, the crop production water footprint in 2015 and 2019 was hundreds of times greater than the consumption footprint. As major crop-exporting areas in northwest China, Xinjiang and Inner Mongolia produce sufficient grain, economic crops, and fruit products to meet local needs and generate surplus for export.

Table 35.1 Agricultural water scarcity in Northwestern provinces from 2000 to 2019 (unit: %)

Region	2000	2005	2010	2015	2019
Inner Mongolia	43.3	32.1	41.7	39.7	35.3
Shaanxi	86.6	60.2	56.7	97.5	65.3
Gansu	89.5	59.2	67.3	89.3	59.7
Ningxia	60.7	70.9	72.7	88.6	75.6
Qinghai	6.0	3.5	4.6	6.7	4.4
Xinjiang	14.3	14.4	15.2	14.4	17.1
Northwest China	36.2	27.7	31.1	36.7	29.5

Table 35.2 Self-sufficiency rate of water resources in Northwest provinces from 2000 to 2019

Degree of self-sufficiency (%)	2000	2005	2010	2015	2019
Inner Mongolia	100.0	100.0	100.0	100.0	100.0
Shaanxi	97.1	100.0	100.0	95.2	92.1
Gansu	74.6	95.9	96.1	95.7	93.1
Ningxia	100.0	100.0	100.0	100.0	99.3
Qinghai	52.6	66.54	56.51	43.95	43.4
Xinjiang	100.0	100.0	100.0	100.0	100.0
Northwest China	100.0	100.0	100.0	100.0	100.0

In contrast, Shaanxi Province, Gansu Province, and Ningxia Hui Autonomous Region face higher levels of agricultural water scarcity, sometimes exceeding 90%, with water self-sufficiency rates fluctuating around 96%. Local crop consumption exceeds production, and the available water resources can no longer meet local demands, leaning more towards reliance on external crop imports to alleviate local water shortages.

As shown in Fig. 35.2, the eastern parts of the northwest region have higher agricultural water resource scarcity than the western parts, and the northern parts have higher water self-sufficiency rates than the southern parts. Central Inner Mongolia, southern Gansu, and the Guanzhong area in Shaanxi Province are all regions with concentrated high water scarcity, especially Xi'an, where agricultural water scarcity has increased annually, rising to 192.4% by 2019. The increasing water scarcity in Xi'an indicates severe water shortages, with local crop production unable to meet people's needs. For arid inland areas, the growth in water scarcity is a serious warning in the development process, necessitating improvements in regional water utilization rates and increased import trade to maintain a balance of water supply and demand. Other western areas, except for Urumqi and Karamay, have agricultural water scarcities less than 100%, indicating that their water resources are sufficient to meet local agricultural consumption needs. Most regions in Xinjiang, Ningxia, and Inner Mongolia have high water self-sufficiency rates reaching 100%, which not only ensures the local demand for crops but also allows for increased production scale for export trade. Additionally, some areas have seen a decline in water self-sufficiency rates, such as the Haixi Mongol and Tibetan Autonomous Prefecture, which decreased from 58.9% in 2000 to 31.7% in 2019, increasingly relying on crop imports from other regions to meet local needs.

35.3.2 Crop Production Water Resource Pressure

The ratio of the crop production water footprint to available water resources serves as an indicator for analyzing water resource pressure. From 2000 to 2015, the water resource pressure index in the northwest region showed an increasing trend

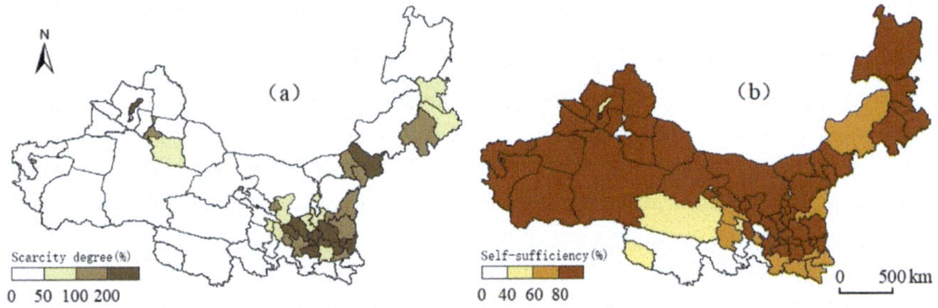

Fig. 35.2 Spatial distribution of agricultural water resources scarcity in Northwest China (2000–2019 mean value)

(Table 35.3), rising from 0.37 in 2000 to 0.50 in 2015, shifting from moderate pressure to severe pressure. By 2019, the water resource pressure index decreased to 0.44, indicating some relief in water resource pressure in the northwest region, although the relationship between crop production and water resources remained in a state of severe pressure.

Among the provinces, Qinghai Province was in a state of no pressure due to its high availability of water resources and lower crop production, thus facing minimal water resource pressure. In contrast, Xinjiang Uyghur Autonomous Region was under moderate pressure; by 2019, its water footprint pressure index exceeded 0.4, indicating significant agricultural water pressure. This was primarily due to little change in the water demand for crop production, but a reduction of 4.745 billion m^3 in available water resources, thus increasing water resource pressure. Furthermore, the continuous increase in crop exports from Xinjiang consumed a large amount of local water resources, thereby increasing agricultural water pressure. The other four provinces had water pressure indices ranging from 0.4 to 1, with relationships between crop production and water resources generally in a state of severe pressure. Crop production consumed substantial water, while the regional available water resources were lower. If these areas appropriately reduced local crop production and opted to import crops from other provinces to meet local consumption needs, it would alleviate the current water resource pressure.

Table 35.3 Water pressure index of Northwest provinces from 2000 to 2019

Pressure	2000	2005	2010	2015	2019
Inner Mongolia	0.58	0.56	0.83	0.70	0.82
Shaanxi	0.84	0.64	0.57	0.93	0.60
Gansu	0.67	0.57	0.65	0.85	0.56
Ningxia	0.63	0.86	0.99	0.93	0.75
Qinghai	0.03	0.02	0.03	0.03	0.02
Xinjiang	0.18	0.22	0.32	0.40	0.41
Northwest China	0.37	0.34	0.42	0.50	0.42

Quantifying the water resource pressure of 63 cities (prefectures, leagues) in the northwest region (Fig. 35.3) revealed that high values indicate unsustainable crop production processes, facing significant water security risks and environmental pressures. The spatial distribution of water resource pressure and agricultural water scarcity generally aligned, showing a pattern of high in the east and low in the west. In 2000, the water resource pressure indices in Qinghai Province, Xinjiang Uyghur Autonomous Region, northern Gansu Province, and southern Shaanxi Province were all below 1. The indices for 14 cities including Dingxi, Yan'an, Guyuan, and Ulanqab ranged between 1 and 5, while Wei'nan and Xianyang had ratios of 5.6 and 9.99, respectively, far exceeding the local available water resources. By 2019, some cities in the northwest region experienced growth in their water pressure indices compared to 2015. Specifically, Xing'an League, Urumqi, and Changji Hui Autonomous Prefecture saw their indices surpass 1, with total increases in their crop water footprints of 1.329 billion m^3, 1.622 billion m^3, and − 0.18 billion m^3, respectively. The rise in water resource pressure in Changji was due to a decrease of 0.584 billion m^3 in available water resources. Xianyang had the highest water resource pressure index at 12.08 among the 63 areas, placing it in an extreme pressure state. The significant agricultural water pressure in this area was partly due to Xianyang's water scarcity limiting the available water for agriculture and partly because an increase in crop production led to higher agricultural water demand.

35.3.3 Sustainability of Agricultural Water Resources in Production

The use of water for agricultural production in the Northwest region is not conducive to the sustainable use of local water resources. From 2000 to 2019, the provinces in the Northwest region experienced continual population growth, an increase in the agricultural water footprint of residents, a rise in virtual water flow, and a corresponding increase in the crop production water footprint. Therefore, establishing sensible crop trade policies could help alleviate regional water resource pressures, while failing to do so could further increase regional water security risks.

Looking at regional differences, Inner Mongolia Autonomous Region and Xinjiang Uyghur Autonomous Region, as the major virtual water export areas for crops in the Northwest, saw increases in their crop virtual water exports of 281.4% and 571.8% respectively over the 20-year period (Table 35.4). By 2019, the virtual water volume of crops exported from Xinjiang was equivalent to 1.41 times the crop consumption water footprint of the autonomous region for that year. Similarly, the virtual water volume of crops exported from Inner Mongolia in 2019 was equivalent to 1.33 times the region's crop consumption water footprint for that year. It is evident that the crop production in Xinjiang and Inner Mongolia far exceeds local demands, making them important strategic resource bases in China. The other four provinces are virtual water import areas for crops. Over 20 years, Shaanxi

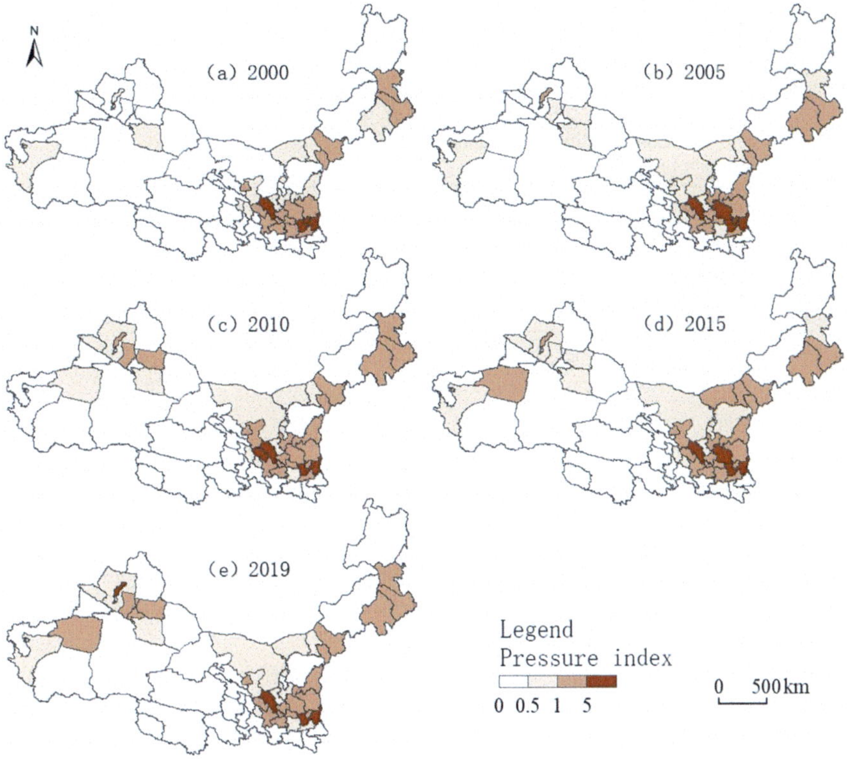

Fig. 35.3 Distribution of water resources stress in Northwest China from 2000 to 2019

Province saw a 184.9% increase in crop virtual water imports, with the ratio of virtual water imports to the consumption footprint increasing from 2.9% in 2000 to 7.9% in 2019; Qinghai Province's virtual water imports grew by 537 million m^3, exceeding 50% of the consumption footprint; Gansu Province saw a decrease in crop virtual water imports by 4.605 billion m^3, with the ratio of virtual water imports to the consumption footprint dropping from 25.4% in 2000 to 6.9% in 2019; Ningxia Hui Autonomous Region transitioned from being a virtual water export region in 2000 to an import region in 2019, with imported virtual water accounting for 0.7% of local consumption.

Cities like Tongliao, Chifeng, and Xing'an League in Inner Mongolia, along with Aksu District, Changji Hui Autonomous Prefecture, and Urumqi in Xinjiang, are virtual water export areas facing significant agricultural water resource pressure. The long-term increase in crop virtual water exports has further intensified local water resource pressures, posing threats to regional water security and the sustainable development of crop production. Conversely, cities such as Xianyang, Weinan, Yinchuan, Wuzhong, and Zhongwei are virtual water import areas. Due to increased crop imports, there is a continual rise in virtual water imports. Moreover, these

Table 35.4 Consumption water footprint and trade virtual water in the Northwest territories 2000 and 2019 (billion m^3)

Region	2000		2019	
	Water consumption footprint	Virtual water flow momentum	Water consumption footprint	Virtual water flow momentum
Inner Mongolia	196.87	66.76	191.46	254.61
Shaanxi	307.02	– 8.95	322.41	– 25.50
Gansu	234.16	– 59.54	194.79	– 13.49
Ningxia	52.21	1.63	53.94	– 0.40
Qinghai	36.94	– 17.52	40.46	– 22.89
Xinjiang	121.50	31.06	148.68	209.19
Northwest China	948.70	13.44	951.74	401.52

areas also import significant amounts of grey water footprints from other regions, which further strains the relationship between crop production and water resources. However, the import of virtual water to some extent alleviates the water resource pressures in these areas, easing the tension between local crop production and water resources.

35.4 Conclusion

This study conducted a detailed examination of spatial-temporal variations in agricultural water resource scarcity, water self-sufficiency, and water resource pressure indices in the Northwest region from 2000 to 2019. Our findings can be summarized as follows:

Imbalance in distribution and agricultural structure. The Xinjiang Uyghur Autonomous Region and the Inner Mongolia Autonomous Region, notable for exporting virtual water, achieve a 100% water self-sufficiency rate, indicating minimal agricultural water scarcity. This capacity fulfills local agricultural demands and reduces water pressure in virtual water-importing regions such as Shaanxi Province, Gansu Province, and Ningxia Hui Autonomous Region, which experience significant scarcity and have self-sufficiency rates around 96%.

The Guanzhong region in Shaanxi, central Inner Mongolia, and southern Gansu face acute water shortages, necessitating enhanced regional water utilization and increased crop imports. Conversely, areas like Tacheng in Xinjiang and significant parts of Ningxia and Inner Mongolia exhibit high self-sufficiency, supporting local and export agricultural demands.

The Northwest region continues to face significant water deficits, impacting agricultural water availability and intensifying pressures, especially in areas with large-scale crop production. Virtual water imports and strategies such as improving crop efficiency, developing new varieties, promoting water-saving irrigation, and reducing high-water-use crop production are essential to alleviate these pressures and support sustainable agricultural development.

References

1. Wei YH, Li H, Yue WZ (2017) Urban land expansion and regional inequality in transitional China. Landscape Urban Plan 163:17–31
2. Zhang JX, Deng MJ, Li P et al (2022) The pattern and regulation of agricultural water resources security in Northwest China from the perspective of virtual flow. Chin Eng Sci 24(1):131–140
3. Jiang S, Wang JH, Zhao Y et al (2017) Sustainability of water resources for agriculture considering grain production, trade and consumption in China from 2004 to 2013. J Cleaner Prod 149:1210–1218
4. China's 2020 Water Resources Bulletin Released. Yangtze Institute for Conservation and Development. 2020. Ministry of Water Resources of the People's Republic of China
5. Wu P (2015) Warning and strategy of north-to-south water transfer project. Prog Water Conservancy Hydropower Sci Technol 35(5):121–123+180
6. Sun SK, Yin YL, Wu PT et al (2019) Geographical evolution of agricultural production in china and its effects on water stress, economy and the environment: the virtual water perspective. Water Resour Res 55(5):4014–4029
7. Aldaya MM, Chapagain AK, Hoekstra AY et al (2011) The water footprint assessment manual: setting the global standard. Earthscan, London
8. Sun SK, Li C, Wang YB et al (2020) Evaluation of the mechanisms and performances of major satellite-based evapotranspiration models in Northwest China. Agric Forest Meteorol 291:108056

Chapter 36
Study on the Change of Microbial Community Structure and Its Response to Environmental Factors in Heavily Polluted Rivers Repaired by Biological Wax

Lei Dong, Xingrui Qi, Yueqi Cao, and Li Lin

Abstract This paper discusses the application of biological wax remediation technology in heavily polluted rivers, especially its response to the diversification of microbial community structure and environmental factors. After 7 months of treatment, the contents of ammonia nitrogen, total phosphorus and chemical oxygen demand in the treatment group decreased by 38%–79%,7%–45% and 14%–33%, respectively, when compablue to the control group. Most of the ammonia nitrogen in the group receiving treatment was class I or II water quality, and the total phosphorus and chemical oxygen demand were basically maintained in class I water quality. The structure of microbial community changed significantly. Based on the analysis of the α diversity of microbial community in each group, it was found that the dominant taxa and classes at phylum and class level were similar before and after treatment, but the dominant bacteria were different at genus level. It shows that biological wax has great potential to enhance the ecological restoration of heavily polluted rivers.

Keywords Biological wax remediation · Heavily polluted river · Microbial community structure · Environmental factors

L. Dong · Y. Cao · L. Lin (✉)
Basin Water Environmental Research Department, Changjiang River Scientific Research Institute, Wuhan, P.R. China
e-mail: linli1229@hotmail.com

Key Lab of Basin Water Resource and Eco-Environmental Science in Hubei Province, Wuhan, P.R. China

L. Dong · L. Lin
Innovation Team for Basin Water Environmental Protection and Governance of Changjiang Water Resources Commission, Wuhan, P.R. China

X. Qi
School of Chemistry, Chemical Engineering and Life Sciences, Wuhan University of Technology, Wuhan, P.R. China

K. Y. Foo et al. (eds.), *Water Nexus Approach for Sustainable Development*, Environmental Science and Engineering, https://doi.org/10.1007/978-3-031-89238-7_36

36.1 Introduction

Heavily polluted rivers are a serious problem facing the world, which not only damages river biological resources and environmental quality of river water, but also endangers human health and sustainable development of society. The degree of pollution changes with time, and spreads quickly, and the influence of pollution is great. The treatment of river pollution is a complex system engineering [1]. At present, the main treatment technologies of polluted rivers at home and abroad are mainly categorized into chemical treatment [2], physical treatment [3] and biological treatment [4–6]. Among them, physical and chemical methods will consume too much energy and materials, and easy to produce secondary pollution can not eradicate the problem of river pollution [7], while Bioremediation technology has gained significant popularity due to its environmentally friendly nature and high cost-effectiveness [8].

At present, there are many biological treatment methods. Peng immobilized natural photosynthetic bacteria (PSB) to treat seriously polluted river water and improve its performance in low-carbon, low-dissolved oxygen (DO), high-nitrogen and high-phosphorus water. This method is not efficient, and the operation of immobilized bacteria is tedious and not easy to be used on a large scale [9]. Wu and others directly add home-grown microorganisms and microbial promoters to water from the river and sediments by unique vaccination process that does not involve removing sediments or using artificial aeration. Following this procedure, pollution can be significantly decreased, the river ecosystem can be revitalized, and water quality can be enhanced, but this method can not be recycled and requires continuous application of microbial agents [10]. However, as a new type of bioremediation material, biofilm can be recovered and reused with high treatment efficiency [11]. It can significantly improve the indexes such as COD, NH_3–N and TP, and can effectively improve the ecological balance of rivers. Its application in river restoration has been widely concerned by researchers [12].

At present, there is little research and application about biological wax, which is mainly used in coating technology to cover the surface of various products in order to increase the products' performance. Jo et al invented a multi-layer barrier coating technology, which uses cationic and anionic NFC and biological wax to produce three-layer barrier coated paper, which is used as paper that serves a functional purpose in packaging [13, 14] with high barrier properties. As a typical urban river, the Xunsi River is located at the southern end of the urban area of Wuhan, and flows into the Yangtze River through Jiangxia, Hongshan and Wuchang districts of Wuhan. Because of the massive release of household wastewater and the lack of nearby sewage treatment plants, the pollution problem is becoming more and more serious. In addition, less than two kilometers downstream of the river estuary is the intake of the Baishazhou Waterworks. Therefore, it is particularly important to solve the problem of sewage treatment in the inspector river as soon as possible.

In this study, Australian TWC biological wax was used to treat sewage from the Xunsi River, to improve the microbial community of heavily polluted rivers by biological wax remediation, and to explore its response to environmental factors. This paper evaluates the application effect of biological wax remediation technology in heavily polluted rivers, explores the structure changes of microbial communities in the process of biological wax repair, and analyzes the response of microbial community to different environmental factors. The experimental results are anticipated that technical support and a theoretical foundation will be provided for the remediation and treatment of heavily polluted rivers and the ecological assessment of river water quality health.

36.2 Materials and Methods

36.2.1 Area of Study and Collection of Samples

As the largest open discharge channel in Wuhan city, Xunshi River is the main discharge place of domestic sewage. In this paper, the Wutai gate section of Xunsi River was selected as the research object, and the microbial samples and water samples were collected and analyzed regularly before and after the application of biological wax.

In this paper, using TWC biological wax water environment treatment technology, and from October 2022 to May 2023, three sampling sites suitable for demonstration projects were selected around Wutai Sluice (treatment group), and downstream 1 km sampling sites were selected as the control group, and the incoming water quality was compared and analyzed. A water sample from 0 m to 0.5 m was collected with a water sampler, and 2 L was collected from each sampling point, which was placed in a sampling bottle and transported to the labouratory. One section of the water body was used to determine total nitrogen (TN), total phosphorus (TP), ammonia nitrogen (NH_3–N) and chemical oxygen demand (COD), and the other part was used to measure the microbiological index.

36.2.2 Microbial Community Analysis

The microbial DNA in the collected samples was extracted and sequenced by 16S high-throughput sequencing technology, and the change of microbial community structure was analyzed.

The water samples stored at – 80 °C were melted at low temperature, and 1 L water samples were extracted and filtered onto 0.22 μm water system filter membrane. The filter membrane was quickly put into the 5 mL centrifuge tube and frozen in the refrigerator at – 80 °C for DNA extraction.

The MagenHipure Soil DNA Kit extraction kit was utilized for extraction. The phytoplankton DNA from the cryopreserved 0.22 μm filter membrane. The specific procedures were conducted in accordance with the kit's instructions, and the DNA concentration was detected by Qubit ®ds DNA HS Assay Kit. Illumina Mi Seq PE300 high-throughput sequencing platform was used for sequencing. PCR and sequencing were completed by Suzhou Jinweizhi Biotechnology Co., Ltd.

36.2.3 Data Analysis

Multivariate statistical analysis focused on examining the relationship between alterations in microbial environmental factors and community structure. The analysis of the relationship between community structure and environmental factors was conducted using Pearson correlation of SPSS 20.0 software.

36.3 Results and Discussion

36.3.1 Repair Effect of Biological Wax

The total nitrogen (TN), ammonia nitrogen (NH_3–N), total phosphorus (TP) and chemical oxygen demand (COD) of the treatment group, the incoming water quality and the control group were analyzed respectively, and the modification in the water quality of Xunsi River was examined both before and after the use of biological wax. As shown in Fig. 36.1, the total nitrogen of incoming water quality increased in part, but did not increase or decrease steadily with time; except for 2023.2 and 2023.4, the content of total nitrogen in other months was more in the treatment group than in the control group. In most cases, the ammonia nitrogen concentration in the treatment group was much less than that in the control group, which decreased by about 38%–79% (average 60.75%); and except for 2023.2, the ammonia nitrogen concentration in the treatment group was basically higher than the incoming water quality in each month. The concentration of total phosphorus in the 2022.10, 2023.3 and 2023.5 treatment groups decreased significantly relative to the incoming water quality, by about 25%–65% (mean 52%). In most cases, the total phosphorus concentration in the treatment group was dramatically lower than that in the control group, reducing about 7%–45% (mean 26.4%). The content of COD in the treatment group augmented first and then decreased over time, and decreased by 25% after 7 months. Compared with the control group (2022.10, 2023.4, 2023.5), the content of COD in the treatment group decreased by 14%–33% (mean 22%).

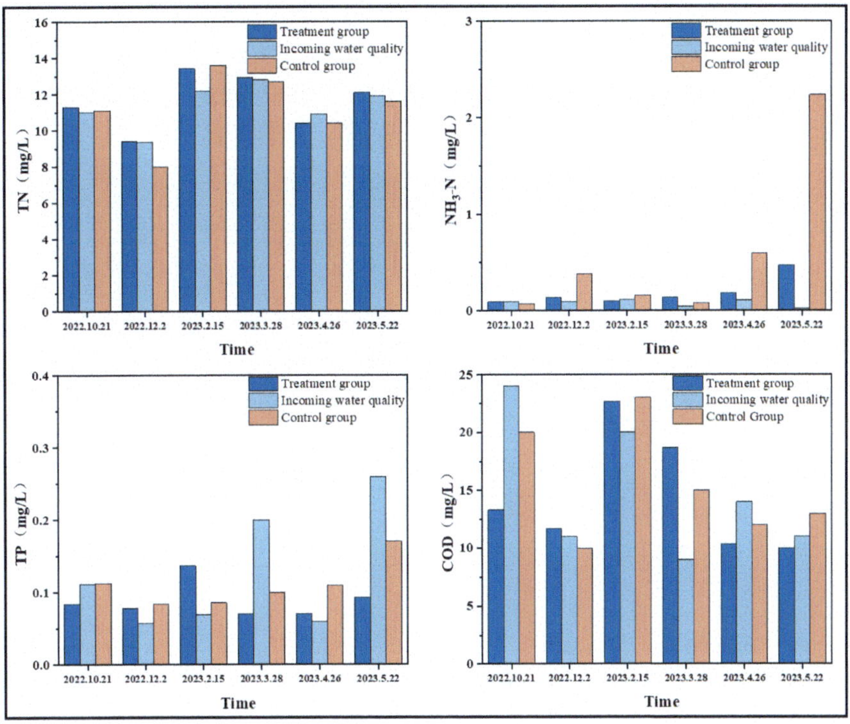

Fig. 36.1 Changes of environmental factors in different groups

36.3.2 Research Area and Sample Collection

High-throughput sequencing results.

High-throughput sequencing of microorganisms [15, 16] from all sample sites in different periods in the demonstration area was divided into five groups: W-TWC (biological wax treated water), CK (control water), TWC (biological wax), MPs (PVC plastic plate group) and BR (brick group). The effective sequence, OTU number and sequencing coverage were obtained, and it was found that more than 16,000 effective sequences were obtained in all samples. The obtained high-quality sequences are divided into OTU according to 97% similarity level, and about 1000–5000 OTUs are obtained.

The number of unique and common bacteria in each group was analyzed, and the results were shown in Fig. 36.2. The total number of bacteria in 5 groups was 2010. In each material, there were 1052 unique bacteria in TWC group, 899 unique bacteria in MPs group and 1093 unique bacteria in BR group. Comparatively speaking, the number of unique bacteria in CK group is bigger than that in W-TWC group, and the quantity of unique bacteria in each material is: TWC $\approx$ BR > MPs. At the same time, the number of unique bacteria in water is higher than that in each material. During

Fig. 36.2 Wayne diagram of unique and common strains in each group

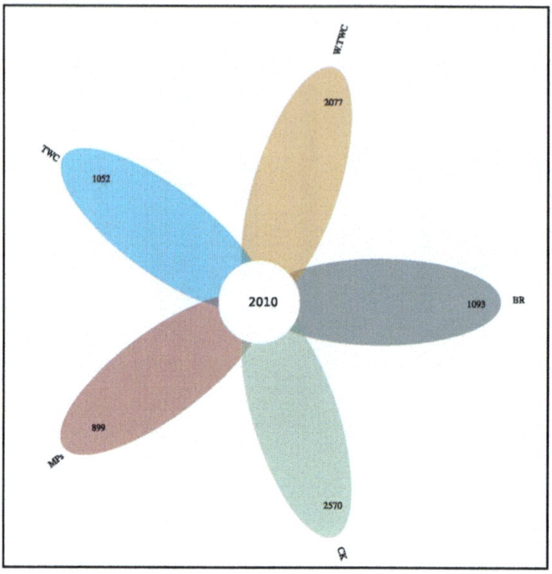

the treatment process (7 months), the number of unique bacteria in water and each material in the treatment group decreased compared with that in the control group.

Carry on the species classification and annotation to the different OTUs, thus get the community structure composition of each sample point in the demonstration area, and count the community composition at each level [17–19]. Results showing in Table 36.1, CK represents the control group, W-TWC represents the water body treated with TWC biological wax, TWC represents the TWC biological wax group, MPs represents the PVC plastic plate group, and BR represents the brick group. Among them, the average number of microorganisms in the control group (CK) included 43 phylums, 444 genus and 310 species, and after 7 months, the species of phyla and classes had little difference, and the quantity of species and genera increased gradually. The average number of microorganisms in TWC biological wax treatment group (W-TWC) was 40 phylums, 433 genus and 307 species after 7 months, the species of phylums and classes decreased slightly, but the number of species and genus increased gradually. Comparing the microbial community composition at various levels between the W-TWC group and the CK group, it was found that there was little difference in the phylum and class species between the treatment group and the control group, and the number of species and genus in the treatment group was slightly less than that in the control group.

By comparing the community composition of microorganisms at different levels on different materials, it was found that the average number of microorganisms in TWC biowax group (TWC) included 25 phylums, 240 genus and 137 species, and after 7 months, the quantity of species changed little, but the species' number and genus showed an increasing trend, while in PVC plastic sheet group (MPs), there were an average of 298 genus and 176 species in 29 phylums, and after 7 months,

Table 36.1 The number of species at different levels under OTU classification

Sample point	Kingdom	Phylum	Class	Order	Family	Genus	Species
W-TWC0	1	41	92	194	247	372	251
W-TWC1	1	43	96	195	252	399	261
W-TWC2	1	39	91	183	246	436	309
W-TWC3	1	40	85	194	280	470	343
W-TWC4	1	39	86	180	279	493	358
W-TWC5	1	37	86	166	252	429	319
CK0	1	44	102	192	241	379	249
CK1	1	43	98	198	252	423	303
CK2	1	39	89	173	252	420	280
CK3	1	49	105	203	279	466	319
CK4	1	40	92	194	276	514	372
CK5	1	42	89	179	270	465	340
TWC1	1	24	65	128	181	224	119
TWC2	1	28	65	132	177	235	134
TWC3	1	27	60	133	181	274	148
TWC4	1	21	45	90	136	197	116
TWC5	1	27	55	113	176	269	166
MPs1	1	32	70	156	212	295	165
MPs2	1	30	72	153	209	300	171
MPs3	1	32	64	145	200	286	162
MPs4	1	24	56	115	172	272	157
MPs5	1	29	70	137	198	339	226
BR1	1	30	76	152	218	315	186
BR2	1	36	82	171	222	323	201
BR3	1	37	82	166	216	321	204
BR4	1	20	54	101	155	244	139
BR5	1	27	65	129	191	325	194

the number of phylums and species decreased slightly, while the number of species and genera showed an increasing trend. The average number of microorganisms in the brick group (BR) included 30 phylums, 306 genus and 185 species, and after 7 months, the number of phyla decreased slightly, while the number of species and genera increased slightly. In comparison, the composition of microorganisms, the number of phylums, genus and species in each group was BR > MPs > TWC. At the same time, the number of phylums, genus and species of microorganisms in the water body is larger than that of each material.

Analysis of microbial community composition of each group.

Chao1 index was selected to analyze the α diversity of microbial communities in each group. The chao1 index is used to calculate the amount of OTU present in the sample. The larger the chao index is, the more the OTU number is, indicating that the species richness of the sample is higher [20]. The results of chao1 index of different sample points are shown in Fig. 36.3. The chao1 index of W-TWC microorganisms in TWC biological wax treatment group increased at first and then decreased after 7 months treatment, while the chao1 index of CK microorganisms in the control group had no obvious change trend after 7 months. In comparison, the Chao1 index did not differ significantly between the CK group and W-TWC group, indicating a minimal variance in bacterial species richness between the treatment and control groups.

Comparing the microbial chao1 index of different materials, it was found that the microbial chao1 index of TWC biological wax group (TWC), PVC plastic sheet group (MPs) and brick group (BR) increased slightly after 7 months (Fig. 36.3a). In comparison, the relationship of chao1 index on different materials is as follows: BR > MPs > TWC, indicating that the species richness of bacteria is the highest on bricks, the second on PVC plastic sheets and the lowest on TWC. In addition, the chao1 index of microorganisms in water is higher than that in all materials, that is, CK > W-TWC > BR > MPs > TWC (Fig. 36.3b), indicating that the species richness of bacteria in water is higher than that in different materials.

36.3.3 Environmental Factor Response

In order to analyze the main environmental factors affecting the arrangement of bacterial populations in each group [21–24], the correlation between bacterial communities and environmental factors was analyzed by CCA method, and the main environmental factors affecting bacterial colonies in the demonstration area were analyzed.

According to Fig. 36.4, the arrows represent different water environmental factors, and the dots represent bacterial communities at different sample points. The longer the arrow, the greater the impact of the environmental factor on the bacterial community. The CCA results illustrated that the arrowheads of various environmental factors were short, indicating that they had little effect on the bacterial community, and only some sample sites had a strong correlation with environmental factors. Among them, W-TWC4 sample sites are greatly affected by TP and NH_3–N, W-TWC3 and CK3 sites are more affected by TN, W-TWC2 sites are more affected by COD, while other sites are far away from environmental factors, indicating that they are less affected by environmental factors, which may be related to fewer environmental factors selected.

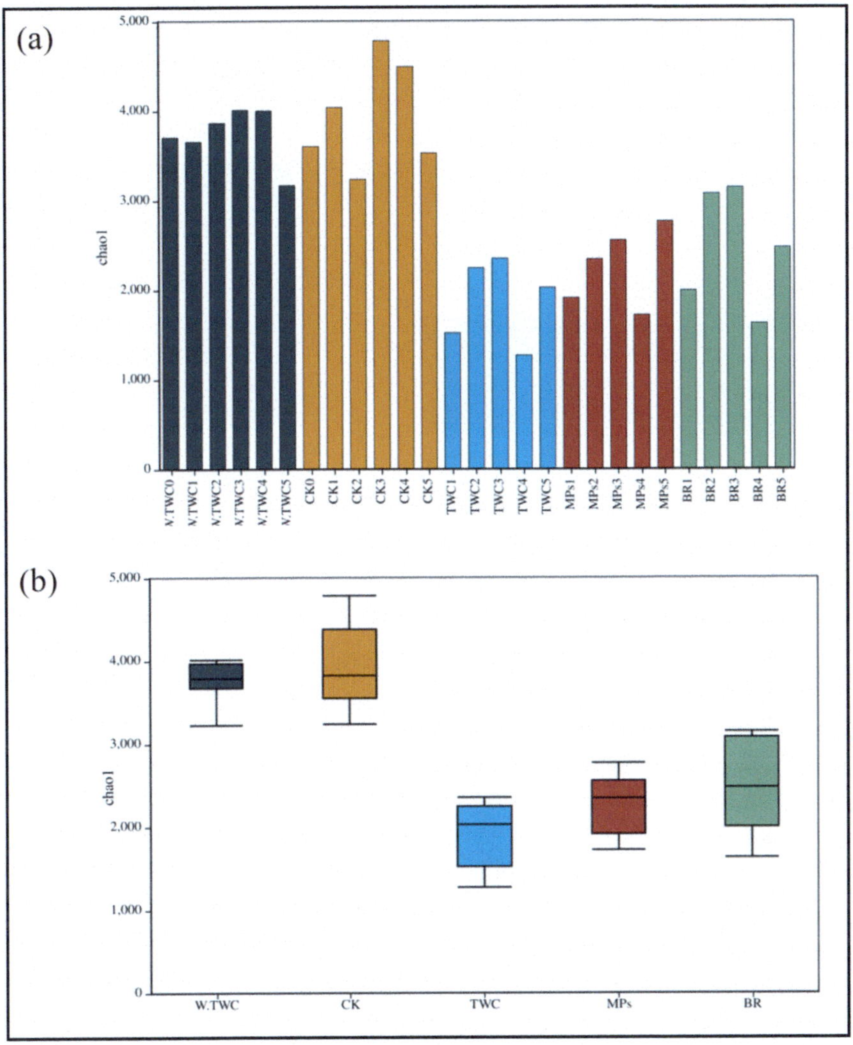

Fig. 36.3 Distribution of microbial chao1 index **a** and its differences among groups **b**

36.4 Conclusion

(1) The content of TN in the demonstration area of Xunsi River is relatively high, which all exceeds the surface water quality standard V, while NH_3–N, TP and COD are all lower than the surface water quality standard of Class IV. After 7 months of treatment, the content of TN in the treatment group remained unchanged as that in the control group and the incoming water group, showing fluctuating changes. In comparison to the control group, the contents of NH_3–N, TP and COD in the

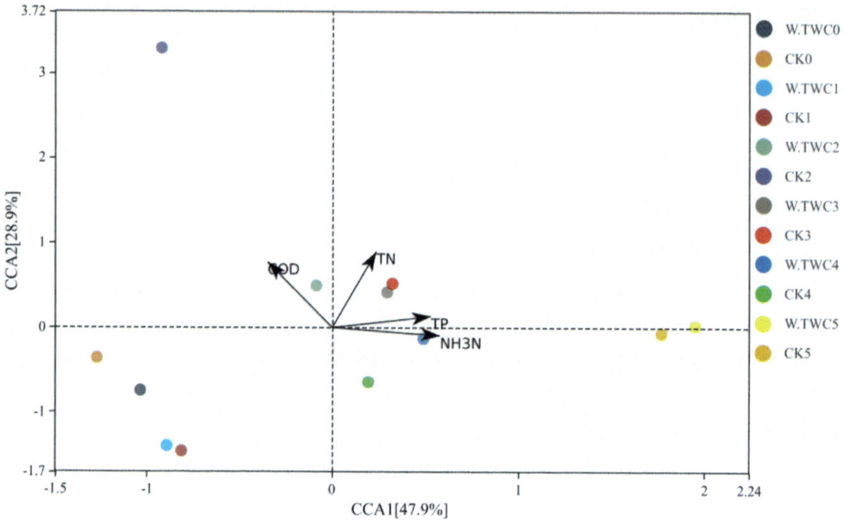

Fig. 36.4 CCA analysis of microbial community and main environmental factors in treatment group and control group

treatment group decreased by 38%–79%, 7%–45% and 14%–33%, respectively, while compared with the incoming water quality, the TP and COD in the treatment group generally decreased by 25%–65% and 9%–45%, respectively.

(2) There was no remarkable difference in chao1 index between the treatment group and the control group, indicating that there was no notable discrepancy in the species diversity of bacteria between the treatment group and the control group. At the same time, although the richness of bacteria in different materials decreased compared with that in water, the diversity and evenness of species were higher. In addition, the dominant taxa and classes of each group are similar at the phylum and class level, but the dominant genera are different at the genus level.

(3) The impact of different environmental factors on the bacterial community of both the control group and the treatment group was minimal, and there was only a strong correlation between the specific bacteria and environmental factors in some periods of the sampling site.

Acknowledgements This work was supported by the Strategic Priority Research Program of the Chinese Academy of Sciences (Grant No. XDA23040303), and the National Natural Science Foundation of China (NO. 52179057).

References

1. Li W, Li X, Su J et al (2014) Sources and mass fluxes of the main contaminants in a heavily polluted and modified river of the North China Plain. Environ Sci Pollut Res 21(8):5678–5688
2. Chang J, Jia W, Qu Y et al (2019) Improving treatment performance of biofilters for heavily polluted surface water using combined organic/inorganic substrates and exogenous carbon source of wood pieces. Desalin Water Treat 169:191–197
3. Song C, Shi Y, Gao H et al (2022) Enhanced removal of dissolved S^{2-} from a river by the pump-and-treat ex-situ process with enriched consortium. Aqua Water Infrastruct Ecosyst Soc 71 (8):926–937
4. Hui Z (2009) Demonstration work of inter-cycle biological contact oxidation process for purification of Daqing river in Dianchi Lake watershed. Chin J Environ Eng 3(2):199–204
5. Quan X, Gu L, Qian Y et al (2013) Characterization of nitrification performance and microbial community in a MBBR and integrated GBBR-MBBR treating heavily polluted river water. Environ Eng Manage J 12(7):1335–1344
6. Hong G, Yue X, Sarfraz H et al (2018) Application of microbial technology used in bioremediation of urban polluted river: a case study of Chengnan River, China. Water 10(5):643
7. Lin Y, Wang L, Xu K et al (2021) Revealing taxon-specific heavy metal-resistance mechanisms in denitrifying phosphorus removal sludge using genome-centric metaproteomics. Microbiome 9:67
8. Jia H, Dong N, Ma H (2010) Evaluation of aquatic rehabilitation technologies for polluted urban rivers and the case study of the Foshan channel. Front Environ Sci Eng China 4:213–220
9. Peng M, Xu H, Yang G et al (2021) Purifying heavily polluted river water using immobilized native photosynthetic bacteria. J Environ Eng 147(8):04021021
10. Wu X, Xie Y (2012) The field application of microbial technology used in bioremediation of urban polluted river. Adv Mater Res 518–523:2906–2911
11. Wan J, Wang P, Qian X et al (2020) Bioinspired paper-based nanocomposites enabled by biowax-mineral hybrids and proteins. ACS Sustain Chem Eng 8(26):9906–9919
12. Olarte G, Garzón L, Sarmiento J et al (2023) Biowax production from the hydrotreatment of refined palm oil (RPO). Processes 11(5):1372
13. Jo HM, Kim DH, Lee SH et al (2022) Multi-layer barrier coating technology using nano-fibrillated cellulose and a hydrophobic coating agent. BioResources 17(4):6222–6233
14. Li G, Xia X, Yang Z (2006) Distribution and sources of polycyclic aromatic hydrocarbons in the middle and lower reaches of the yellow river, China. Environ Pollut 144(3):985–993
15. Lu Q, Song Y, Mao G et al (2020) Spatial variation in bacterial biomass, community composition and driving factors across a eutrophic river. Ecotoxicol Environ Saf 205:111113
16. Lekunberri I, Gasol JM, Acinas SG et al (2014) The phylogenetic and ecological context of cultured and whole genome-sequenced planktonic bacteria from the coastal NW Mediterranean Sea. Syst Appl Microbiol 37:216–228
17. Zhang T, Wang P (2022) The response of bacterial communities to organic matter in the surface sediment of the Pearl River estuary. Plankon Benthos Res 17(1):45–56
18. Spain A, Krumholz L, Elshahed M (2009) Abundance, composition, diversity and novelty of soil proteobacteria. ISME J 3:992–1000
19. Guo B, Yu N, David G et al (2021) Effects of micro-aeration on microbial niches and antimicrobial resistances in blackwater anaerobic digesters. Water Res 196:117035
20. Chao A (1984) Nonparametric estimation of the number of classes in a population. Scand J Stat 11:265–270
21. Vilmi A, Zhao W, Picazo F et al (2020) Ecological processes underlying community assembly of aquatic bacteria and macroinvertebrates under contrasting climates on the Tibetan Plateau. Sci Total Environ 702:134974
22. Luef B, Aspetsberger F, Hein T et al (2010) Impact of hydrology on free-living and particle-associated microorganisms in a river floodplain system (Danube, Austria). Freshw Biol 52(6):1043–1057

23. Liu K, Hu A, Hou L et al (2022) Effects of land-use patterns on the biogeography of the sediment bacteria in the Yarlung Tsangpo River. FEMS Microbiol Ecol 98:1–10
24. Chen X, Lu J, Zhu J et al (2020) Characteristics of denitrifying bacteria in different habitats of the Yongding River wetland, China. J Environ Manage 275:111273

Chapter 37
Synthesis and Adsorption Properties of Ternary-Grafted Starch-Based Hydrogel

Haihua Li, Yuhua Gao, Meifang Yan, and Na Li

Abstract The biomass-based hydrogels can be used as the adsorbents in sewage water, and they have good adsorption capacity for heavy ions and dyes. The starch-based hydrogel was prepared with cornstarch (St) and sodium carboxymethyl cellulose (CMC) as raw materials, acrylamide (AM), acrylic acid (AA) and 2-acrylamide-2-methylpropane sulfonic acid (AMPS) as grafting monomers through grafting and cross-linking reactions and named St/CMC/P(AM-AA-AMPS). The synthesis conditions and the adsorption performances for Cu^{2+} were studied. The results showed that the suitable synthesis conditions were m(St):m(CMC) = 1:0.5, m(St):m(AM):m(AA):m(AMPS) = 1:3:5:2, m(St):m(APS) = 1:0.06, reaction temperature 75 °C, and reaction time 3 h. The results of adsorption experiments showed that the hydrogel had good adsorption properties for water and Cu^{2+}. The swelling ratio was 126.77 g/g after immersed in water for 3 h. When the hydrogel was used to adsorb Cu^{2+} ($\rho_0 = 30$ mg/L) in the simulated water, the removal rate of Cu^{2+} was 93.3%, and the adsorption capacity was 80.51 mg/g.

Keywords Starch · Hydrogel · Heavy ion · Synthesis · Adsorption

37.1 Introduction

Excessive copper ions in soil and water can lead to crop diseases, the decline in the self purification ability of water bodies, and toxic effects on aquatic organisms [1]. Copper ions can also enter the organism through the food chain, causing a series of physiological and metabolic disorders. The main treatment methods for Cu^{2+} in water include chemical precipitation, displacement, adsorption, and ion exchange [2]. Among them, adsorption method is widely used, and activated carbon, zeolite

H. Li (✉) · Y. Gao · M. Yan
Institute of Energy Sources, Hebei Academy of Sciences, Shijiazhuang, Hebei, China
e-mail: 873756647@qq.com

N. Li
Hebei Sunwater Water Treatment Co., Ltd, Shijiazhuang, Hebei, China

K. Y. Foo et al. (eds.), *Water Nexus Approach for Sustainable Development*,
Environmental Science and Engineering, https://doi.org/10.1007/978-3-031-89238-7_37

molecular sieve, fly ash, minerals, etc. can all be used as adsorption materials for Cu^{2+}.

Hydrogels are the new polymers with three-dimensional network structures, and these three-dimensional structures often contain many hydrophilic groups, such as $-OH$, $-CONH_2$, $-NH_2$ and $-COOH$, which can store a large amount of water and maintain the complete structure in water. Therefore, some hydrogels can be used for water purification and have great advantages [3]. Starch-based hydrogels can not only reduce the cost of hydrogels, but also have good biocompatibility and biodegradability, which is of great research value [4]. Vinyl monomers such as acrylamide (AM) and acrylic acid (AA) that react with starch to form hydrogels have been studied most [5]. In this kind of starch-based hydrogel, the hydrophilic functional group is mainly carboxyl group, which can be used as the adsorbent for heavy metal ions. However, due to the single type of functional group, the adsorption capacity of starch-based hydrogel is generally low.

In this paper, a ternary-graft starch-based hydrogel was prepared from cornstarch (St) and sodium carboxymethyl cellulose (CMC) as raw materials, AM, AA and 2-acrylamido-2-methylpropanesulfonic acid (AMPS) as graft monomers using graft and cross-linking reactions and the product was named St/CMC/P(AM-AA-AMPS). The molecular structure was characterized by infrared spectrum (ATR/FTIR) and thermogravimetric analysis (TGA), and the adsorption performances of the product were studied.

37.2 Materials and Methods

37.2.1 Reagents

St, CMC, AM, AA and AMPS were bought from Aladdin Reagent Company (Shanghai, CHN); N.N-Methylenebisacrylamide (NMBA) was bought from McLean Reagent Company (Shanghai, CHN); ammonium persulfate (APS) was purchased from Yongda Reagent Company (Tianjin, CHN). The reagents used in the experiments are all analytical grade, except St, St is regent grade. All reagents are used directly.

37.2.2 Synthesis of Hydrogel

1.0 g of St and 15 mL of deionized water were put into beaker 1 to form a suspension. Beaker 1 was put into a 90 °C water bath to gelatinize for 15 min, and then took out to cool down. In beaker 2, 0.5 g of CMC was dissolved in 30 mL of deionized water to form a transparent viscous solution. 5 mL of AA was added into beaker 3, 11.7 g of sodium hydroxide solution (15 wt%) was dripped into the beaker under stirring to achieve a neutralization degree of 60% for AA. Then the mixed powder of AM (3.0 g), AMPS (1.0 g), and NMBA (0.2 g, as the cross-linking agent) was dissolved in the AA solution. 0.06 g of initiator APS was weighed with a weighing boat and dissolved in 5 mL of deionized water. The solutions in beaker 2 and 3 were poured into beaker 1 and stirred evenly, and then the initiator solution was added into the precursor solution to form a uniform mixture. To eliminate air, N_2 was filled into the beaker for 1 min, and then the beaker mouth was sealed by a preservative film. The beaker 1 was placed in a 75 °C water bath and the mixture was stood to react for 5 h to obtain a kind of creamy white gel. The hydrogel was broken into pieces, and then the pieces were soaked with deionized water and ethanol. The product was dried under vacuum at 60 °C to get white particles, which were the target product.

37.2.3 Characterization of Hydrogel

The molecular structure of the starch-based hydrogel was determined by ATR/FTIR (Frontier, PE, USA). The thermal stability of hydrogel was analyzed by TGA (STA449F3, NETZSCH, GER).

37.2.4 Water–adsorption Experiment

The swelling ratio (*Sr*, g/g) indicates the water-absorption performance of the hydrogel. The larger the swelling ratio, the better the water absorption performance. The swelling ratio of hydrogel was measured by the gravimetric method, 0.2 g of the hydrogel was weighed accurately and packaged with a tea bag, and then the bag was immersed into sufficient deionized water. Every 0.5 h, the bag was taken out from water and weighed after cleaning the outside of tea bag using pieces of tissue. *Sr* can be calculated with the follow formula:

$$Sr = \frac{m_2 - m_1}{m_0} \tag{37.1}$$

Where m_0 is the initial mass of hydrogel, g; m_1 is the initial mass of hydrogel and tea bag, g; m_2 is the weight of hydrogel and tea bag after water absorption, g.

37.2.5 Adsorption Experiment

0.01–0.05 g of starch-based hydrogel was put into a simulate aqueous solution containing 30 mg/L of Cu^{2+}, and the initial concentration of Cu^{2+} was denoted by ρ_0, mg/L. The hydrogel adsorbent adsorbed Cu^{2+} from the simulated solution under magnetic stirring of 400 rpm for 1 h, following by static adsorption for 4 h. Then used a TU-1900 dual-beam ultraviolet visible spectrometer (UV-Vis, Beijing Puxi General Instrument Co., Ltd, CHN), the equilibrium concentration of Cu^{2+} in the supernatant was determined and denoted by ρ_e, mg/L. The removal rate ($E\%$) and the equilibrium adsorption capacity of Cu^{2+} (q_e, mg/g) are calculated by the Formulas 37.2 and 37.3, respectively.

$$E\% = (1 - \rho_e/\rho_0) \times 100\% \quad (37.2)$$

$$q_e = \frac{(\rho_0 - \rho_e) \times V}{m} \quad (37.3)$$

Where m is the dose of adsorbent, g; V is the volume of the simulated aqueous solution, L.

37.3 Results and Discussion

37.3.1 Synthesis Conditions of Hydrogel

Effect of initiator dosage on adsorption capacity.

Fixed m(St):m(AM):m(AA):m(AMPS) = 1:3:5:1, m(St):m(CMC) = 1:0.5, the effect of initiator APS dose on adsorption capacity of hydrogel was investigated. It can be seen in Fig. 37.1, the adsorption capacity rapidly increases with the increase of initiator dose, but when the initiator dose exceeds 6% of cornstarch mass, the adsorption capacity shows a downward trend. This is mainly because when the amounts of initiator are too small, the amounts of free radicals generated are less, and the probability of reaction between monomers is smaller. However, when those of initiator are too large, too many free radicals are generated, which hinders chain growth reaction and produces many short chain products. In addition, it also leads to an increase in the self polymerization of active monomers, which produces more by-products, thus affecting the adsorption performance of hydrogels.

Effect of CMC dose on adsorption capacity.

Fixed m(St):m(AM):m(AA):m(AMPS) = 1:3:5:1, m(St):m(APS) = 1:0.06, Fig. 37.2 shows the effect of CMC dose on adsorption capacity. As the content of CMC in raw materials increases, the adsorption capacity of the product increases rapidly. When the CMC content is greater than 50% of the starch content, the increase in adsorption capacity slows down. CMC can greatly increase the adsorption performance of hydrogels, mainly because CMC contains many hydrophilic groups $-COO^-$, which can greatly increase the number of effective groups in hydrogels.

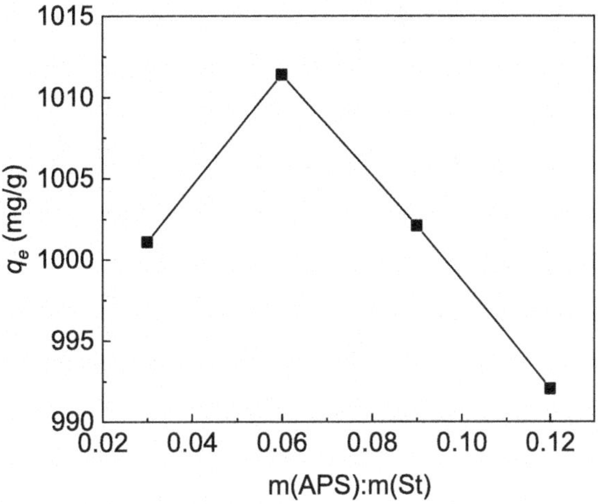

Fig. 37.1 Effect of initiator dose on adsorption capacity

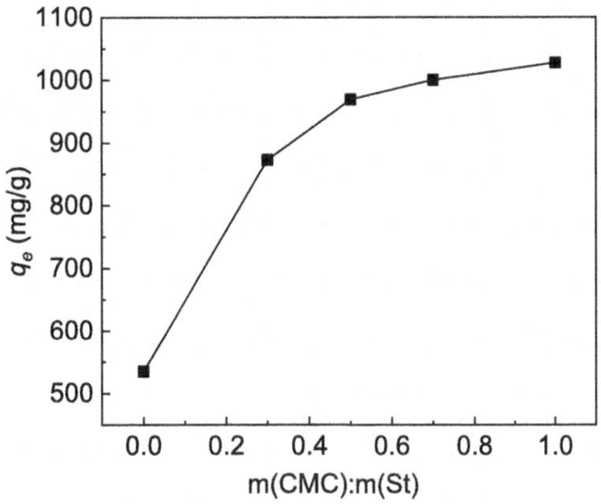

Fig. 37.2 Effect of CMC dose on adsorption capacity

Effect of AMPS dose on adsorption capacity.

Fixed m(St):m(AM):m(AA) = 1:3:5, m(St):m(CMC) = 1:0.5, m(St):m(APS) = 1:0.06, Fig. 37.3 shows that the effect of AMPS dose on adsorption capacity. As the content of AMPS in grafting monomers increases, the adsorption capacity gradually increases. This is mainly because with the increase of AMPS content in monomers, the amount of $-SO_3H$ introduced into the hydrogel through the graft reaction increases, and there is an electrostatic interaction between the deprotonated sulfonic acid group and the MB molecule, thus improving the adsorption performance of the hydrogel. Taking into account both cost and performance factors, it is more appropriate to choose m(St):m(AMPS) = 1:2.

In conclusion, the optimal synthesis conditions of starch-based hydrogel were m(St):m(CMC) = 1:0.5, m(St):m(AM):m(AA):m(AMPS) = 1:3:5:2, m(St):m(APS) = 1:0.06, reaction temperature 75 °C, and reaction time 3 h.

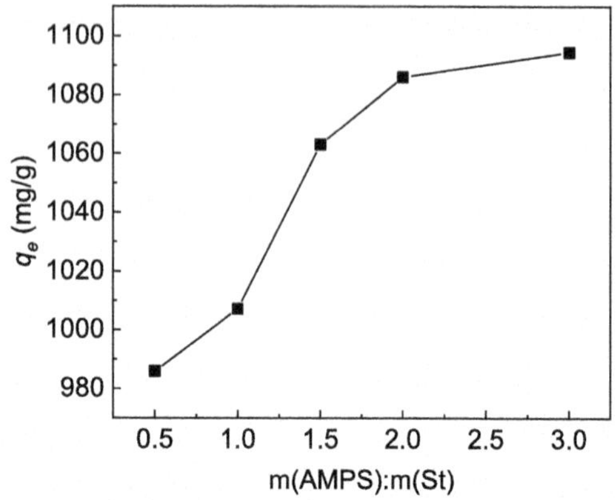

Fig. 37.3 Effect of AMPS dose on adsorption capacity

37.3.2 Characterization of Hydrogel

The structure of starch-based hydrogel prepared under the optimal conditions was characterized by ATR/FTIR, and the result is shown in Fig. 37.4. In the spectrum of the hydrogel, the primary amide exhibits bimodal peaks at 3330 cm^{-1} and 3200 cm^{-1} [6]. The vibration absorption peak of carbonyl group is located at 1657 cm^{-1}. The bending vibration of N–H occurs at 1552 cm^{-1}, and symmetric and asymmetric stretching vibrations of $-COO^-$ at 1449 and 1402 cm^{-1} [7]. The characteristic absorption peaks of sulfonic acid groups are at 1187 cm^{-1} and 1040 cm^{-1}. The new peaks in the spectrum demonstrate that the molecular chains of the product contain the functional groups of carbonyl, amino, carboxyl and sulfonic.

The TGA test result of the same hydrogel is shown in Fig. 37.5. The main water loss stage of the hydrogel is between 320–390 °C and 440–540 °C. The mass loss of

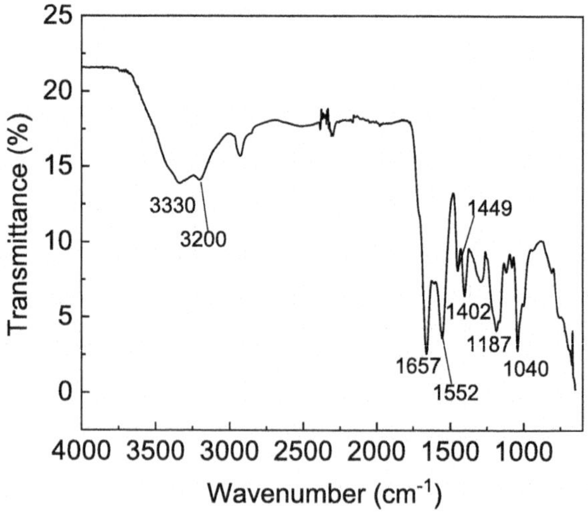

Fig. 37.4 Spectrum of hydrogel

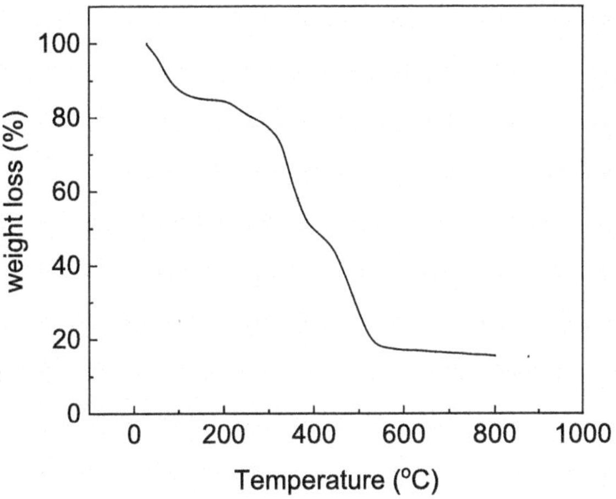

Fig. 37.5 TGA cuver of starch-based hydrogel

the previous stage is mainly because of the fracture of the St molecular chain and the disintegration of the cross-linking structure in the hydrogel. The latter is due to the decomposition of the grafted side chain. The mass of the hydrogel residue is 15.58%. Compared with that of cornstarch, the mass loss rate and total mass loss of hydrogel are smaller, which indicates that hydrogel has higher thermal stability.

37.3.3 Water-Adsorption Performance of Hydrogel

The changes of the swelling ratio over the immersion time in water are shown in Fig. 37.6. The swelling ratio of hydrogel increases rapidly with the immersion time, and changes slightly after the immersion time in water exceeds 1 h, which indicates that the water-absorption rate of the hydrogel is very fast. When the immersion time was 3 h, the swelling ratio was 126.77 g/g.

37.3.4 Adsorption Performance of Hydrogel

The effect of the hydrogel dose on the removal rate ($E\%$) and the adsorption capacity (q_e) of Cu^{2+} are shown in Fig. 37.7. With the increase of dose, the removal rate of Cu^{2+} increases rapidly. When the dose of hydrogel was 400 mg/L, $E\% = 93.3\%$ and $q_e = 80.51$ mg/g. Continue to increase the dose, $E\%$ does not increased significantly, while the adsorption capacity continued to decrease. This is mainly because $E\%$ and q_e are positivity correlated with the numbers of active adsorption sites provided by the adsorbent. With the increase of the hydrogel dose, more heavy ions can be adsorbed, thus greatly increasing the removal rate. However, it can be seen from the Formula (37.3), in addition to the amount of Cu^{2+} adsorbed, the adsorption capacity is also

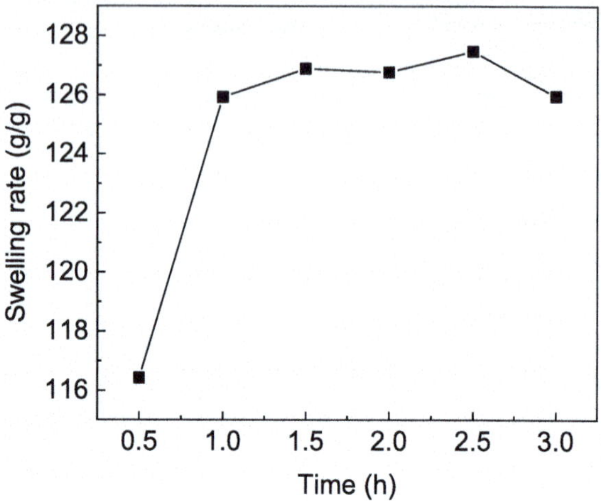

Fig. 37.6 Variation of swelling ratio of hydrogel over immersion time in water

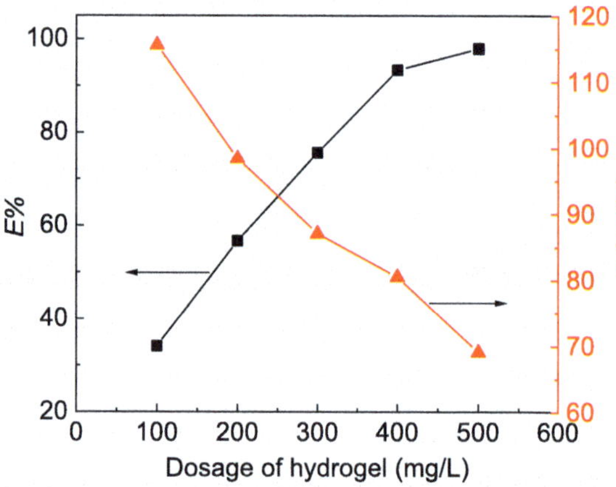

Fig. 37.7 Effect of dose of starch-based hydrogel on adsorption

negatively related to the mass of the adsorbent added. When the dose of starch-based hydrogel exceeds 400 mg/L, all of Cu^{2+} ions have been adsorbed already by a part of the active adsorption sites, and the occupied proportion of the adsorption site of hydrogel is low, and the adsorption capacity continues to decline.

37.4 Conclusions

The synthetic conditions of starch-based hydrogel were screened. The suitable synthesis conditions of the hydrogel were m(St):m(AM):m(AA):m(AMPS) = 1:3:5:2, m(St):m(CMC) = 1:0.5, m(St):m(APS) = 1:0.06, reaction temperature 75 °C, reaction time 3 h.

The adsorption performances of the hydrogel for water and Cu^{2+} were tested, and the results showed that the hydrogel had good adsorption performances for water and Cu^{2+}. After immersing in water for 3 h, the swelling ratio of the hydrogel was 126.77 g/g. When the initial concentration of Cu^{2+} was 30 mg/L and the hydrogel dose was 400 mg/L, the removal rate of Cu^{2+} was 93.3%, and the adsorption capacity was 80.51 mg/g.

Acknowledgements The research was sponsored by Fundamental Research Fund System Pilot Project of Hebei Academy of Sciences (2024PF03).

References

1. Shrestha R, Ban S, Devkota S et al (2021) Technological trends in heavy metals removal from industrial wastewater: a review. J Environ Chem Eng 9(4):105688
2. Yuan YW, Li RY, Peng SJ (2022) Research progress on chemical modification of waste biomass cellulose to prepare heavy metal adsorbents. Polym Bull 80(11):11671–11700
3. Hafida FH, Tayeb B, Amina S (2021) Development of functional chitosan-based superabsorbent hydrogel nanocomposites for adsorptive removal of basic red 46 textile dye. Polym Bull 79(8):1–32
4. Warunee T, Jate P, Sayant S (2018) Novel biodegradable hydrogel based on natural polymers: synthesis, characterization, swelling/reswelling and biodegradability. Eur Polymer J 112:678–687
5. Chai WS, Cheun JY, Kumar PS et al (2021) A review on conventional and novel materials towards heavy metal adsorption in wastewater treatment application. J Clean Prod 296:126589
6. Azizi S, Shaki H, Shafeeyan MS et al (2024) Chitosan-polyethylene oxide nanofiber/polyacrylamide-co-acrylic acid hydrogel nanocomposite: a copper ion adsorbent of aqueous solutions. Inorg Chem Commun 159:111682
7. Yin XC, Ke T, Zhu H et al (2023) Efficient removal of heavy metals from aqueous solution using licorice residue-based hydrogel adsorbent. Gels 9(7):907059

The manufacturer's authorised representative in the EU is Springer Nature Customer Service Centre GmbH, Europaplatz 3, 69115 Heidelberg, Germany. If you have any concerns regarding our products, please contact ProductSafety@springernature.com

Printed and bound by CPI Group (UK) Ltd, Croydon, CR0 4YY
05/10/2026
02228127-0013